Synthesis Lectures on Mathematics & Statistics

Series Editor

Steven G. Krantz, Department of Mathematics, Washington University, Saint Louis, MO, USA

This series includes titles in applied mathematics and statistics for cross-disciplinary STEM professionals, educators, researchers, and students. The series focuses on new and traditional techniques to develop mathematical knowledge and skills, an understanding of core mathematical reasoning, and the ability to utilize data in specific applications.

Mario Petrich · Norman R. Reilly

Completely Regular Semigroup Varieties

A Comprehensive Study with Modern Insights

 Springer

Mario Petrich
Bol, Brac, Croatia

Norman R. Reilly
Department of Mathematics
Simon Fraser University
Burnaby, BC, Canada

ISSN 1938-1743   ISSN 1938-1751 (electronic)
Synthesis Lectures on Mathematics & Statistics
ISBN 978-3-031-42893-7   ISBN 978-3-031-42891-3 (eBook)
https://doi.org/10.1007/978-3-031-42891-3

This Springer imprint is published by the registered company Springer Nature Switzerland AG
The registered company address is: Gewerbestrasse 11, 6330 Cham, Switzerland

Paper in this product is recyclable.

We dedicate this book
to the memory of Libor Polák (1950–2020)

Preface

This book is a natural continuation of the authors' text "Completely Regular Semigroups" [PR1]. However, in order to read this volume, all that is required is a general background in algebra and varieties of algebras as well as a basic knowledge of semigroup theory. Where results from the previous volume are required, they are meticulously cited so that the reader may easily refer to them.

It was natural that the first volume should deal mainly with the elementary features of the theory such as the structure of completely regular semigroups, characterization of congruences and the lattice of congruences on completely regular semigroups. "Completely Regular Semigroups" contains the beginnings of the treatment of the rich topic of varieties of completely regular semigroups and the lattice of such varieties. Inevitably, the focus was on building the basics, a necessary foundation for a deeper investigation of the lattice of varieties of completely regular semigroups. The treatment culminated with the introduction of the Mal'cev product, a concept that will figure prominently in this volume.

As well as filling a practical role and vital reference, "Completely Regular Semigroups" exposed the need for a second volume of the same general kind to keep up with the growth of the theory studied and cover in greater depth topics that were merely touched upon in the published volume. This volume inherits the general form of presentation from its predecessor with an array of topics more advanced than in the predecessor, such as the basic theory of varieties, a modern and unified treatment of varieties of bands, and Polák's theorem. Building on the solid foundation provided in the earlier volume, the subjects covered here are the most important parts of the core theory of varieties of completely regular semigroups. The overall goal of this text is to give a unified treatment of the most important developments about varieties of completely regular semigroups and related topics in order to make them more accessible to non-specialists and future students.

The text is written with complete proofs except when the relevant statements belong to other branches of algebra and would require serious digressions to include in detail. Several indices are provided and specialists in algebra should be able to pinpoint what they want fairly easily. Abstract algebra is the main basic background for reading this book. The sources of the material are copiously noted at the end of each section, as well as broader references enabling the reader to pursue further reading on all of the topics discussed.

The authors deeply appreciate the comments, corrections and advice received from Jiří Kad'ourek. The authors benefitted greatly from the careful reading of the manuscript by Maria Szendrei and Edmond Lee. Antonio Cegarra was also helpful. The authors cannot thank Dragan Blagojević enough for the hours that he spent applying his masterful editorial skills to the text. The second author also wishes to thank the Department of Mathematics, Simon Fraser University for its support.

Bol, Croatia   Mario Petrich
Burnaby, Canada   Norman R. Reilly

Introduction

It was a great loss to the mathematical community when Mario Petrich died on September 15, 2021. He was a leader in his field and an inspiration to many others. He was variously a friend, colleague and co-author of the second author for over forty years and is greatly missed.

The theory of completely regular semigroups sits in a sweet spot between the long established theory of groups and the general theory of semigroups, finite or infinite. The structure of completely regular semigroups has significant levels of complexity, even when it is assumed that we know everything that we might wish to know regarding the structure of groups and their varieties. Completely regular semigroups exhibit a wealth of interesting structures. Although the characterization of a congruence on a completely regular semigroup by kernel and trace is a good deal more complicated than the situation that pertains for congruences in groups by normal subgroups, that very fact leads to a rich theory not enjoyed by groups. When viewed as algebras with respect to the binary operation of multiplication and the unary operation of inversion, which we do throughout this text, completely regular semigroups constitute a variety $\mathcal{CR}$ for which the lattice $\mathcal{L}(\mathcal{CR})$ of subvarieties displays many beautiful substructures. The fact that completely regular semigroups are unions of groups provides enough leverage to provide a rich deep diverse theory, many aspects of which suggest avenues of development in more general classes of semigroups.

The focus of this text is on varieties of completely regular semigroups, especially the lattice of varieties of completely regular semigroups. There are two possible general approaches to the problem of finding a description of $\mathcal{L}(\mathcal{CR})$. One, which may be thought of as the "local" approach, is to study in detail various intervals starting at the bottom of the lattice and working up the lattice with the goal of obtaining a complete detailed description in each case. The second approach, which may be termed the "global" approach, consists of studying various decompositions of $\mathcal{L}(\mathcal{CR})$ with the goal of obtaining an isomorphic copy of $\mathcal{L}(\mathcal{CR})$ as a direct or subdirect product that can then be the starting point for even deeper investigations both locally and globally.

The first approach reveals many interesting facts but is mainly restricted to the lattice of subvarieties of varieties that are known to be joins of smaller and, hopefully, better understood varieties. Nevertheless there are some exceptions to this limitation at the very bottom of this lattice such as the lattice of varieties of completely simple semigroups and the lattice of varieties of bands (considered in Chap. 3). This approach, though important, cannot possibly lead to a complete description of $\mathcal{L}(\mathcal{CR})$.

The second approach attacks the structure of the entire lattice $\mathcal{L}(\mathcal{CR})$ and, in view of the complexity of $\mathcal{L}(\mathcal{CR})$, must necessarily entail some imprecise features. The core of this approach consists of various decompositions of $\mathcal{L}(\mathcal{CR})$ by means of a number of relations which may be successfully defined on $\mathcal{L}(\mathcal{CR})$. These make use of the kernel and trace relations defined on the lattice of fully invariant congruences on the free completely regular semigroup of countably infinite rank transferred to $\mathcal{L}(\mathcal{CR})$ under the familiar antiisomorphism. This effort achieves an apex in an isomorphic copy of $\mathcal{L}(\mathcal{CR})$ due to Polák.

In order to set the stage, we focus in Chap. 1 on the model of the free completely regular semigroup CR_X on a countably infinite set X first investigated by A. H. Clifford. This has the form of U/ζ where U is the free unary semigroup on X and ζ is the congruence on U generated by the relations $u = u(u)^{-1}u$, $u = (u^{-1})^{-1}$, $u(u)^{-1} = (u)^{-1}u$. This characterization facilitates the descriptions of Green's relations with the help of a few basic operators on words. From there we proceed to the fundamental study of fully invariant congruences on CR_X and a family of congruences associated with each such fully invariant congruence. The critical component in this chapter is the beginning of the study of the restrictions of the kernel and trace relations $K, K_\ell, K_r, T, T_\ell, T_r$ introduced in [PR1] to the lattice of fully invariant congruences on CR_X. Each of these relations is a complete congruence. Since the classes of these relations are intervals, each of these relations associates to each fully invariant congruence two new congruences, namely the least and greatest congruence in its class. Several of the most important of these congruences can be characterized in terms of the initial congruence with the help of appropriate operators on words. The chapter ends by establishing the very important result that $\mathcal{L}(\mathcal{CR})$ is modular.

In Chap. 2, the standard antiisomorphism between the lattice of fully invariant congruences and the lattice of varieties is used to translate the six complete congruences $K, K_\ell, K_r, T, T_\ell, T_r$ on the lattice of fully invariant congruences to six complete congruences on $\mathcal{L}(\mathcal{CR})$. These complete congruences are then employed in almost every chapter in the analysis of various aspects of $\mathcal{L}(\mathcal{CR})$. Chapter 2 is devoted to a detailed study of the classes of each of the relations in $\Sigma = \{K, K_\ell, K_r, T, T_\ell, T_r\}$ on $\mathcal{L}(\mathcal{CR})$. Since each $P \in \Sigma$ is a complete congruence, it follows that each class of P is an interval of the form $[\mathcal{V}_P, \mathcal{V}^P]$ for some $\mathcal{V} \in \mathcal{L}(\mathcal{CR})$. This leads us to a whole host of questions concerning the varieties of the form $\mathcal{V}_P$, or $\mathcal{V}^P$, such as bases of identities and structural relationships between elements of $\mathcal{V}$ and elements of $\mathcal{V}_P$ or $\mathcal{V}^P$ and to questions about the mappings $\mathcal{V} \mapsto \mathcal{V}_P, \mathcal{V} \mapsto \mathcal{V}^P$. All of these questions are thoroughly explored in Chap. 2.

In Chap. 3 we turn our attention to the lattice of varieties of bands. Every band S can be viewed as a completely regular semigroup in which the unary operation is the identity mapping, $(x)^{-1} = x$, for all $x \in S$, so that the basic theory of bands can be dealt with by treating them purely as semigroups. As a consequence, the first important interval in $\mathcal{L}(\mathcal{CR})$, outside of the lattice of varieties of groups, to be completely described was the principal ideal generated by the variety $\mathcal{B}$ of bands, that is, the lattice $\mathcal{L}(\mathcal{B})$ of all varieties of bands. This major result was achieved independently by Birjukov, Fennemore and Gerhard. There is a vast literature on the topic of bands. Since this is a book about completely regular semigroups, it seems appropriate to enrich the theory of bands with the specific insights that can be derived from that more general theory. With this in mind, we provide a new proof of these results using the tools provided by the theory of completely regular semigroups and exhibit the relations that lie behind the commonly displayed diagram of $\mathcal{L}(\mathcal{B})$. The sublattice $\mathcal{L}(\mathcal{B})$ is a single K-class in $\mathcal{L}(\mathcal{CR})$, but the T_ℓ and T_r relations restricted to $\mathcal{L}(\mathcal{B})$ reveal its inner partitions. Since the mapping $\mu_\mathcal{B} : \mathcal{V} \mapsto \mathcal{U} \cap \mathcal{B}$ is a complete retraction of $\mathcal{L}(\mathcal{CR})$ onto $\mathcal{L}(\mathcal{B})$, it follows that the relation $\mathbf{B}$ induced by $\mu_\mathcal{B}$ on $\mathcal{L}(\mathcal{CR})$ is a complete congruence on $\mathcal{L}(\mathcal{CR})$ and the classes of $\mathbf{B}$ are intervals that reach deeply into the lattice $\mathcal{L}(\mathcal{CR})$. In other words, $\mathcal{L}(\mathcal{B})$ is much more than just an interesting sublattice of $\mathcal{L}(\mathcal{CR})$, the structure of $\mathcal{L}(\mathcal{B})$ is imprinted over the whole of the lattice $\mathcal{L}(\mathcal{CR})$. Our approach enables us to reveal the critical role, indeed the central role, of $\mathbf{B}$ in Polák's theorem in the next chapter. This approach also leads us to the meet subsemilattice Υ consisting of the upper ends of the intervals in $\mathbf{B}$. This is a fascinating topic which we will revisit in [PR3]. Chapter 3 concludes with descriptions of the free band on a countably infinite set.

Chapter 4 could be considered the jewel in the crown of the theory of $\mathcal{L}(\mathcal{CR})$. After some preliminaries, we present the theorem due to Polák. This remarkable theorem represents the lattice $\mathcal{L}(\mathcal{CR})$ as a subdirect product of copies of a lattice consisting of $\{\mathcal{V}_K \mid \mathcal{V} \in \mathcal{L}(\mathcal{CR})\}$ augmented by three elements at the bottom. It might appear that we are exchanging one mysterious structure for another. But the general notion is to describe each variety in terms of other varieties that are, one hopes, smaller and better known or easier to work with than the initial variety. Indeed, it may be difficult to identify the image of an element of $\mathcal{L}(\mathcal{CR})$ in the subdirect product or to gain further insight from the representation especially, for instance, if $\mathcal{V}K = \{\mathcal{V}\}$ and there are indeed such varieties, as we shall see in the next volume. However, in other situations Polák's theorem was a great leap forward. For one thing it shed a bright light on the structure of much larger intervals than could be considered before, such as the lattice of varieties of orthodox or locally orthodox varieties. For another, it can be a great help in evaluating the join of two varieties—whose solution can often be quite elusive. In addition the technique of working with (Polák) ladders can be used to describe certain intervals in $\mathcal{L}(\mathcal{CR})$ in complete detail. In [PR3] we will illustrate the application of Polák's theorem to general intervals of the form $[\mathcal{U}, \mathcal{V}]$ where $\mathcal{U}_K = \mathcal{U}$, $\mathcal{V}_K = \mathcal{V}$ and $\mathcal{V}$ covers $\mathcal{U}$. This can be used to give a complete description of the K-classes of the form $\mathcal{A}_p K$, where $\mathcal{A}_p$ denotes the variety of abelian groups of prime exponent p.

This story is far from over. A key element in the above development is the class of those varieties that are minimal in their K-classes. We know about important cases within the lattice of varieties of completely simple semigroups and we will see in the next part that all varieties of the form $H\mathcal{U}$, consisting of all completely regular semigroups with subgroups lying in the nontrivial group variety $\mathcal{U}$ are minimal in their K-classes. It is likely that much more can be said about this topic.

This volume is sufficiently advanced to be primarily for use as the basis for a graduate course or to assist researchers interested in recent developments in the theory of varieties of completely regular semigroups. We are confident that the rich structures described in this text will also inspire researchers in neighbouring fields.

Contents

Fully Invariant Relations

We started a systematic study of varieties of completely regular semigroups and the associated lattice $\mathcal{L}(\mathcal{CR})$ of varieties of completely regular semigroups, in [PR1, Chapter VIII] by considering the varieties of completely simple semigroups. Ibid Chap. IX on Mal'cev products serves as an introduction to the study of general varieties of completely regular semigroups. The present chapter aims to resume this introduction by considering fully invariant congruences on a free unary semigroup as well as on a free completely regular semigroup.

This chapter begins with a characterization of Green's relations on a free completely regular semigroup. The relationship between fully invariant congruences and varieties is then used to transfer results concerning the lattice of congruences on the free unary semigroup U or the free completely regular semigroup CR_X to the lattice of varieties of completely regular semigroups. As the arsenal of knowledge of congruences includes the kernel and trace relations as its primary tool, it is most natural to transfer some of this onto the lattice of varieties via the usual lattice antiisomorphism. By means of the apparatus characteristic to words over a set, such as the content, the length, and so on, a relation $\overline{\rho}$ derived from a fully invariant congruence ρ on U is proved to be itself a fully invariant congruence and that in $\mathcal{L}(\mathcal{CR})$, $\overline{\rho}$ corresponds to $\mathcal{V}^K$ where $\mathcal{V}$ corresponds to ρ. Any fully invariant congruence ρ on U gives rise to a relation ρ^s which, with some restrictions, also turns out to be itself a fully invariant congruence, now corresponding to $\mathcal{V}_{T_r}$ where $\mathcal{V}$ corresponds to ρ. The formulations of $\overline{\rho}$ and ρ^s are critical in establishing that the mappings $\mathcal{V} \mapsto \mathcal{V}_K$ and $\mathcal{V} \mapsto \mathcal{V}^{T_r}$ are both complete endomorphisms of $\mathcal{L}(\mathcal{CR})$.

It is a rare occurrence in semigroups that we find a large lattice of commuting congruences. So it may come as a surprize that fully invariant congruences on a free completely simple semigroup on a countably infinite set commute. Even more astonishing is the fact that fully invariant congruences under $\mathcal{D}$ on a free completely regular semigroup on a countably infinite set also commute. It is, however, not the case that on the latter semigroup all fully

© The Author(s), under exclusive license to Springer Nature Switzerland AG 2024
M. Petrich and N. R. Reilly, *Completely Regular Semigroup Varieties*, Synthesis Lectures on Mathematics & Statistics, https://doi.org/10.1007/978-3-031-42891-3_1

invariant congruences commute. Nonetheless the lattice of fully invariant congruences on a free completely regular semigroup on a countably infinite set is arguesian and therefore $\mathcal{L}(\mathcal{CR})$ is also arguesian.

Fully invariant congruences play their most important role in their relationship to varieties. This fact begins to emerge in this chapter but will be fully evident in the next.

Throughout this chapter, let $X = \{x_1, x_2, x_3, \ldots\}$ **denote a fixed countably infinite set,** $P = X \cup \{(,)^{-1}\}$, **where (and** $)^{-1}$ **are new variables and** $U = U_X$ **denote the free unary semigroup on** X, as described in [PR1, Section I.10].

1.1   Free Completely Regular Semigroups

The frame that we usually impose on a semigroup of unknown structure is that of Green's relations. They give us a first inkling of the structure of the semigroup, or at least a rough idea of the divisibility relations between elements. In the spirit of [PR1, Section I.10], a free completely regular semigroup on X can be described as a quotient of U modulo a suitable congruence. We employ this to obtain a characterization of Green's relations.

We refer the reader back to [PR1, Sections I.9 and I.10] for much of the notation relating to words in U that we will be using in this chapter. In particular the definitions, for any $u \in U$, of $\sigma(u)$, the last variable from X to appear in u for the first time, $\varepsilon(u)$, the first variable from X to appear for the last time, $i(u)$, the word obtained from u by retaining only the first occurrence of each variable from X, $f(u)$, the word obtained from u by retaining only the last occurrence of each variable from X and $\#(u)$, the number of variables from X that appear in u, remain unchanged in meaning and *refer only to elements of* X. However, we need some new notation and we also must adapt some of the earlier notation to the context of U where the parenthetical variables (and $)^{-1}$ require special treatment.

Notation 1.1.1 For any $u \in P^+$, let

$$u^0 = u(u)^{-1},$$

$\overline{u} =$ the word obtained from u by deleting all unmatched parentheses,

$h(u) =$ the *head* of u is the first variable from X occurring in u,

$t(u) =$ the *tail* of u is the last variable from X occurring in u,

$c(u) =$ the *content* of u is the set of variables from X appearing in u,

$\#(u) = |c(u)|$

$|u| =$ the *length* of u is the number of variables from X

appearing in u counting multiplicities,

$|u|_P =$ the *P-length* of u is the number of variables from $X \cup \{(,)^{-1}\}$

appearing in u counting multiplicities,

$s(u) = \overline{w}$, where w is the longest initial segment of u containing all but one
of the variables in $c(u)$,

$e(u) = \overline{w}$, where w is the longest final segment of u containing all but one
of the variables in $c(u)$,

$\gamma(u) = \overline{w}$, where w is the shortest initial segment of u with $c(u) = c(w)$,

$\delta(u) = \overline{w}$, where w is the shortest final segment of u with $c(u) = c(w)$.

The words $\gamma(u)$ and $\delta(u)$ are the *left* and *right indicators* of u, respectively.

The application of the operator $s(-)$ to elements of U requires special care in regard to the treatment of the parentheses " ("and")$^{-1}$". Consider the element $w = ((x)^{-1}y)^{-1}z \in U$ where $x, y, z \in X$. Then $c(w) = \{x, y, z\}$ and the longest initial segment of w containing all but one of the variables in $c(w)$ is $((x)^{-1}y)^{-1}$. Since this contains no unmatched parentheses, we have

$$s(w) = \overline{((x)^{-1}y)^{-1}} = ((x)^{-1}y)^{-1}.$$

On the other hand, $c(s(w)) = \{x, y\}$ so that the longest initial segment of $s(w)$ containing all but one of the variables in $c(s(w))$ is $((x)^{-1}$ which contains an unmatched parenthesis. Hence

$$s^2(w) = s(s(w)) = \overline{((x)^{-1}} = (x)^{-1}.$$

Generally speaking, we will simply write u^{-1} for $(u)^{-1}$ and, in particular, x^{-1} for $(x)^{-1}$. It is easy to see that for $p \in P^+, u \in U$,

$$c(p) = c(\overline{p}),$$
$$\gamma(u) = s(u)\,\sigma(u), \quad \delta(u) = \varepsilon(u)\,e(u), \quad \gamma^2 = \gamma, \quad \delta^2 = \delta.$$

A subset A of (respectively, a relation ρ on) a semigroup or unary semigroup S is *fully invariant* if $A\theta \subseteq A$ (respectively, $a\,\rho\,b \Rightarrow a\theta\,\rho\,b\theta$) for all endomorphisms θ of S. We denote by $\Gamma(S)$ the set of fully invariant (unary) congruences on S. In any group G, congruences correspond to normal subgroups. Consequently, in that context we denote the lattice of fully invariant subgroups of G by $\Gamma(G)$.

Since we will be considering U as a unary semigroup, unless we explicitly say otherwise, a congruence on U will always mean a *unary congruence* on U and a homomorphism of U will always mean a unary homomorphism of U.

Theorem 1.1.2 (i) *The semigroup congruence ζ on U generated by the set of all pairs of the form*

$$\left(u, uu^{-1}u\right), \quad \left(uu^{-1}, u^{-1}u\right), \quad \left(u, \left(u^{-1}\right)^{-1}\right) \quad (u \in U) \tag{1.1.1}$$

is the least unary completely regular congruence on U. Moreover, $(u^{-1})\zeta = (u\zeta)^{-1}$. *Consequently,* $(U/\zeta, \iota)$, *where* $\iota : x \mapsto x\zeta$ $(x \in X)$, *is a free completely regular semigroup on X.*

(ii) ζ *is a fully invariant congruence on U.*

Proof (i) For any $u \in U$, we have from (1.1.1) that

$$u\zeta = (u\zeta)(u^{-1}\zeta)(u\zeta), \quad (u\zeta)(u^{-1}\zeta) = (u^{-1}\zeta)(u\zeta),$$

and also that

$$(u^{-1})\zeta = (u^{-1}\zeta)((u^{-1})^{-1}\zeta)(u^{-1}\zeta) = (u^{-1}\zeta)(u\zeta)(u^{-1}\zeta).$$

Thus $(u^{-1})\zeta$ is an inverse of $u\zeta$ that commutes with $u\zeta$. Hence every element of U/ζ is completely regular, whence U/ζ is itself completely regular. By [PR1, Lemma II.2.1], we also have $(u^{-1})\zeta = (u\zeta)^{-1}$. It follows that ζ is indeed a unary congruence on U.

On the other hand, if ρ is any unary congruence on U such that U/ρ is completely regular then, for all $u \in U$, we must have that $(u^{-1})\rho = (u\rho)^{-1}$, which yields $(uu^{-1}u)\rho = (u\rho)(u\rho)^{-1}(u\rho) = u\rho$ so that $u \rho uu^{-1}u$. Similarly, $uu^{-1} \rho u^{-1}u$ and $u \rho (u^{-1})^{-1}$. Hence $\zeta \subseteq \rho$ and ζ must be the least unary completely regular congruence on U. Thus, in the notation of [PR1, Section I.8], $\zeta = \rho_{\mathcal{CR}}$. The final assertion is a consequence in [PR1, Proposition I.8.15].

(ii) Clearly the class $\mathcal{U}$ of all unary semigroups, that is semigroups endowed with a unary operation $a \to a^{-1}$, is a variety containing $\mathcal{CR}$ as a subvariety and with U as the free object on X in $\mathcal{U}$. Consequently, by [PR1, Theorem I.8.14] and part (i), ζ is fully invariant. □

Notation 1.1.3 Henceforth we denote by ζ the least completely regular unary congruence on U and write $\mathrm{CR}_X = U/\zeta$. We also write CR_n for $\mathrm{CR}_{\{x_1,...,x_n\}}$.

This characterization of CR_X provides some latitude in how we define varieties since it is clear that if $u, u', v, v' \in U$ are such that $u \zeta u'$ and $v \zeta v'$, then the identities $u = v$ and $u' = v'$ define the same varieties; that is, $[u = v] = [u' = v']$.

Any (unary) homomorphism of U into a unary semigroup V is uniquely determined by the image of the free generators (of U) in X and any homomorphism of CR_X into a completely regular semigroup is likewise uniquely determined by the image of the free generators (of CR_X) in X. Because CR_X is a unary quotient of U, it is a straightforward exercise to show that every (unary) homomorphism (respectively, endomorphism) of U induces a homomorphism (respectively, endomorphism) of CR_X. Conversely, every endomorphism of CR_X is induced by an endomorphism of U.

Corollary 1.1.4 $CR_{\{x\}}$ *is an infinite cyclic group generated by* x.

Proof The proof is left as an exercise.   □

The next observation is elementary but fundamental.

Lemma 1.1.5 *Let* $u, v \in U$ *and* $u \zeta v$. *Then* $c(u) = c(v)$.

Proof For any $w \in U$ we clearly have

$$c(w) = c(ww^{-1}w) = c(ww^{-1}) = c(w^{-1}w) = c\big((w^{-1})^{-1}\big).$$

The claim then follows from the description of the semigroup congruence generated by a relation in [PR1, Lemma I.5.2].   □

A simple refinement of the argument in Theorem 1.1.2 yields a useful characterization of the least congruence on U corresponding to any subvariety $\mathcal{V}$ of $\mathcal{CR}$.

Lemma 1.1.6 *For any* $\mathcal{V} = [u_\alpha = v_\alpha]_{\alpha \in A} \in \mathcal{L}(\mathcal{CR})$, *the least unary* $\mathcal{V}$-*congruence on* U *is the semigroup congruence generated by all pairs of the form*

$$\big(u, uu^{-1}u\big), \quad \big(uu^{-1}, u^{-1}u\big), \quad \big(u, (u^{-1})^{-1}\big), \quad \big(u_\alpha\varphi, v_\alpha\varphi\big)$$

where $u \in U$, $\alpha \in A$, $\varphi \in \mathcal{E}(U)$, *where* $\mathcal{E}(U)$ *denotes the semigroup of unary endomorphisms of* U.

Proof The proof is left as an exercise.   □

Lemma 1.1.7 *Let* w *be an initial segment of* $u \in U$.

(i) $u \zeta \overline{w}v$ *for some* $v \in U$.
(ii) $u \zeta s(u)\sigma(u)v$ *for some* $v \in U$.
(iii) $u \zeta \overline{w}(\overline{w})^{-1}u$.

Proof (i) Let $u = wt$ where $t \in P^*$. If $w \in U$, then $\overline{w} = w$ and, by [PR1, Lemma I.10.3], the claim follows. So suppose that $w \notin U$. By [PR1, Lemma I.10.1], this means that w must contain an unmatched parenthesis (. Let $w = p (q$ where p contains no unmatched parentheses so that $p \in U \cup \{\emptyset\}$, $q \in P^+$. Since $u \in U$, there must be a parenthesis $)^{-1}$ to match the displayed parenthesis. In other words, there is an initial segment $r)^{-1}$ of t, ending in a parenthesis $)^{-1}$ which matches the unmatched parenthesis in $p (q$. Thus $p(qr)^{-1}$ is an initial segment of u where $qr \in U$. Let $s \in P^*$ be such that $t = r)^{-1}s$.

Then $u = p(qr)^{-1}s \zeta pqr(qr)^{-2}s$, where pq is obtained from $w = p (q$ by deleting one unmatched parenthesis. Clearly, $pqr(qr)^{-2}s \in U$. Hence by induction on the number of unmatched parentheses in w, there exists $v \in P^+$ such that $u \zeta \overline{w}v$ and $\overline{w}v \in U$. Since $\overline{w} \in U$, it follows that v is also an element of U and the claim is established.

(ii) This follows from part (i) by choosing w to be the shortest initial segment of u containing all the variables. Then $\overline{w} = s(u)\sigma(u)$.

(iii) By part (i), $u \zeta \overline{w}v$ for some $v \in U$ so that $u \zeta \overline{w}v \zeta \overline{w(\overline{w})}^{-1}\overline{w}v \zeta \overline{w(\overline{w})}^{-1}u$. □

An important feature of the operators s and σ is that when they are applied to ζ-related elements they yield elements that are also ζ-related.

Lemma 1.1.8 *Let $u, u' \in U$ and be such that $u \zeta u'$. Then $\sigma^k(u) = \sigma^k(u')$ and $s^k(u) \zeta s^k(u')$ for all positive integers k.*

Proof Assume first that $k = 1$. Let $\sigma(u) = x$ and $s(u) = w$. It suffices to establish the claim when u' is obtained from u by a single elementary ζ-transition involving one of the pairs $(v, vv^{-1}v)$, $(vv^{-1}, v^{-1}v)$, $(v, (v^{-1})^{-1})$ for some $v \in U$.

Let $u = ab$ where $a, b \in P^*$ and a is the shortest initial segment of u with $c(a) = c(u)$. Then $a = a_1x$ for some $a_1 \in P^*$ and $s(u) = \overline{a_1x} = \overline{a_1}x$.

Let u' be obtained from u by means of a substitution of the form $v \to vv^{-1}v$.

First consider the case where v is a subword of a not involving x. Then v must be a subword of a_1 and we have $a_1 = a_2va_3$ for some $a_2, a_3 \in P^*$. All parentheses in v are matched. Hence $s(u) = \overline{a_1} = a_2^*va_3^*$ for some $a_2^*, a_3^* \in P^*$.

Under the substitution $v \to vv^{-1}v$ we have

$$u = ab = a_2va_3xb \to u' = a_2vv^{-1}va_3xb.$$

All parentheses in $v^{-1}v$ are matched so that $s(u') = \overline{a_2vv^{-1}va_3} = a_2^*vv^{-1}va_3^*$. Thus $s(u')$ is also obtained from $s(u)$ by the substitution $v \to vv^{-1}v$. Therefore $s(u) \zeta s(u')$. We also see from the above discussion that $\sigma(u) = \sigma(u') = x$ in this case.

If v is a subword of b or overlaps a and b, then the left indicator remains unchanged. A similar discussion applies if the substitution is of the form $vv^{-1}v \to v$.

The only additional observation required to make the same argument apply to substitutions of the form $vv^{-1} \leftrightarrow v^{-1}v$ and $v \leftrightarrow (v^{-1})^{-1}$ is that for any initial segment d of v, the word $(d$ is an initial segment of $v^{-1} = (v)^{-1}$ and $((d$ is an initial segment of $((v)^{-1})^{-1}$ where $\overline{d} = \overline{(d} = \overline{((d}$. The only initial segment of $(v)^{-1}$ not obtained in this way is $(v)^{-1}$ and the only initial segments of $((v)^{-1})^{-1}$ not obtained are $((v)^{-1}$ and $((v)^{-1})^{-1}$.

The remainder of the proof of the case $k = 1$ then follows the same pattern as that for the first case and is left as an exercise. The case $k > 1$ follows by an obvious induction argument. □

One consequence of Lemma 1.1.8 is that we can view s as an operator on CR_X by defining $s(u\zeta) = s(u)\zeta$ for any $u \in U$. In the same spirit, if $u, u', v, v' \in U$ and $u\,\zeta\,u'$, $v\,\zeta\,v'$ then $s^k(u)\,\zeta\,s^k(u')$ and $s^k(v)\,\zeta\,s^k(v')$ so that the identities $s^k(u) = s^k(v)$ and $s^k(u') = s^k(v')$ define the same varieties; that is, $[s^k(u) = s^k(v)] = [s^k(u') = s^k(v')]$. Consequently, we do not need to be overly particular whether we apply s to an element of U or the corresponding class in CR_X.

We now come to the characterizations of Green's relations on CR_X. For such a complicated structure, these turn out to be relatively simple.

Theorem 1.1.9 *Let $u, v \in U$.*

(i) *$u\zeta\ \mathcal{D}\ v\zeta$ if and only if $c(u) = c(v)$.*
(ii) *$u\zeta\ \mathcal{L}\ v\zeta$ if and only if $\varepsilon(u) = \varepsilon(v)$ and either $e(u)\ \zeta\ e(v)$ or $e(u) = \emptyset = e(v)$.*
(iii) *$u\zeta\ \mathcal{R}\ v\zeta$ if and only if $\sigma(u) = \sigma(v)$ and either $s(u)\ \zeta\ s(v)$ or $s(u) = \emptyset = s(v)$.*

Proof (i) *Necessity*: We have

$$u\zeta\ \mathcal{D}\ v\zeta \implies \text{there exist } p, q \in U \text{ such that } u\zeta = p\zeta v\zeta q\zeta = (pvq)\zeta$$
$$\implies c(u) = c(pvq) \quad \text{by Lemma 1.1.5}$$
$$\implies c(v) \subseteq c(u).$$

Similarly, $c(u) \subseteq c(v)$ and equality prevails.

Sufficiency: Let $u, v \in U$ and $c(u) = c(v)$. Since $\mathcal{D}$ is a semilattice congruence on CR_X, we have, for all $x, y \in U$, that

$$x\zeta\ \mathcal{D}\ x^2\zeta, \quad (x\zeta)(y\zeta)\ \mathcal{D}\ (y\zeta)(x\zeta), \quad (x)^{-1}\zeta\ \mathcal{D}\ x\zeta$$

from which it clearly follows that $u\zeta\ \mathcal{D}\ v\zeta$.

(ii) and (iii): These cases are dual to each other. So we consider only (iii).

Necessity: By part (i), we have $c(u) = c(v)$. Let $t \in U^1$ be such that $u\zeta = (v\zeta)(t\zeta) = (vt)\zeta$. Then, by Lemma 1.1.5 $c(vt) = c(u) = c(v)$ so that by Lemma 1.1.8, $\sigma(u) = \sigma(vt) = \sigma(v)$ and if $s(u) \neq \{\emptyset\}$ then $s(u)\ \zeta\ s(vt) = s(v)$.

Sufficiency: If $s(u) = s(v) = \emptyset$, then $c(u) = c(v) = \{x\}$ for some $x \in X$ and $u\ \mathcal{H}\ v$ by Corollary 1.1.4. So suppose that $s(u), s(v) \in U$ and that $s(u)\ \zeta\ s(v)$. Since $c(u) = c(s(u)\sigma(u))$, we have $u\zeta\ \mathcal{D}\ s(u)\sigma(u)\zeta$ by part (i). By Lemma 1.1.7(ii) this implies that $u\ \zeta\ s(u)\sigma(u)(s(u)\sigma(u))^{-1}u$ so that, by [PR1, Corollary II.4.3(ii)], $u\zeta\ \mathcal{R}\ (s(u)\sigma(u))\zeta$. Similarly, $v\zeta\ \mathcal{R}\ (s(v)\sigma(v))$. Thus, by the hypothesis, we have $u\zeta\ \mathcal{R}\ v\zeta$. □

One consequence of Theorem 1.1.9 is that, if $|c(u)| > 1$, then $D_{u\zeta}$ has infinitely many distinct $\mathcal{L}$- and $\mathcal{R}$-classes.

Corollary 1.1.10 *Let $u, v \in U$ be such that $c(u) = c(v)$. Let $s = \gamma(u)$ and $t = \delta(v)$. Then*

$$H_{(uv)\zeta} = \{(swt)\zeta \mid w \in U, c(w) \subseteq c(u)\}. \tag{1.1.2}$$

In particular, for any $x \in X$, we have $H_{x\zeta} = \{w\zeta \mid c(w) = \{x\}\}$.

Proof Let A denote the set on the right hand side of (1.1.2). By Theorem 1.1.9(ii)(iii), $A \subseteq H_{(uv)\zeta}$. Conversely, since $H_{(uv)\zeta}$ is a subgroup with identity $((uv)\zeta)^0 = (uv(uv)^{-1})\zeta = ((uv)^{-1}uv)\zeta$, it follows that, for any $w\zeta \in H_{(uv)\zeta}$, we have

$$w\zeta = ((uv)^0 w (uv)^0)\zeta = (uv(uv)^{-1} w (uv)^{-1} uv)\zeta \in A. \qquad \square$$

The characterizations of Green's relations on CR_X above already provide sufficient information regarding the structure of CR_X to enable us to derive some important basic information concerning congruences on CR_X.

Corollary 1.1.11 *On CR_X, we have $\mu = \mathcal{H}^0 = \mathcal{L}^0 = \mathcal{R}^0 = \tau = \epsilon$.*

Proof By duality and since $\mu = \mathcal{H}^0 = \mathcal{L}^0 \cap \mathcal{R}^0$, it suffices to show that $\mathcal{L}^0 = \epsilon$. Suppose that $u, v \in U$ are such that $u\zeta \, \mathcal{L}^0 \, v\zeta$. Then $c(u) = c(v) = \{x_1, \ldots, x_n\}$, say. Let $z \in X \setminus c(u)$, $u' = zu$ and $v' = zv$. Since $\mathcal{L}^0$ is a congruence, we have $(u'\zeta, v'\zeta) \in \mathcal{L}^0 \subseteq \mathcal{L}$ so that, by Theorem 1.1.9, $u = e(u') \, \zeta \, e(v') = v$. Thus $\mathcal{L}^0 = \epsilon$. The proof of the claim that $\tau = \epsilon$, will follow from Theorem 2.4.3(xii) which is to come below. $\qquad \square$

Notation 1.1.12 Let η denote the least semilattice congruence on U.

Clearly η/ζ will then be the least semilattice congruence on CR_X. It should cause no confusion if we also write η for the least semilattice congruence on CR_X.

Corollary 1.1.13 *Let $u, v \in U$. Then $u \, \eta \, v \Leftrightarrow c(u) = c(v)$.*

Proof This follows from [PR1, Corollary II.1.5] and Theorem 1.1.9 (i). $\qquad \square$

The existence of the free completely regular semigroup on a set X was first established by McAlister [1]. The main results in this section are due to Clifford [5]. Corollary 1.1.11 is due to Pastijn and Trotter [1]. Alternative models for CR_X were obtained by Gerhard [7,8] and Trotter [9]. Clifford [5] gives explicit descriptions of the free completely regular semigroups on one and two generators and also describes the free objects in the variety of completely simple semigroups. Ajan [1] studied presentations of completely regular semigroups by generators and relations; see also Ajan and Pastijn [1]. Casimiro and Skapinakis [1] present alternative bases of identities for various subvarieties of $\mathcal{CR}$ such as $\mathcal{CR}$ itself as well as the subvarieties of cryptogroups and orthogroups employing $x \leftrightarrow x^0$ as an additional unary operation. Jones and Trotter [1] study the Howson property for free objects in varieties of completely regular semigroups.

1.2   Fully Invariant Congruences

We gather here some elementary observations concerning fully invariant congruences related to equivalence relations in certain ways.

Definition 1.2.1 We will refer to the antiisomorphism $\pi : \mathcal{V} \mapsto \rho_{\mathcal{V}}$ between the lattice $\mathcal{L}(\mathcal{CR})$ and $\Gamma(\mathrm{CR}_X)$ (see [PR1, Theorem I.8.14]) as the *standard antiisomorphism* . This correspondence will play a major role in the discussions to follow in this and subsequent chapters.

From Theorem 1.1.2(ii), we know that ζ is a fully invariant congruence and clearly ω (the universal congruence) and ι (the identity congruence) are fully invariant congruences on U and $E(S)$ is a fully invariant subset of any completely regular semigroup S.

Notation 1.2.2 Let Δ denote the interval $[\zeta, \omega]$ in $\Gamma(U)$ and let $\Gamma = \Gamma(\mathrm{CR}_X)$.

The correspondence between Δ and Γ is of fundamental importance.

Lemma 1.2.3 *The mapping $\rho \mapsto \rho/\zeta$ $(\rho \in \Delta)$ is a lattice isomorphism of Δ onto Γ.*

Proof The proof is left as an exercise.   □

Lemma 1.2.4 *Let $S \in \mathcal{CR}$ and $\rho \in \mathcal{C}(S)$.*

(i) *ρ is fully invariant if and only if both $\ker \rho$ and $\mathrm{tr}\, \rho$ are fully invariant.*
(ii) *$\Gamma(S)$ is a complete sublattice of $\mathcal{C}(S)$.*

Proof (i) This follows immediately from [PR1, Lemma VI.3.1].
(ii) The proof is left as an exercise.   □

As might be expected, some of the most natural relations on a semigroup are fully invariant.

Lemma 1.2.5 *For any $S \in \mathcal{CR}$ the binary relations $\leq$, $\leq_\ell$, $\leq_r$, $\leq_d$ and the relations $\mathcal{H}$, $\mathcal{L}$, $\mathcal{R}$ and $\mathcal{D}$ are fully invariant.*

Proof This follows easily from their respective definitions in [PR1, Sections II.4 and I.7]. $\square$

Recall that for any relation ρ on a semigroup, ρ^* denotes the least congruence containing ρ while, if ρ is an equivalence relation, then ρ^0 is the largest congruence contained in ρ.

Lemma 1.2.6 *Let ρ be a fully invariant relation on a semigroup S. Then $\rho^* \in \Gamma(S)$.*

Proof Let $a, b \in S$, $a \rho^* b$ and φ be an endomorphism of S. Then there exist $c_i, d_i \in S^1$, $p_i, q_i \in S$, with $p_i \rho q_i$ or $q_i \rho p_i$ for $i = 1, \ldots, n$ and

$$a = c_1 p_1 d_1, \ c_1 q_1 d_1 = c_2 p_2 d_2, \ldots, c_n q_n d_n = b.$$

Consequently, we have

$$a\varphi = (c_1\varphi)(p_1\varphi)(d_1\varphi)$$
$$(c_1\varphi)(q_1\varphi)(d_1\varphi) = (c_2\varphi)(p_2\varphi)(d_2\varphi)$$
$$\vdots$$
$$(c_n\varphi)(q_n\varphi)(d_n\varphi) = b\varphi$$

where either $(p_i\varphi) \rho (q_i\varphi)$ or $(q_i\varphi) \rho (p_i\varphi)$ since ρ is fully invariant. Therefore, $a\varphi \ \rho^* \ b\varphi$ and ρ^* is fully invariant. $\square$

Corollary 1.2.7 *For any semigroup S and any of Green's relation $\mathcal{K}$, the relation $\mathcal{K}^*$ is a fully invariant congruence.*

Proof This is an immediate consequence of Lemmas 1.2.5 and 1.2.6. $\square$

The next result will be important when we consider congruences of the form ρ^K (as defined in [PR1 Section VII.i]).

Lemma 1.2.8 *Let ρ be a fully invariant equivalence relation on a semigroup S. Let S have the following property:*

> *for any endomorphism φ of S and any $a, b \in S$,*
> *there exists an epimorphism φ^* of S onto S*
> *with $a\varphi = a\varphi^*$ and $b\varphi = b\varphi^*$.*   (*)

Then ρ^0 is fully invariant.

Proof Let $a, b \in S, a \rho^0 b$ and φ be any endomorphism of S. By condition $(*)$, there exists an epimorphism φ^* of S onto S with $a\varphi = a\varphi^*, b\varphi = b\varphi^*$. If S lacks an identity, then extend φ^* to an endomorphism of S^1 by defining $1\varphi^* = 1$. For any elements $s, t \in S^1$ there exist, by the hypothesis on φ^*, elements $s', t' \in S$ such that $s'\varphi^* = s, t'\varphi^* = t$. Since $a \rho^0 b$, we have that $s'at' \rho s'bt'$. But ρ is fully invariant and so

$$s(a\varphi)t = (s'\varphi^*)(a\varphi^*)(t'\varphi^*) = (s'at')\varphi^* \ \rho \ (s'bt')\varphi^*$$
$$= (s'\varphi^*)(b\varphi^*)(t'\varphi^*) = s(b\varphi)t.$$

Therefore $a\varphi \rho^0 b\varphi$ and ρ^0 is also fully invariant.   □

The most important applications of Lemma 1.2.8 will come in the context of relatively free objects.

Lemma 1.2.9 *Let* $\mathcal{V} \in \mathcal{L}(\mathcal{CR})$ *and* F *be a free object in* $\mathcal{V}$ *on* X. *Then* F *satisfies the condition* $(*)$ *of Lemma* 1.2.8.

Proof Let $a, b \in F$ and φ be any endomorphism of F. Since F is generated by X, there exists a finite subset Y of X such that a and b are elements of the completely regular subsemigroup of F generated by Y. Let φ^* be any bijection of $X \setminus Y$ onto X. Extend φ^* to a mapping of X into F by prescribing that $y\varphi^* = y\varphi$ for all $y \in Y$. Then φ^* extends to a unique endomorphism of F. Since φ^* agrees with φ on Y, it follows that $a\varphi^* = a\varphi$ and $b\varphi^* = b\varphi$. In addition φ^* maps $X \setminus Y$ onto X and X generates F. Hence φ^* is an epimorphism.   □

Corollary 1.2.10 *Let* $\mathcal{V} \in \mathcal{L}(\mathcal{CR})$, F *be a free object in* $\mathcal{V}$ *on* X *and* $\mathcal{K}$ *be any of Green's relations. Then* $\mathcal{K}^0$ *is a fully invariant congruence on* F.

Proof This follows immediately from Lemmas 1.2.5, 1.2.8, and 1.2.9.   □

Theorem 1.2.11 *Let* ρ *be a fully invariant congruence on a completely regular semigroup* S, $P \in \{K, K_\ell, K_r, T, T_\ell, T_r\}$ *and* $Q \in \{T, T_\ell, T_r\}$. *Then* ρ_P *is fully invariant. If* S *satisfies the condition* $(*)$ *in Lemma 1.2.8, then* ρ^Q *is fully invariant. In particular, for any fully invariant congruence* ρ *on* CR_X, *we have that* ρ_P *and* ρ^Q *are also fully invariant congruences.*

Proof By Lemmas 1.2.4 and 1.2.5, $\mathcal{H}$, $\leq_\ell$ and $\leq_r$ and tr ρ are all fully invariant. It follows that $\rho \cap \mathcal{H}$, $\rho \cap \leq_\ell$, $\rho \cap \leq_r$ are also fully invariant. By [PR1, Theorems VII.1.2, VII.2.2 and VII.4.2 and its dual], we have

$$\rho_K = (\rho \cap \mathcal{H})^*, \quad \rho_T = (\operatorname{tr} \rho)^*, \quad \rho_{T_\ell} = (\rho \cap \leq_r)^*, \quad \rho_{T_r} = (\rho \cap \leq_\ell)^*.$$

By Lemma 1.2.6, each of these is fully invariant. The assertions concerning K_ℓ and K_r now follow from Lemma 1.2.4(ii) and [PR1, Lemma VII.4.13].

Now let S satisfy the condition (∗). Since $\mathcal{H}$, $\mathcal{L}$ and $\mathcal{R}$ are fully invariant relations, it follows easily that $\rho \vee \mathcal{H}$, $\rho \vee \mathcal{L}$ and $\rho \vee \mathcal{R}$ are also fully invariant. By [PR1, Theorem I.2.8], it now follows that ρ^T, ρ^{T_ℓ}, ρ^{T_r} are all fully invariant. □

Notable absentees from the list in Theorem 1.2.11 are ρ^K, ρ^{K_ℓ} and ρ^{K_r}. These omissions are partially addressed in the next result.

Theorem 1.2.12 *Let $\lambda \in \Delta$, $\rho = \lambda/\zeta$, and $u, v \in U$. Then the following are equivalent.*

(i) $u\zeta \, \rho^K \, v\zeta$.
(ii) $(puq(pvq)^{-1})\zeta \in \ker \rho$ *for all $p, q \in U$.*
(iii) $wxuxw \, \lambda \, wxvxw$ *for all (some) $w \in U$ with $c(uv) \subseteq c(w)$ and for $x \in X \setminus c(w)$.*
(iv) $txuxt \, \lambda \, txvxt$ *where $t = (yx_1 \ldots x_n y)^0 y$, $c(uv) = \{x_1, \ldots, x_n\}$, $x \neq y$ and $x, y \in X \setminus c(uv)$.*

Proof (i) *implies* (ii). For any $p, q \in U$ we have

$$u\zeta \, \rho^K \, v\zeta \implies (puq)\zeta \, \rho^K \, (pvq)\zeta$$
$$\implies (puq)(pvq)^{-1}\zeta \in \ker \rho^K = \ker \rho = \ker \rho.$$

(ii) *implies* (iii). We will show that (ii) implies the stronger version of (iii) with the formulation "for all". Let $w \in U$ be such that $c(uv) \subseteq c(w)$ and let $x \in X \setminus c(w)$. Then with $p = wx, q = xw$ we can deduce from (ii) that

$$(wxuxw(wxvxw)^{-1})\zeta \in \ker \rho. \tag{1.2.1}$$

From the hypotheses on w and x, we have

$$\sigma(wxuxw) = x = \sigma(wxvxw), \qquad \varepsilon(wxuxw) = x = \varepsilon(wxvxw),$$
$$s(wxuxw) = s(w) = s(wxvxw), \quad e(wxuxw) = e(w) = e(wxvxw).$$

By Theorem 1.1.9(ii)(iii), $(wxuxw)\zeta \, \mathcal{H} \, (wxvxw)\zeta$ so that

$$((wxuxw)\zeta)^0 = ((wxvxw)\zeta)^0$$

and trivially

$$\left(((wxuxw)\zeta)^0, ((wxvxw)\zeta)^0\right) \in \operatorname{tr} \rho. \tag{1.2.2}$$

By (1.2.1) and (1.2.2) we now have that $(wxuxw)\zeta \; \rho \; (wxvxw)\zeta$. Equivalently

$$wxuxw \; \lambda \; wxvxw,$$

as required.

The condition in part (iv) is clearly a special case of the strong condition in part (iii). On the other hand, for any $w \in U$ with $c(uv) \subseteq c(w)$, the substitution $y \to w$ transforms the condition in part (iv) into an instance of the weak condition in part (iii).

(iii) *implies* (i). (We will use the weaker version of (iii) with the formulation "for some".) Let τ denote the relation on U defined by: for $a, b \in U$,

$$a \; \tau \; b \; \Longleftrightarrow \; wxaxw \; \lambda \; wxbxw \quad \text{for some}$$
$$w \in U \quad \text{with} \quad c(ab) \subseteq c(w) \quad \text{and for some} \quad x \in X \setminus c(w).$$

We will show that τ is a congruence on U such that τ/ζ is K-related to ρ. Clearly τ is reflexive and symmetric. To see that τ is transitive, let $a, b, d \in U$ be such that $a \; \tau \; b \; \tau \; d$, say,

$$pxaxp \; \lambda \; pxbxp, \tag{1.2.3}$$

$$qybyq \; \lambda \; qydyq, \tag{1.2.4}$$

for some $p, q \in U$, $x, y \in X$ such that $c(ab) \subseteq c(p)$, $c(bd) \subseteq c(q)$, $x \notin c(p)$, $y \notin c(q)$. Let $r \in U$, $z \in X$ be such that $c(abd) \subseteq c(r)$, $z \notin c(abd)$. Define an endomorphism φ of U which maps all variables in $c(abd)$ identically and maps all other variables to zr^2z. Then from (1.2.3) and the fact that λ is fully invariant, we have $p'zrzazrzp' \; \lambda \; p'zrzbzrzp'$ for some $p' \in U$ with $c(p') \subseteq c(r) \cup \{z\}$. Premultiplying by $rz(p'zrz)^{-1}$ and postmultiplying by $(zrzp')^{-1}zr$, we obtain

$$rz(p'zrz)^{-1}(p'zrz)a(zrzp')(z \; rzp')^{-1}zr$$
$$\lambda \; rz(p'zrz)^{-1}(p'zrz)b(zrzp')(zrzp')^{-1}zr$$

or

$$rz(p'zrz)^0a(zrzp')^0zr \; \lambda \; rz(p'zrz)^0b(zrzp')^0zr$$

which, by Theorem 1.1.9(i)(ii), reduces to $rzazr \; \lambda \; rzbzr$. Applying the same argument to (1.2.4) we obtain $rzbzr \; \lambda \; rzdzr$. By the transitivity of λ we then have $rzazr \; \lambda \; rzdzr$ where $c(ad) \subseteq c(r)$ and $z \in X \setminus c(ad)$. Thus $a \; \tau \; d$ and τ is transitive.

Now let a, b, p, x be as above and $d \in U$. Substituting dz for x in (1.2.3), where $z \in X \setminus c(abd)$, we obtain $pdzadzp \; \lambda \; pdzbdzp$ which, multiplied on the right by d, yields $(pd)zadz(pd) \; \lambda \; (pd)zbdz(pd)$ where $c(abd) \subseteq c(pd)$, $z \notin c(abd)$. Therefore $ad \; \tau \; bd$. Dually, $da \; \tau \; db$. Thus τ is a congruence.

By hypothesis $\zeta \subseteq \lambda$ and it is straightforward to see that $\lambda \subseteq \tau$, so that $\zeta \subseteq \tau$.

Next we wish to show that $\ker \tau/\zeta = \ker \rho$. Since $\lambda \subseteq \tau$ we have $\lambda/\zeta \subseteq \tau/\zeta$ and $K = \ker \rho = \ker \lambda/\zeta \subseteq \ker \tau/\zeta$. So let $g\zeta \in \ker \tau/\zeta$. We have

$$g\zeta \in \ker \tau/\zeta \implies g^{-1}\zeta \in \ker \tau/\zeta \implies g^{-1}\zeta \; \tau/\zeta \; g^{-2}\zeta$$

so that there exist $w \in U$, $x \in X$ such that $c(g) \subseteq c(w)$, $x \notin c(w)$ and

$$wxg^{-1}xw \; \lambda \; wxg^{-2}xw.$$

Hence

$$(gw)^{-1}gwxg^{-1}x(wg)(wg)^{-1} \; \lambda \; (gw)^{-1}gwxg^{-2}x(wg)(wg)^{-1}$$

or

$$(gw)^0 xg^{-1}x(wg)^0 \; \lambda \; (gw)^0 xg^{-2}x(wg)^0. \tag{1.2.5}$$

Now let φ denote the endomorphism of U defined by

$$\varphi(y) = \begin{cases} y & \text{if } y \in c(g) \\ g & \text{if } y = x \\ g^0 & \text{otherwise.} \end{cases}$$

Since λ is fully invariant, we get from (1.2.5) that

$$(gw')^0 gg^{-1}g(w'g)^0 \; \lambda \; (gw')^0 gg^{-2}g(w'g)^0$$

where $w' = \varphi(w)$ and $c(w') \subseteq c(g)$. By Theorem 1.1.9(ii)(iii), this yields $g\zeta \; \lambda/\zeta \; g^0\zeta$ so that $g\zeta \in \ker \lambda/\zeta$. Therefore we have $\ker \tau/\zeta \subseteq \ker \lambda/\zeta$ and equality follows:

$$\ker \rho = \ker \lambda/\zeta = \ker \tau/\zeta.$$

Hence $(\tau/\zeta)^K = \rho^K$. Finally assume that u and v satisfy the weaker form of (iii). Then $u \; \tau \; v$ so that $u\zeta \; (\tau/\zeta) \; v\zeta$ which implies that $u\zeta \; (\tau/\zeta)^K \; v\zeta$ and therefore $u\zeta \; \rho^K \; v\zeta$ and (i) holds. $\qquad\square$

Corollary 1.2.13 *Let $\rho \in \Gamma(\mathrm{CR}_X)$. Then ρ^K, ρ^{K_ℓ} and ρ^{K_r} are fully invariant.*

Proof Let $\lambda \in \Delta$ be such that $\rho = \lambda/\zeta$ and let τ be the congruence defined in the proof that (iii) implies (i) in Theorem 1.2.12. By the equivalence of the statements (i) and (iii) in Theorem 1.2.12 we have, for any $u, v \in U$, that

$$u\zeta \; \rho^K \; v\zeta \iff u \; \tau \; v \iff u\zeta \; \tau/\zeta \; v\zeta.$$

In other words $\rho^K = \tau/\zeta$. It is easily seen that τ is fully invariant from which it follows that ρ^K is also fully invariant.

The claims concerning ρ^{K_ℓ} and ρ^{K_r} now follow from the facts that ρ^K (just proved) and ρ^{T_r} (Theorem 1.2.11) are fully invariant together with [PR1, Lemmas VII.4.13, 2.4(ii)]. □

The equivalence of parts (i) and (ii) in Theorem 1.2.12, the full invariance of ρ_K, ρ_T (Theorem 1.2.11) as well as Corollary 1.2.13 are due to Pastijn and Trotter [1]. The equivalence of parts (i) and (iii) in Theorem 1.2.12 is due to Polák [1]. The rest of the section is mainly folklore. Fully invariant congruences play an important role in Pastijn [4], Pastijn and Trotter [1], Petrich and Reilly [1], Polák [1, 2, 3] and Reilly [2].

1.3   Kernel and Trace Relations

Here we transfer the upper and lower operators associated with the relations K, K_ℓ, K_r, T, T_ℓ and T_r from $\Gamma(\mathrm{CR}_X)$ to $\mathcal{L}(\mathcal{CR})$ via the standard lattice antiisomorphism of these two lattices. It is a general phenomenon that the more special the structure, the richer the properties. The main result asserts that the relation K on $\Gamma(\mathrm{CR}_X)$ is a complete congruence. Recalling from [PR1, Theorems VII.2.2 and VII.4.2] that T, T_ℓ and T_r are complete congruences on $\mathcal{C}(S)$ for any (completely) regular semigroup S and that K, K_ℓ and K_r are generally not $\vee$-congruences makes the case at hand appear quite remarkable. These relations play a seminal role in the theory of congruences on completely regular semigroups. We shall see in this section that their counterparts for varieties also play a fundamental role.

Notation 1.3.1 For any $\rho \in \Delta$ (respectively, $\rho \in \Gamma(\mathrm{CR}_X)$), let $[\rho]$ denote the variety of completely regular semigroups corresponding to ρ. If $\rho \in \Delta$, then clearly $[\rho] = [\rho/\zeta]$. For any $\mathcal{V} \in \mathcal{L}(\mathcal{CR})$, let $\zeta_\mathcal{V}$ (respectively, $\rho_\mathcal{V}$) denote the fully invariant congruence in Δ (respectively, $\Gamma(\mathrm{CR}_X)$) corresponding to $\mathcal{V}$. Note that, in particular, $\zeta_\mathcal{S} = \eta$.

In [PR1, Chapter VII], we introduced various kernel and trace relations on the congruence lattice of an arbitrary completely regular semigroup and developed their basic properties. From [PR1, Theorem I.8.14] we know that the mappings $\rho \mapsto [\rho]$, $\mathcal{V} \mapsto \zeta_\mathcal{V}$ are mutually inverse antiisomorphisms between Δ and $\mathcal{L}(\mathcal{CR})$ and that the mappings $\rho \to [\rho]$, $\mathcal{V} \to \rho_\mathcal{V}$ are mutually inverse antiisomorphisms between $\Gamma(\mathrm{CR}_X)$ and $\mathcal{L}(\mathcal{CR})$. We take advantage of the latter correspondence and the fact that any relation on $\mathcal{C}(\mathrm{CR}_X)$ restricts to a relation on $\Gamma(\mathrm{CR}_X)$. These antiisomorphisms make it natural to introduce the following symbolism.

Notation 1.3.2 Let $\Sigma = \{K, K_\ell, K_r, T, T_\ell, T_r\}$ where K, K_ℓ, K_r, T, T_ℓ, T_r denote the usual relations on $\mathcal{C}(\mathrm{CR}_X)$. Let $P \in \Sigma$. We will also denote by P:

(i) the restriction of P to $\Gamma(\mathrm{CR}_X)$,
(ii) the relation defined by

$$\mathcal{U} \ P \ \mathcal{V} \ \text{ if } \ \rho_{\mathcal{U}} \ P \ \rho_{\mathcal{V}} \quad (\mathcal{U}, \mathcal{V} \in \mathcal{L}(\mathcal{CR})).$$

(iii) the relation defined by

$$\lambda \ P \ \rho \ \text{ if } \ (\lambda/\zeta) \ P \ (\rho/\zeta) \quad (\lambda, \rho \in \Delta), \tag{1.3.1}$$

and will use the notation Σ for the sets of corresponding relations on Δ, $\Gamma(\text{CR}_X)$, and $\mathcal{L}(\mathcal{CR})$, as appropriate. The precise meaning should be clear from the context.

In contrast to the situation for arbitrary completely regular semigroups (see [PR1, Exercise VII.1.12(iii)]), the lower and upper operators associated with K are both order preserving on $\Gamma(\text{CR}_X)$.

Lemma 1.3.3 *Let* $\lambda, \rho \in \Gamma(\text{CR}_X)$ *and* λ_K, ρ_K, λ^K, ρ^K *be defined as in [PR1, Theorem VII.1.2]. Then* $\ker \lambda \subseteq \ker \rho \implies \lambda_K \subseteq \rho_K$, $\lambda^K \subseteq \rho^K$.

Proof The assertion relating to the first inclusion follows from [PR1, Corollary VII.1.3] and that concerning the second containment from the equivalence of parts (i) and (ii) in Theorem 1.2.12.   □

The next result is of major importance in the study of $\mathcal{L}(\mathcal{CR})$.

Theorem 1.3.4 *The relations in* Σ *are complete congruences on* Δ, $\Gamma(\text{CR}_X)$ *and* $\mathcal{L}(\mathcal{CR})$.

Proof Since, by Lemma 1.2.3, $\Gamma(\text{CR}_X)$ and Δ are isomorphic and $\Gamma(\text{CR}_X)$ and $\mathcal{L}(\mathcal{CR})$ are antiisomorphic, it suffices to establish the result for $\Gamma(\text{CR}_X)$. By [PR1, Theorems VII.2.2, VII.4.2 and its dual], we know that T, T_ℓ and T_r are complete congruences on $\mathcal{C}(\text{CR}_X)$ and it follows easily that their restrictions to $\Gamma(\text{CR}_X)$ must also be complete congruences.

With respect to K, let $\rho_\alpha \in \Gamma(\text{CR}_X)$ for $\alpha \in A$. Then $\rho_\alpha \subseteq \bigvee_{\alpha \in A} \rho_\alpha$ for $\alpha \in A$ and therefore, by Lemma 1.3.3, $(\rho_\alpha)^K \subseteq \left(\bigvee_{\alpha \in A} \rho_\alpha \right)^K$ $(\alpha \in A)$. Hence

$$\bigvee_{\alpha \in A} \rho_\alpha \subseteq \bigvee_{\alpha \in A} (\rho_\alpha)^K \subseteq \left(\bigvee_{\alpha \in A} \rho_\alpha \right)^K$$

from which we conclude that $\left(\bigvee_{\alpha \in A} \rho_\alpha \right) K \left(\bigvee_{\alpha \in A} \rho_\alpha^K \right)$ and therefore that K respects arbitrary joins. That K respects arbitrary meets was established in [PR1, Theorem VII.1.2]. This also implies the claim concerning K_ℓ and K_r.   □

In the light of Theorem 1.3.4, for any $\rho \in \Delta$ (respectively, $\rho \in \Gamma(\mathrm{CR}_X)$), $\mathcal{V} \in \mathcal{L}(\mathcal{CR})$ and $P \in \Sigma$, the classes ρP and $\mathcal{V}P$ are intervals in Δ, (respectively, $\Gamma(\mathrm{CR}_X)$) and $\mathcal{L}(\mathcal{CR})$, respectively. This permits the following symbolism.

Notation 1.3.5 For $\rho \in \Delta$ or $\rho \in \Gamma(\mathrm{CR}_X)$, $\mathcal{V} \in \mathcal{L}(\mathcal{CR})$ and $P \in \Sigma$, we define $\rho_P, \rho^P, \mathcal{V}_P$ and $\mathcal{V}^P$ by the requirements that $\rho P = [\rho_P, \rho^P]$, $\mathcal{V}P = [\mathcal{V}_P, \mathcal{V}^P]$.

It is important to notice that, by virtue of Theorem 1.2.11 and Corollary 1.2.13, for any $\rho \in \Gamma(\mathrm{CR}_X)$, the congruences ρ_P and ρ^P are the same whether we take them in $\Gamma(\mathrm{CR}_X)$ or in $\mathcal{C}(\mathrm{CR}_X)$. In terms of the Notations 1.3.2 and 1.3.5, we have

$$(\zeta_\mathcal{V})_P = \zeta_{\mathcal{V}^P}, \quad (\zeta_\mathcal{V})^P = \zeta_{\mathcal{V}_P} \text{ for all } P \in \Sigma.$$

The congruences and free objects corresponding to varieties of the form $\mathcal{V}_P$ can be characterized as follows.

Lemma 1.3.6 *Let $\mathcal{V} \in \mathcal{L}(\mathcal{CR})$ and $V = U/\zeta_\mathcal{V} = F\mathcal{V}_X$.*
(i) *We have*

$$\zeta_{\mathcal{V}_K}/\zeta_\mathcal{V} = \tau_V, \quad \zeta_{\mathcal{V}_T}/\zeta_\mathcal{V} = \mu_V, \quad \zeta_{\mathcal{V}_{T_r}}/\zeta_\mathcal{V} = \mathcal{R}_V^0, \quad \zeta_{K_r}/\zeta_\mathcal{V} = (\tau \cap \mathcal{R}^0)_V.$$

(ii) *We have*

$$F\mathcal{V}_K \cong F\mathcal{V}/\tau_V, \ F\mathcal{V}_T \cong F\mathcal{V}/\mu_V, \ F\mathcal{V}_{T_r} \cong F\mathcal{V}/\mathcal{R}_V^0, \ F\mathcal{V}_{K_r} \cong F\mathcal{V}/(\tau \cap \mathcal{R}^0)_V.$$

Proof (i) This follows immediately from [PR1, Propositions VII.1.5, VII.2.5] and the duals of [PR1, Propositions VII.4.5 and VII.4.14] while part (ii) follows immediately from part (i). $\square$

The characterizations of kernel and trace can be used to obtain bases for the varieties of the form $\mathcal{V}^K$, $\mathcal{V}^T$, etc.

Lemma 1.3.7 *Let $\rho \in \Gamma(\mathrm{CR}_X)$.*

(i) $[\rho_K] = [\rho]^K = [u = u^0 \mid u\zeta \, \rho \, (u\zeta)^0]$.
(ii) $[\rho_T] = [\rho]^T = [u^0 = v^0 \mid (u\zeta)^0 \, \rho \, (v\zeta)^0]$.
(iii) $[\rho_{T_r}] = [\rho]^{T_r} = [u^0 = v^0 \mid (u\zeta)^0 \, \rho \, (v\zeta)^0$ *and* $(u\zeta)^0 = (u\zeta)^0(v\zeta)^0]$.
(iv) $[\rho_{K_r}] = [\rho]^{K_r} = [\rho_K \vee \rho_{T_r}] = [\rho]^K \cap [\rho]^{T_r}$.

Proof These claims follow immediately from the facts that $\rho_K = (\rho \cap \mathcal{H})^*$ [PR1, Theorem VII.1.2], $\rho_T = (\mathrm{tr} \, \rho)^*$ [PR1, Theorem VII.2.2], $\rho_{T_r} = (\rho \cap \leq_r)^*$ (the dual of

[PR1, Theorem VII.4.2]) and since $\rho_{K_r} = \rho_K \vee \rho_{T_r}$ [PR1, Lemma VII.4.13], respectively, and the antiisomorphism between the lattices of fully invariant congruences and varieties.   □

The following characterizations of $\mathcal{D}$-equivalent elements that are related under ρ^P for $P \in \{K, T, T_\ell, T_r\}$ can be helpful.

Lemma 1.3.8 *Let $\rho \in \Gamma(\mathrm{CR}_X)$ and $u, v \in \mathrm{CR}_X$ be such that $u \mathcal{D} v$.*

(i) $u \rho^K v \Leftrightarrow xuy(xvy)^{-1} \in \ker \rho$, *for all $x, y \in \mathrm{CR}_X^1$*,
(ii) $u \rho^T v \Leftrightarrow (xuy)^0 \rho (xvy)^0$, *for all $x, y \in \mathrm{CR}_X^1$*,
(iii) $u \rho^{T_r} v \Leftrightarrow (xuy)^0 \rho \mathcal{R} (xvy)^0 \rho$, *for all $x, y \in \mathrm{CR}_X^1$*,
(iv) $u \rho^{K_r} v \Leftrightarrow (xuy)(xvy)^{-1} \in \ker \rho$ *and $(xuy)^0 \rho \mathcal{R} (xvy)^0 \rho$, for all $x, y \in \mathrm{CR}_X^1$*,

Proof (i) We have

$$u \rho^K v \Longleftrightarrow u\rho \left((\rho^K/\rho) \cap \mathcal{D}\right) v\rho$$
$$\Longleftrightarrow u\rho \left(\tau \cap \mathcal{D}\right) v\rho \qquad \text{by [PR1, Proposition VII.1.5]}$$
$$\Longleftrightarrow (xuy)(xvy)^{-1}\rho \in E(\mathrm{CR}_X/\rho) \qquad \text{for all } x, y \in \mathrm{CR}_X^1$$
$$\qquad \text{by [PR1, Lemma VI.1.14]}$$
$$\Longleftrightarrow (xuy)(xvy)^{-1} \in \ker \rho \qquad \text{for all } x, y \in \mathrm{CR}_X^1 \,.$$

(ii) We have

$$u \rho^T v \Longleftrightarrow u \left(\rho \vee \mathcal{H}\right)^0 v \qquad \text{by [PR1, Theorem VII.2.2]}$$
$$\Longleftrightarrow xuy \left(\rho \vee \mathcal{H}\right) xvy \qquad \text{for all } x, y \in \mathrm{CR}_X^1$$
$$\Longleftrightarrow (xuy)\rho \, \mathcal{H} \, (xvy)\rho \qquad \text{for all } x, y \in \mathrm{CR}_X^1$$
$$\qquad \text{by [PR1, Theorem VI.5.1(ii)]}$$
$$\Longleftrightarrow (xuy)^0 \rho (xvy)^0 \qquad \text{for all } x, y \in \mathrm{CR}_X^1 \,.$$

(iii) This follows in a similar fashion to part (ii), using the dual of [PR1, Theorem VII.4.2] and [PR1, Theorem VI.5.1(ii)].

(iv) This follows from parts (i), (iii) and [PR1, Lemma VII.4.13].   □

Lemmas 1.3.3 and 1.3.8 are due to Pastijn–Trotter [1]. Theorem 1.3.4 was proved by Pastijn [2]. The relations K_ℓ and K_r on $\mathcal{L}(\mathcal{CR})$ were first considered by Reilly [2].

1.4   Congruences ρ_K and ρ^K

In this section we shall see how to every fully invariant congruence on a free object we can associate two such congruences playing an important role relative to the K-relation. These congruences will be constructed on the basis of the very structure of free completely regular semigroups. For every congruence ρ on U, we define a relation $\overline{\rho}$ by means of a simple inductive procedure involving the invariants s and e of a word in U. This ingenious concept produces a wealth of interesting results. After some preparation, we show that $\overline{\rho}$ is a congruence. For $\rho \in \Delta$, that is when ρ is fully invariant and contains ζ, the principal result concerning $\overline{\rho}$ asserts that $\overline{\rho}/\zeta = (\rho/\zeta)_K$ where, from the results already established, we know that $(\rho/\zeta)_K$ is fully invariant. It then follows easily that $\overline{\rho}$ is also fully invariant. Note that $(\rho/\zeta)_K$ has the usual meaning of $(\ )_K$ for congruences on a completely regular semigroup.

Notation 1.4.1 For any $\rho \in \mathcal{C}(U)$, define a relation $\overline{\rho}$ on U by induction on $\#(u)$ by

$$u \,\overline{\rho}\, v \quad \text{if} \quad c(u) = c(v), \quad u \,\rho\, v, \quad s(u) \,\overline{\rho}\, s(v), \quad e(u) \,\overline{\rho}\, e(v).$$

where $\emptyset \,\overline{\rho}\, \emptyset$.

Clearly $\overline{\rho} \subseteq \rho$. A useful characterization of $\overline{\rho}$ is given by the following.

Lemma 1.4.2 *For any $\rho \in \mathcal{C}(U)$ and $u, v \in U$, we have*

$$u \,\overline{\rho}\, v \iff w(u) \,(\rho \cap \eta)\, w(v) \quad \text{for all } w \in \{s, e\}^*.$$

Proof The proof follows easily from the definitions of $\overline{\rho}$ and η by induction on $\#(u)$. □

Whenever $\rho \subseteq \eta$, the condition in Lemma 1.4.2 simplifies to:

$$u \,\overline{\rho}\, v \iff w(u) \,\rho\, w(v) \quad \text{for all } w \in \{s, e\}^*.$$

On the other hand, if we take $\rho = \omega$, then the condition reduces to:

$$u \,\overline{\rho}\, v \iff w(u) \,\eta\, w(v) \quad \text{for all } w \in \{s, e\}^*.$$

In the next two results we gather some simple technical observations about $\overline{\rho}$.

Lemma 1.4.3 *Let $\rho \in \mathcal{C}(U)$, $u, v \in U$ and $u \,\overline{\rho}\, v$.*

(i) $\sigma^k(u) = \sigma^k(v), \quad 1 \le k \le \#(u)$.

(ii) $i(u) = i(v)$.
(iii) $w(u) \, \overline{\rho} \, w(v)$ *for all* $w \in \{s, e\}^*$.

Proof (i) This is a consequence of the definition of $\overline{\rho}$ and Lemma 1.4.2.
(ii) This is a reformulation of part (i).
(iii) This follows from the definition of $\overline{\rho}$ by means of a simple induction argument. □

Lemma 1.4.4 *Let* $\lambda, \rho \in \mathcal{C}(U)$. *We have*

$$\text{(i)}\ \overline{\overline{\rho}} = \overline{\rho} \subseteq \eta \cap \rho. \qquad \text{(ii)}\ \textit{If}\ \lambda \subseteq \rho,\ \textit{then}\ \overline{\lambda} \subseteq \overline{\rho}.$$

Proof (i) This is an immediate consequence of the definition of $\overline{\rho}$.
(ii) This follows by a straightforward induction argument from the definitions of $\overline{\lambda}$ and $\overline{\rho}$. □

As we shall see, the relation $\overline{\rho}$ has some remarkable properties, beginning with the fact that it is a congruence.

Lemma 1.4.5 *Let* $\rho \in \mathcal{C}(U)$, *then* $\overline{\rho} \in \mathcal{C}(U)$.

Proof Let $\rho \in \mathcal{C}(U)$. It follows immediately from the definition of $\overline{\rho}$ that it is an equivalence relation. We wish to show that, for any x, y, $z \in U$, if $x \, \overline{\rho} \, y$, then $zx \, \overline{\rho} \, zy$ and $xz \, \overline{\rho} \, yz$. By duality, it suffices to show that $zx \, \overline{\rho} \, zy$. From the fact that $x \, \overline{\rho} \, y$, we have $c(x) = c(y)$ and $c(zx) = c(zy)$. In addition, $x \, \overline{\rho} \, y$ implies that $x \, \rho \, y$ and therefore $zx \, \rho \, zy$ since ρ is a congruence. We argue by induction on $\#(zx)$.

Since x, y, $z \in U$, we have $\#(x), \#(y), \#(z) \geq 1$. Also, $c(zx) = c(zy)$ so that $\#(zx) = \#(zy)$. If $\#(zx) = \#(zy) = 1$, then $z = x = y$, so that $s(zx) = \emptyset = s(zy)$ and $e(zx) = \emptyset = e(zy)$. If $z \neq x$, then $s(zx) = z = s(zy)$ and $e(zx) = x = y = e(zy)$. Either way, $s(zx) \, \overline{\rho} \, s(zy)$ and $e(zx) \, \overline{\rho} \, e(zy)$. For $k \geq 2$ we have $s^k(zx) = \emptyset = s^k(zy)$ and $e^k(zx) = \emptyset = e^k(zy)$. Therefore $zx \, \overline{\rho} \, zy$.

Now let $\#(zx) = \#(zy) \geq 2$ and that the conclusion $zx \, \overline{\rho} \, zy$ holds for similar situations with a smaller number of variables. Consider any $w \in \{s, e\}^*$. If $w = \emptyset$, then trivially $w(zx) = zx \, (\rho \cap \eta) \, zy = w(zy)$. So assume that $w \in \{s, e\}^+$.

Case: $w = us$ *for some* $u \in \{s, e\}^*$. If $c(x) = c(y) \subseteq c(z)$, then $s(zx) = s(z) = s(zy)$ so that $w(zx) = us(zx) = us(zy) = w(zy)$. Suppose that $c(x)$ is not contained in $c(z)$. Then $s(zx) = zs^m(x)$, where m is the least positive integer such that $\sigma^m(x) \notin c(z)$. By Lemma 1.4.3(i), we have $\sigma^r(x) = \sigma^r(y)$ for $1 \leq r \leq \#(x)$ and so $s(zy) = zs^m(y)$. Since, by Lemma 1.4.3(iii), $s^m(x) \, \overline{\rho} \, s^m(y)$ and, in addition, $\#(zs^m(x)) = \#(zx) - 1$, we have from the induction assumption that $zs^m(x) \, \overline{\rho} \, zs^m(y)$. Hence

$$w(zx) = us(zx) = u(zs^m(x)) \, (\rho \cap \eta) \, u(zs^m(y)) = us(zy) = w(zy).$$

Case: $w = ue$ for some $u \in \{s, e\}^*$. If $c(z) \subseteq c(x)$ then, by the dual of Lemma 1.4.3(ii), $e(zx) = e(x) \; \overline{\rho} \; e(y)$ from which it follows that

$$w(zx) = ue(zx) = ue(x) \; (\rho \cap \eta) \; ue(y) = ue(zy) = w(zy).$$

Suppose that $c(z)$ is not contained in $c(x)$. Then $e(zx) = e^m(z)x$ where m is the least positive integer such that $\varepsilon^m(z) \notin c(x)$. Then $\#(e^m(z)x) = \#(zx) - 1$ so that, by the induction hypothesis, $e^m(z)x \; \overline{\rho} \; e^m(z)y$. Hence

$$w(zx) = ue(zx) = u(e^m(z)x) \; (\rho \cap \eta) \; u(e^m(z)y) = ue(zy) = w(zy).$$

Thus in all cases $w(zx) \; (\rho \cap \eta) \; w(zy)$ so that $cx \; \overline{\rho} \; cy$. Therefore $\overline{\rho}$ is a congruence. $\square$

Corollary 1.4.6 *Let $\zeta \subseteq \rho \in \mathcal{C}(U)$. Then*

(i) $\zeta \subseteq \overline{\rho} \subseteq \rho$. *In particular $\zeta = \overline{\zeta}$.*
(ii) *If $\zeta \subseteq \rho$, $u \; \rho \; v$ and $u\zeta \; \mathcal{H} \; v\zeta$, then $u \; \overline{\rho} \; v$.*
(iii) $(\rho/\zeta) \cap \mathcal{H} = (\overline{\rho}/\zeta) \cap \mathcal{H}$.

Proof (i) Clearly $w(p) = w(q)$ for any $p, q \in \{u, uu^{-1}u, uu^{-1}, u^{-1}u, (u^{-1})^{-1}\}$ and any $w \in \{s, e\}$ so that the first containment is an immediate consequence of the definitions of ζ and $\overline{\rho}$, Lemmas 1.4.2 and 1.4.5. The second containment follows easily from Lemma 1.4.4.

(ii) Let $\zeta \subseteq \rho$, $u \; \rho \; v$ and $u\zeta \; \mathcal{H} \; v\zeta$. Since $\mathcal{H} \subseteq \mathcal{D}$, we have $u\zeta \; \mathcal{D} \; v\zeta$ and therefore by Theorem 1.1.9(i), we have $c(u) = c(v)$. Also, by Theorem 1.1.9(ii)(iii), $s(u) \; \zeta \; s(v)$ and $e(u) \; \zeta \; e(v)$, while by part (i), $\zeta \subseteq \overline{\rho}$. Hence we see that $u \; \overline{\rho} \; v$.

(iii) Combining part (ii) with the observation that $\overline{\rho} \subseteq \rho$, we obtain the desired result. $\square$

In certain induction arguments based on the size of the content of words it will be useful to have the following.

Notation 1.4.7 For $n \geq 1$, let

$$U_n = \{u \in U \mid \#(u) = n\}, \quad U_n^* = \{u \in U \mid \#(u) \leq n\}.$$

Note that $U_1^* = U_1$ and that if $\rho \in \mathcal{C}(U)$ and $\rho \subseteq \eta$, then $u \; \rho \; v$ implies that $c(u) = c(v)$ so that $\rho = \bigcup_{n=1}^{\infty} \rho|_{U_n}$.

Lemma 1.4.8 *If $\rho \in \mathcal{C}(U)$ and $\rho \subseteq \eta$, then $\rho|_{U_1^*} = \overline{\rho}_{U_1}$.*

Proof If $u, v \in U_1^*$ with $c(u) = c(v) = \{x\}$ and $u \; \rho \; v$, then $s(u) = e(u) = s(v) = e(v) = \emptyset$ so that $u \; \overline{\rho} \; v$. $\square$

The operator $\rho \to \overline{\rho}$ is extremely important in the study of fully invariant congruences and varieties due to its remarkable connection with the operator $(\ )_K$ on the lattice of congruences.

Theorem 1.4.9 *For every* $\rho \in \Delta$*, the equality* $(\rho/\zeta)_K = \overline{\rho}/\zeta$ *holds.*

Proof By Corollary 1.4.6(i), we have $\zeta \subseteq \overline{\rho}$. We will obtain the desired conclusion by showing that $\ker(\overline{\rho}/\zeta) = \ker(\rho/\zeta)_K$, $\mathrm{tr}(\overline{\rho}/\zeta) = \mathrm{tr}(\rho/\zeta)_K$.

Since $\overline{\rho} \subseteq \rho$, we have $\ker(\overline{\rho}/\zeta) \subseteq \ker(\rho/\zeta)$. Let $u \in U$ be such that $u\zeta \in \ker(\rho/\zeta)$. Then $u \, \rho \, u^0$. Also $s(u) = s(u^0)$ and $e(u^0) = e(u)$ so that, for any $w \in \{s, e\}^+$, $w(u) = w(u^0)$. Since $c(u) = c(u^0)$, it follows from Lemma 1.4.2 that $u \, \overline{\rho} \, u^0$ and therefore $u\zeta \in \ker(\overline{\rho}/\zeta)$. Consequently $\ker(\rho/\zeta) \subseteq \ker(\overline{\rho}/\zeta)$ and equality follows.

By the minimality of $(\rho/\zeta)_K$, we have $(\rho/\zeta)_K \subseteq \overline{\rho}/\zeta$. In order to establish the reverse inclusion, let $u, v \in U$ be such that $u\zeta \, \overline{\rho}/\zeta \, v\zeta$. Then $u \, \overline{\rho} \, v$ so that $c(u) = c(v)$. In order to show that $u\zeta \, (\rho/\zeta)_K \, v\zeta$, we proceed by induction on $\#(u)$.

First suppose that $\#(u) = 1$. Then $c(u) = c(v) = \{x\}$ for some element $x \in X$, whence $u^0 = uu^{-1} \zeta \, xx^{-1} \zeta \, vv^{-1} = v^0$ and $(u\zeta)^0 = (v\zeta)^0$. Trivially, this implies that $(u\zeta)^0$ $(\rho/\zeta)_K \, (v\zeta)^0$. Also, by the first part of the proof,

$$(u\zeta)(v\zeta)^{-1} \in \ker(\overline{\rho}/\zeta) = \ker(\rho/\zeta) = \ker(\rho/\zeta)_K.$$

Hence, by [PR1, Lemma VI.3.1], $u\zeta \, (\rho/\zeta)_K \, v\zeta$.

Now let $\#(u) > 1$ and, as our induction hypothesis, let us assume that for all $p, q \in U$ with $p\zeta \, \overline{\rho}/\zeta \, q\zeta$ and $\#(p) < \#(u)$ we have $p\zeta \, (\rho/\zeta)_K \, q\zeta$. From the hypothesis that $u\zeta \, \overline{\rho}/\zeta \, v\zeta$, we know that $u^0\zeta \, \overline{\rho}/\zeta \, v^0\zeta$ so that $u^0 \, \overline{\rho} \, v^0$. By the definition of $\overline{\rho}$, we have $s(u) = s(u^0) \, \overline{\rho} \, s(v^0) = s(v)$ so that $s(u)\zeta \, (\rho/\zeta)_K \, s(v)\zeta$ by the induction hypothesis. Again since $u^0 \, \overline{\rho} \, v^0$ we have $c(u) = c(v)$, $\sigma(u) = \sigma(v) = x$, say, and $c(s(u)) = c(s(v))$. Hence by Theorem 1.1.9(iii), we obtain

$$u\zeta \, \mathcal{R} \, (s(u)x)\zeta \, (\rho/\zeta)_K \, (s(v)x)\zeta \, \mathcal{R} \, v\zeta$$

so that $u\zeta \, ((\rho/\zeta)_K \vee \mathcal{R}) \, v\zeta$. By [PR1, Theorem VI.5.1(ii)], we have

$$(u\zeta)(\rho/\zeta)_K \, \mathcal{R} \, (v\zeta)(\rho/\zeta)_K.$$

By duality, also $(u\zeta)(\rho/\zeta)_K \, \mathcal{L} \, (v\zeta)(\rho/\zeta)_K$ and consequently

$$(u\zeta)(\rho/\zeta)_K \, \mathcal{H} \, (v\zeta)(\rho/\zeta)_K.$$

Therefore $(u^0\zeta)(\rho/\zeta)_K = (v^0\zeta)(\rho/\zeta)_K$ so that $u^0\zeta \, (\rho/\zeta)_K \, v^0\zeta$. But the hypothesis that $u\zeta \, \overline{\rho}/\zeta \, v\zeta$ and the fact that $\ker(\overline{\rho}/\zeta) = \ker(\rho/\zeta)_K$ imply that

$$(u\zeta)(v\zeta)^{-1} \in \ker(\overline{\rho}/\zeta) = \ker(\rho/\zeta)_K .$$

Hence, by [PR1, Lemma VI.3.1], we have $u\zeta \, (\rho/\zeta)_K \, v\zeta$ and the proof is complete. $\qquad\square$

In terms of Notation 1.3.5, we can write Theorem 1.4.9 as $\overline{\rho} = \rho_K$ ($\rho \in \Delta$).

Corollary 1.4.10 *For any $\rho \in \Delta$, $\overline{\rho}$ is fully invariant.*

Proof Clearly ρ/ζ is a fully invariant congruence on CR_X and therefore, by Theorem 1.2.11, so also is $(\rho/\zeta)_K$. By Theorem 1.4.9, $\overline{\rho}/\zeta$ is fully invariant and from Lemma 1.2.3 it follows easily that $\overline{\rho}$ is fully invariant. $\qquad\square$

From [PR1, Corollary VII.1.3] we know that the mapping $\rho \to \rho_K$ is a lower closure operator on $\mathcal{C}(S)$ for any completely regular semigroup S. Within $\Gamma(CR_X)$ we can now state a stronger result.

Theorem 1.4.11 *The mapping $\rho \mapsto \rho_K$ ($\rho \in \Gamma(CR_X)$) is a complete endomorphism of $\Gamma(CR_X)$.*

Proof By Lemma 1.2.3 and Theorem 1.4.9, it suffices to show that the mapping $\chi : \rho \to \overline{\rho}$ ($\rho \in \Delta$) is a complete endomorphism of Δ. By Theorem 1.3.4, K is a complete congruence on Δ and, by Theorem 1.4.9, $\overline{\rho} = \rho_K$. Hence, by [PR1, Lemma I.2.2], χ respects arbitrary joins.

Let $\rho_\alpha \in \Delta$, $\alpha \in A$, $\lambda = \bigcap_{\alpha \in A} \overline{\rho}_\alpha$ and $\rho = \bigcap_{\alpha \in A} \rho_\alpha$. We wish to show that $\lambda = \overline{\rho}$. In order to do so, it suffices to show that $\lambda|_{U_n^*} = \overline{\rho}|_{U_n^*}$ for all n. We proceed by induction on n. By Lemma 1.4.8, we obtain

$$\overline{\rho}|_{U_1^*} = \rho|_{U_1^*} = \bigcap_{\alpha \in A} \left(\rho_\alpha|_{U_1^*}\right) = \bigcap_{\alpha \in A} \left(\overline{\rho}_\alpha|_{U_1^*}\right) = \left(\bigcap_{\alpha \in A} \overline{\rho}_\alpha\right)\Bigg|_{U_1^*} = \lambda|_{U_1^*}.$$

Now assume that $\lambda|_{U_n^*} = \overline{\rho}|_{U_n^*}$ and consider U_{n+1}^*. Let $u, v \in U_{n+1}^*$. We have

$$
\begin{aligned}
u \, \lambda \, v &\iff u \, \overline{\rho}_\alpha \, v && \text{for all } \alpha \in A \\
&\iff u \, \rho_\alpha \, v, \;\; c(u) = c(v), \;\; s(u) \, \overline{\rho}_\alpha \, s(v), \;\; e(u) \, \overline{\rho}_\alpha \, e(v) \\
&&& \text{for all } \alpha \in A \\
&\iff u \, \rho \, v, \;\; c(u) = c(v), \;\; s(u) \, \lambda \, s(v), \;\; e(u) \, \lambda \, e(v) \\
&\iff u \, \rho \, v, \;\; c(u) = c(v), \;\; s(u) \, \overline{\rho} \, s(v), \;\; e(u) \, \overline{\rho} \, e(v) \\
&&& \text{by the induction hypothesis} \\
&\iff u \, \overline{\rho} \, v.
\end{aligned}
$$

Thus $\lambda|_{U_{n+1}^*} = \overline{\rho}|_{U_{n+1}^*}$ and, by induction, we have $\lambda = \overline{\rho}$, as required. $\qquad\square$

The material for this section is taken from Polák [1, 2]. The formulation in Notation 1.4.1 first appeared in Pastijn [2]. We have used the notation from Gerhard and Petrich [1].

1.5   Congruence ρ^{T_r}

In the preceding section, to every fully invariant congruence ρ, we associated the congruence ρ_K. Here we will construct a congruence ρ^{T_r} relative to the relation T_r. It will turn out that ρ^{T_r} has as strong properties as ρ_K, such as the mapping $\rho \rightarrow \rho^{T_r}$ being a complete endomorphism of the lattice of fully invariant congruences. It has its dual counterpart ρ^{T_ℓ} with the same property, though their intersection $\rho^T = \rho^{T_\ell} \cap \rho^{T_r}$ does not. The construction and properties of ρ^{T_r} depend on the mysterious structure of CR_X.

For any $\rho \in \Delta$, that is, ρ a fully invariant congruence on U containing ζ, we define ρ^s by means of ρ and the invariant s of a word in U. After establishing some basic properties of ρ^s, we then characterize the relation $\mathcal{R}$ on U/ρ in terms of ρ^s. In the main result we show that congruences $\lambda, \rho \in [\zeta, \zeta_{\mathcal{LRB}}]$ are T_r-equivalent if and only if $\lambda^s = \rho^s$. The surprising outcome is that the operator $\rho \rightarrow \rho^s$ aligns nicely with the T_r relation on Δ and $\Gamma(CR_X)$.

Notation 1.5.1 For any binary relation ρ on U and any elements $u, v \in U$, define

$$u \, \rho^s \, v \quad \text{if there exist} \quad w, t \in U \quad \text{with} \quad u = s(w), \; v = s(t) \quad \text{and} \quad w \, \rho \, t.$$

We denote by ρ^e the right-left dual of ρ^s. Any expression of the form $\rho^{\alpha\beta\gamma\cdots}$, where $\alpha, \beta, \gamma, \ldots \in \{s, e\}$, is to be read as $\cdots ((\rho^\alpha)^\beta)^\gamma \cdots$. Where required, we shall also adopt the conventions that $s(\emptyset) = \emptyset = e(\emptyset)$, $\emptyset \, \rho^s \, \emptyset$, and $\emptyset \, \rho^e \, \emptyset$. As before, we also write $\eta = \zeta_{\mathcal{S}}$ for the least semilattice congruence on U. We will also denote by $\hat{\rho}$ the least fully invariant unary congruence on U containing ρ.

First some basic observations. To place them in context, it is helpful to recall that, from [PR1, Corollary IX.5.3(ii)], we know that $\mathcal{L}(C\mathcal{R})$ is the disjoint union of the intervals $[\mathcal{J}, \mathcal{RRO}]$, $[\mathcal{LZ}, L\mathcal{RRO}]$ and $[\mathcal{LRB}, C\mathcal{R}]$.

Lemma 1.5.2 *Let* $\lambda, \rho \in \Delta$.

(i) $\zeta \subseteq \rho \subseteq \rho^s$.

(ii) $\lambda \subseteq \rho \; \rightarrow \; \lambda^s \subseteq \rho^s$.

(iii) $\rho^s \subseteq \rho^{es}$.

(iv) *If* $\zeta_{L\mathcal{RRO}} \subseteq \rho \subseteq \omega$, *then*

$$\widehat{\rho^{ss}} = \widehat{\rho^s} = \begin{cases} \omega & \text{if } \zeta_{\mathcal{RRO}} \subseteq \rho \subseteq \omega, \\ \zeta_{\mathcal{LZ}} & \text{if } \zeta_{L\mathcal{RRO}} \subseteq \rho \subseteq \zeta_{\mathcal{LZ}}. \end{cases}$$

(v) *If $\rho \subseteq \zeta_{\mathcal{LRB}}$, then $\rho^s \in \Delta$ and $\rho^{ss} = \rho^s \subseteq \zeta_{\mathcal{LRB}}$.*

(vi) $\zeta^s_{\mathcal{LRB}} = \zeta_{\mathcal{LRB}}$.

(vii) *If $\mathcal{U} \in [\mathcal{S}, \mathcal{RRO}]$ and $\mathcal{V} \in [\mathcal{LNB}, \mathcal{LRRO}]$, then $(\zeta_{\mathcal{U}})^s = (\zeta_{\mathcal{S}})^s \notin \Delta$ and $(\zeta_{\mathcal{V}})^s = (\zeta_{\mathcal{LNB}})^s \notin \Delta$.*

Proof (i) The first containment is by the definition of Δ. Now let $u, v \in U$ be such that $u \rho v$. Let $z \in X \setminus c(uv)$. Then $uz \rho vz$, $u = s(uz)$, $v = s(vz)$, so that $u \rho^s v$ which establishes the second containment.

(ii) This follows immediately from the definitions of λ^s and ρ^s.

(iii) By the dual of part (i), we have $\rho \subseteq \rho^e$. The claim then follows from part (ii).

(iv) First assume that $\zeta_{\mathcal{RRO}} \subseteq \rho \subseteq \omega$. Let $x, y \in X$. By [PR1, Theorem IX.5.1(iii)], we have that $xy \zeta_{\mathcal{RRO}} y^0 xy$ so that $xy \rho y^0 xy$. Hence $x = s(xy) \rho^s s(y^0 xy) = y^0$, together with the fact that $\widehat{\rho^s}$ is a fully invariant congruence, implies that $\omega \subseteq \widehat{\rho^s}$ and equality follows.

Substituting xz for x in $xy \rho y^0 xy$, where $z \in X \setminus \{x, y\}$ we obtain $xzy \rho y^0 xzy$ so that $xz \rho^s y^0 x$ and $x \rho^{ss} y^0$. Therefore $\widehat{\rho^{ss}} = \omega$ and the claims hold.

Now assume that $\zeta_{\mathcal{LRRO}} \subseteq \rho \subseteq \zeta_{\mathcal{LZ}}$. Let $u, v, w, t \in U$ be such that $u = s(v), v = s(t)$ and $w \rho t$. Then

$$w \rho t \implies w \zeta_{\mathcal{LZ}} t \implies h(w) = h(t) \implies h(u) = h(s(w)) = h(s(t)) = h(v).$$

Hence, for any $u, v \in U$ such that $u \rho^s v$, we must have $h(u) = h(v)$. Thus $\rho^s \subseteq \zeta_{\mathcal{LZ}}$ and therefore also $\widehat{\rho^s} \subseteq \zeta_{\mathcal{LZ}}$. On the other hand, by [PR1, Theorem IX.5.1(iv)], for any $x, y, z \in X$ we have $xyz \zeta_{\mathcal{LRRO}} (xz)^0 xyz$ so that $xyz \rho (xz)^0 xyz$. Hence $xy \rho^s (xz)^0 x$ so that, with the substitution $z \to x^0$, and since $\widehat{\rho^s}$ is fully invariant, we find that $xy \widehat{\rho^s} x$. Hence $\zeta_{\mathcal{LZ}} \subseteq \widehat{\rho^s}$ and equality prevails.

Multiplying $xyz \rho (xz)^0 xyz$ on the right by $w \in X \setminus \{x, y, z\}$, we obtain $xyzw \rho (xz)^0 xyzw$, so that $xyz \rho^s (xz)^0 xyz$ whence $xy \rho^{ss} (xz)^0 x$. Since $\widehat{\rho^{ss}}$ is a fully invariant congruence, it follows that $xy \widehat{\rho^{ss}} x$. Hence $\zeta_{\mathcal{LZ}} \subseteq \widehat{\rho^{ss}}$ and equality prevails, as required.

(v) Let $\rho \subseteq \zeta_{\mathcal{LRB}}$, $u, v \in U$ and $u \rho^s v$. Then there exist $w, t \in U$ with $w \rho t$, $u = s(w)$ and $v = s(t)$. By [PR1, Lemma V.1.4], we have $i(w) = i(t) = x_1 \ldots x_n$, say. Then x_n must be the last variable to appear for the first time in both w and t. Therefore $i(u) = i(s(w)) = x_1 \ldots x_{n-1} = i(s(t)) = i(v)$ and, again by [PR1, Lemma V.1.4], we have $u \zeta_{\mathcal{LRB}} v$. Thus $\rho^s \subseteq \zeta_{\mathcal{LRB}}$.

The next step is to show that ρ^s is a unary congruence. Clearly ρ^s is symmetric. For any $u \in U, z \notin c(u)$, we have $uz \rho uz$ so that $u \rho^s u$ and ρ^s is reflexive.

Next let $u, v, w \in U$ be such that $u \rho^s v \rho^s w$. Since $\rho^s \subseteq \zeta_{\mathcal{LRB}}$ we have $c(u) = c(v) = c(w)$. Then there exist $p, q, r, t \in U$ with

$$p \rho q, \quad r \rho t, \quad u = s(p), \quad s(q) = v = s(r), \quad w = s(t).$$

By hypothesis $\rho \subseteq \zeta_{\mathcal{LRB}}$ so that $i(p) = i(q)$ and $i(r) = i(t)$. Let $\sigma(p) = \sigma(q) = y$, $\sigma(r) = \sigma(t) = z$, and φ be the endomorphism of U defined by $y\varphi = z$ and $x\varphi = x$ for all $x \in X \setminus \{y\}$. Let $p' = p\varphi$ and $q' = q\varphi$. Then $p' \rho q'$ and $\sigma(p') = \sigma(q') = z$ while $s(q)$ is unchanged by φ so that $s(q') = s(q)$. Invoking Theorem 1.1.9(iii), we have

$$p'\rho = q'\rho \; \mathcal{R} \; (s(q')\sigma(q'))\rho = (s(q)zt)\rho = (s(r)z)\rho \; \mathcal{R} \; r\rho = t\rho,$$

whence $p'\rho \; t^0 p'$. Now $u = s(p)$ is unchanged by φ so that $u = s(p')$. Also

$$c(p') = c(p\varphi) = (c(u) \cup \{y\})\varphi = c(u) \cup \{z\} = c(v) \cup \{z\} = c(r) = c(t)$$

which implies that $s(t^0 p') = s(t) = w$. Thus $u \; \rho^s \; w$ and transitivity is established.

To see that ρ^s is a congruence, let $p, u, v \in U$ and $u \; \rho^s \; v$. For suitable $w, t \in U$, we have $u = s(w)$, $v = s(t)$ and $w \; \rho \; t$. Since $\rho, \rho^s \subseteq \zeta_{\mathcal{LRB}}$, we have $c(w) = c(t)$ and $c(u) = c(v)$ and $\sigma(w) = \sigma(t) = y$, say, where $y \in X \setminus c(u)$. Let $z \in X \setminus c(pu)$ and φ be the endomorphism of U defined by $y\varphi = z$ and $x\varphi = x$ for $x \in X \setminus \{y\}$. Since ρ is fully invariant, we must have $w\varphi \; \rho \; t\varphi$ so that $p(w\varphi) \; \rho \; p(t\varphi)$. By the choice of z, we have $pu = ps(w) = ps(w\varphi) = s(p(w\varphi))$ and similarly $pv = s(p(t\varphi))$. Thus $pu \; \rho^s \; pv$. Therefore, ρ^s is a left congruence.

Let ψ be the endomorphism of U defined by $y\psi = pz$ and $x\psi = x$ for $x \in X \setminus \{y\}$. Then $w\psi \; \rho \; t\psi$ where $up = s(w\psi)$ and $vp = s(t\psi)$. Thus $up \; \rho^s \; vp$ and ρ^s is a right congruence on U and therefore a ρ^s is a congruence.

Since $\rho \subseteq \rho^s$ and U/ρ is completely regular and ρ^s/ρ is a semigroup congruence on U/ρ, it follows that ρ^s/ρ respects inverses and therefore that ρ^s respects the unary operation and is a unary congruence.

Next we establish that ρ^s is fully invariant. Let $u, v, w, t \in U$ be such that $u = s(w)$, $v = s(t)$, $w \; \rho \; t$, so that $u \; \rho^s \; v$. Since $\rho \subseteq \zeta_{\mathcal{RRB}}$, we have $i(w) = i(t)$ and, in particular, $\sigma(w) = \sigma(t) = y$, say. Let φ be any endomorphism of U. Let $z \in X \setminus c(u\varphi)$ and define an endomorphism ψ of U by $y\psi = z$ and $x\psi = x\varphi$ for $x \in X \setminus \{y\}$. Then $w\psi \; \rho \; t\psi$ while

$$u\varphi = u\psi = s(w\psi), \quad v\varphi = v\psi = s(t\psi).$$

Thus $u\varphi \; \rho^s \; v\varphi$ and ρ^s is a fully invariant congruence on U so that $\rho^s \in \Delta$.

Now that we know that $\rho^s \in \Delta$ and that $\rho^s \subseteq \zeta_{\mathcal{LRB}}$, we can apply the above arguments to ρ^s instead of ρ to establish that also $\rho^{ss} \in \Delta$ and $\rho^{ss} \subseteq \zeta_{\mathcal{LRB}}$. Next let $a, b \in U$ be such that $a \; \rho^{ss} \; b$. There exist $p, q, u, v \in U$ such that

$$a = s(p), \quad b = s(q), \quad p = s(u), \quad q = s(v), \quad u \; \rho \; v.$$

For suitable $x_1, \ldots x_n \in X$, where $n \geq 3$, we have

$$i(a) = i(b) = x_1 \cdots x_{n-2},$$
$$i(p) = i(q) = x_1 \cdots x_{n-2} x_{n-1},$$
$$i(u) = i(v) = x_1 \cdots x_{n-2} x_{n-1} x_n.$$

Let φ denote the endomorphism of U that maps x_n to x_1 and all other variables identically. Then $u\varphi \, \rho \, v\varphi$ while $a = s(p) = s((u)\varphi)$, $b = s(v) = s((v)\varphi)$ so that $a \, \rho^s \, b$. Thus $\rho^{ss} \subseteq \rho^s$ and the reverse inclusion follows from part (ii).

(vi) This follows immediately from parts (i) and (v).

(vii) We postpone the proof of these claims until Lemma 3.1.9.   □

The close connection between ρ^s and the relation $\mathcal{R}$ on U/ρ is brought out in the next result, which generalizes Theorem 1.1.9(iii) from CR_X to U/ρ.

Lemma 1.5.3 *Let $\rho \in \Delta$, $\rho \subseteq \eta$ and $u, v \in U$. Then*

$$u\rho \, \mathcal{R} \, v\rho \iff c(u) = c(v), \quad s(u) \, \rho^s \, s(v).$$

Proof *Direct part.* Since $u^0 v \, \rho \, v$, $v^0 u \, \rho \, u$ and $\rho \subseteq \eta$, by Theorem 1.1.9(i), we must have $c(u) = c(v)$. Also $s(u) = s(u^0) = s(u^0 v)$, and $u^0 v \, \rho \, v$ so that $s(u) \, \rho^s \, s(v)$.

Converse. The claim is trivial if $\#(u) = 1$. So suppose that $\#(u) > 1$. Since $c(u) = c(v)$ we have, by Theorem 1.1.9(i), that $u\rho \, \mathcal{D} \, v\rho$. By [PR1, Corollary IX.5.3(ii)], the following three cases will cover all the possibilities.

Case: $\zeta_{\mathcal{RRO}} \subseteq \rho \subseteq \omega$. In this case $U/\rho \in \mathcal{RRO}$ and, by the dual of [PR1, Lemma V.3.1], U/ρ is a semilattice of right groups. By Theorem 1.1.9, $u\zeta \, \mathcal{D} \, v\zeta$. Hence $u\rho \, \mathcal{R} \, v\rho$.

Case: $\zeta_{\mathcal{LRRO}} \subseteq \rho \subseteq \zeta_{\mathcal{LZ}}$. By Lemma 1.5.2(iv), $\rho^s \subseteq \zeta_{\mathcal{LZ}}$ so that we must have $h(u) = h(v)$. Let $h(u) = x$. By [PR1, Lemma IX.5.4], we have

$$u\rho = ((x^0)\rho)(u\rho) \, \mathcal{R} \, ((x^0)\rho)(v\rho) = v\rho.$$

Case: $\rho \subseteq \zeta_{\mathcal{LRB}}$. By Lemma 1.5.2(v), $\rho^s \subseteq \zeta_{\mathcal{LRB}}$ so that $c(u) = c(v)$ and $c(s(u)) = c(s(v))$. Hence $\sigma(u) = \sigma(v) = y$, say. By hypothesis there exist $p, q \in U$ such that $s(u) = s(p)$, $s(v) = s(q)$ and $p \, \rho \, q$. Since $\rho \subseteq \zeta_{\mathcal{LRB}}$, we must have $\sigma(p) = \sigma(q) = z$, say. Let φ be the endomorphism of U defined by $z\varphi = y$, $x\varphi = x$ for all $x \in X \setminus \{z\}$. Then $p\varphi \, \rho \, q\varphi$ where

$$s(p\varphi) = s(p) = s(u), \quad \sigma(p\varphi) = y, \quad s(q\varphi) = s(q) = s(v), \quad \sigma(q\varphi) = y = \sigma(v).$$

By Theorem 1.1.9(iii) this yields $u\zeta \, \mathcal{R} \, (p\varphi)\zeta \, (\rho/\zeta) \, (q\varphi)\zeta \, \mathcal{R} \, v\zeta$ so that $u\rho \, \mathcal{R} \, v\rho$.   □

Corollary 1.5.4 *Let $\rho \in \Delta$, $\rho \subseteq \eta$ and $u, v \in U$. Then*

$$u\rho \, \mathcal{H} \, v\rho \iff u \, \eta \, v, \quad s(u) \, \rho^s \, s(v), \quad e(u) \, \rho^e \, e(v).$$

Proof This is an immediate consequence of Lemma 1.5.3 and its dual.   □

It is important now to describe in familiar terms certain T_r-classes that have a special role in the later discussions.

Theorem 1.5.5 (i) *The following intervals are T_r-classes.*

$$[\mathcal{J}, \mathcal{RG}], \quad [\mathcal{LZ}, \mathcal{CS}], \quad [\mathcal{S}, \mathcal{RRO}], \quad [\mathcal{LNB}, \mathcal{LRRO}], \quad [\mathcal{LRB}, \mathcal{R}^*].$$

(ii) *Let $\mathcal{V} \in [\mathcal{S}, \mathcal{CR}]$. If $\mathcal{V}T_r$ is not one of the classes listed above, then $\mathcal{LRB}T_r < \mathcal{V}T_r$ in the lattice $\mathcal{L}(\mathcal{CR})/T_r$ so that the list in* (i) *is a complete list of all T_r-classes containing elements of $\mathcal{R}^*$.*

This result might appear to be more appropriately placed in Chap. II. However, that would play havoc with the the proof of the next theorem, which requires these results.

Proof The proof in each case follows a similar pattern and therefore we will only provide the details for the interval $[\mathcal{LRB}, \mathcal{R}^*]$.

Let $\mathcal{V} \in [\mathcal{LRB}, \mathcal{R}^*]$. By definition, $(\zeta_\mathcal{V})^{T_r}$ is the greatest congruence in $(\zeta_\mathcal{V})T_r$. By the dual of [PR1, Corollary VII.4.9(ii)], this means that $(\zeta_\mathcal{V})^{T_r}$ is the greatest congruence λ on U such that $\lambda/\zeta_\mathcal{V} \subseteq \mathcal{R}$ on $F = U/\zeta_\mathcal{V}$.

By [PR1, Theorem IX.5.1(v)] and the definition of the Malcev product, there exists an $\mathcal{LRB}$ congruence ρ on F over $\mathcal{RG}$. By the dual of [PR1, Lemma VI.4.9], we have $\rho \subseteq \mathcal{R}$. Since $(F/\zeta_\mathcal{V})/(\zeta_{\mathcal{LRB}}/\zeta_\mathcal{V}) \in \mathcal{LRB}$, it follows that $\zeta_{\mathcal{LRB}}/\zeta_\mathcal{V} \subseteq \rho \subseteq \mathcal{R}$. By [PR1, Lemma V.1.2], $\mathcal{R} = \varepsilon$ on any member of $\mathcal{LRB}$. Hence we must have $\zeta_{\mathcal{LRB}}/\zeta_\mathcal{V} = \mathcal{R}$ so that, by the preliminary remarks, $(\zeta_\mathcal{V})^{T_r} = \zeta_{\mathcal{LRB}}$ and, equivalently, $\mathcal{V}T_r = \mathcal{LRB}$. Thus $[\mathcal{LRB}, \mathcal{R}^*] \subseteq \mathcal{LRB}T_r$.

In order to establish the reverse containment, let $\mathcal{V} \in \mathcal{L}(\mathcal{CR})$ be such that $\mathcal{V} T_r \mathcal{LRB}$ and, as before, let $F = U/\zeta_\mathcal{V}$ where we will also identify the latter semigroup with $(U/\zeta)/(\zeta_\mathcal{V}/\zeta)$. Note that $U/\zeta_{\mathcal{LRB}}$ is a left regular band and therefore has trivial $\mathcal{R}$-classes. Hence

$$\mathcal{V} T_r \mathcal{LRB} \implies \zeta_\mathcal{V}/\zeta \; T_r \; \zeta_{\mathcal{LRB}}/\zeta$$
$$\implies (\zeta_\mathcal{V}/\zeta) \vee \mathcal{R} = (\zeta_{\mathcal{LRB}}/\zeta) \vee \mathcal{R}$$
$$\text{by the dual of [PR1, Corollary VII.4.9(iii)]}$$
$$\implies (\zeta_\mathcal{V}/\zeta) \vee \mathcal{R} = \zeta_{\mathcal{LRB}}/\zeta \quad \text{by the preceding remarks}$$
$$\implies \zeta_\mathcal{V} \subseteq \zeta_{\mathcal{LRB}}$$
$$\implies \mathcal{LRB} \subseteq \mathcal{V}.$$

By the dual of [PR1, Corollary VII.4.9(iv)], $(\zeta_{\mathcal{LRB}}/\zeta)/(\zeta_\mathcal{V}/\zeta) \subseteq \mathcal{R}$. This implies that there exists a congruence ρ on $F = U/\zeta_\mathcal{V}$ over right groups such that $F/\rho \in \mathcal{LRB}$, that is, such that $F \in \mathcal{RG} \circ \mathcal{LRB} = \mathcal{R}^*$, by [PR1, Theorem IX.5.1(v)]. Hence $\mathcal{V} \in [\mathcal{LRB}, \mathcal{R}^*]$. Thus $\mathcal{LRB}T_r \subseteq [\mathcal{LRB}, \mathcal{R}^*]$ and equality prevails.

(ii) By [PR1, Corollary IX.5.3], $\mathcal{V}$ must lie in $[\mathcal{LRB}, \mathcal{CR}]$, yet $\mathcal{V} \notin \mathcal{LRB}T_r$. Hence, $\mathcal{LRB}T_r \leq \mathcal{V}T_r$ and $\mathcal{V}T_r \neq \mathcal{LRB}T_r$. The claim follows. □

The importance of the operator $\rho \to \rho^s$ becomes evident in the next result.

Theorem 1.5.6 *Let $\rho \in \Delta$. Then*

$$(\rho/\zeta)^{T_r} = \begin{cases} \omega/\zeta & \text{if } \zeta_{\mathcal{RG}} \subseteq \rho \subseteq \omega, \\ \zeta_{\mathcal{LZ}}/\zeta & \text{if } \zeta_{\mathcal{CS}} \subseteq \rho \subseteq \zeta_{\mathcal{LZ}}, \\ \zeta_{\mathcal{S}}/\zeta & \text{if } \zeta_{\mathcal{RRO}} \subseteq \rho \subseteq \zeta_{\mathcal{S}} = \eta, \\ \zeta_{\mathcal{LNB}}/\zeta & \text{if } \zeta_{L\mathcal{RRO}} \subseteq \rho \subseteq \zeta_{\mathcal{LNB}}, \\ \rho^s/\zeta & \text{if } \rho \subseteq \zeta_{\mathcal{LRB}}. \end{cases}$$

Equivalently, we have

$$\zeta_{\mathcal{V}_{T_r}} = \begin{cases} \omega & \text{if } \mathcal{T} \subseteq \mathcal{V} \subseteq \mathcal{RG}, \\ \zeta_{\mathcal{LZ}} & \text{if } \mathcal{LZ} \subseteq \mathcal{V} \subseteq \mathcal{CS}, \\ \zeta_{\mathcal{S}} & \text{if } \mathcal{S} \subseteq \mathcal{V} \subseteq \mathcal{RRO}, \\ \zeta_{\mathcal{LNB}} & \text{if } \mathcal{LNB} \subseteq \mathcal{V} \subseteq L\mathcal{RRO}, \\ \rho^s & \text{if } \mathcal{LRB} \subseteq \mathcal{V}. \end{cases}$$

Proof By [PR1, Corollary IX.5.3(i)], the listed cases cover all possible values of $\rho \in \Delta$. By Theorem 1.5.5 the intervals

$$[\zeta_{\mathcal{RG}}, \omega], \quad [\zeta_{\mathcal{CS}}, \zeta_{\mathcal{LZ}}], \quad [\zeta_{\mathcal{RRO}}, \zeta_{\mathcal{S}}], \quad [\zeta_{L\mathcal{RRO}}, \zeta_{\mathcal{LNB}}]$$

are T_r-intervals. The claims for these intervals then follow immediately.

So there remains only one case, the one that connects the operator $\rho \to \rho^s$ directly to T_r, namely: $\rho \subseteq \zeta_{\mathcal{LRB}}$.

By Lemma 1.5.2(v), ρ^s is a fully invariant congruence and we have $\rho^s \subseteq \zeta_{\mathcal{LRB}} \subseteq \zeta_{\mathcal{S}} = \eta$. Let $\lambda \in \Delta$ be such that $\lambda/\zeta = (\rho/\zeta)^{T_r}$. We wish to prove that $\lambda = \rho^s$. Note that, by the dual of [PR1, Theorem VII.4.2], $\lambda/\zeta = ((\rho/\zeta) \vee \mathcal{R})^0$.

We show first that $\lambda \subseteq \rho^s$. Since $\rho \subseteq \eta$, we have $\rho/\zeta \subseteq \mathcal{D}$ so that $(\rho/\zeta)^{T_r} \subseteq \mathcal{D}$ which implies that $\lambda \subseteq \eta$. So let $u, v \in U$ and $u \lambda v$. Then $c(u) = c(v)$. For $x \in X \setminus c(u)$, we have $ux \lambda vx$ whence $(ux)\zeta ((\rho/\zeta) \vee \mathcal{R})^0 (vx)\zeta$. By [PR1, Theorem VI.5.1(ii)], we have $((ux)\zeta)(\rho/\zeta) \mathcal{R} ((vx)\zeta)(\rho/\zeta)$ which implies that $(ux)\rho \mathcal{R} (vx)\rho$ in U/ρ. It follows that $(ux)^0 vx \rho vx$ which together with $u = s((ux)^0 vx)$ and $v = s(vx)$ implies that $u \rho^s v$ and thus $\lambda \subseteq \rho^s$.

To consider the reverse containment, let $u, v \in U$ and $u \, \rho^s \, v$. For suitable $w, t \in U$, we have $u = s(w)$, $v = s(t)$ and $w \, \rho \, t$. By the hypothesis that $\rho \subseteq \zeta_{\mathcal{LRB}}$ and [PR1, Lemma V.1.4], we have $c(w) = c(t)$, $c(u) = c(v)$ and $\sigma(w) = \sigma(t) = z$, say. Let $y \in c(u)$ and define an endomorphism φ of U by the stipulation that $z\varphi = y$ and $x\varphi = x$ for $x \in X \setminus \{z\}$. Then $w\varphi \, \rho \, t\varphi$ and $c(w\varphi) = c(u) = c(v) = c(t\varphi)$. Also $s(w\varphi) = s(s(w)) = s(u)$, $s(t\varphi) = s(s(t)) = s(v)$. If we let $p = w\varphi$ and $q = t\varphi$, then

$$p \, \rho \, q, \quad s(u) = s(p), \quad s(v) = s(q), \quad c(p) = c(q) = c(u) = c(v).$$

Hence $\sigma(p) = \sigma(u) = m$ and $\sigma(q) = \sigma(v) = n$, say. By Theorem 1.1.9(iii) we have,

$$u\zeta \, \mathcal{R} \, (s(u)m)\zeta = (s(p)m)\zeta \, \mathcal{R} \, p\zeta = (w\varphi)\zeta \, (\rho/\zeta) \, (t\varphi)\zeta$$
$$= q\zeta \, \mathcal{R} \, (s(q)n)\zeta = (s(v)n)\zeta \, \mathcal{R} \, v\zeta.$$

Thus $u\zeta \, ((\rho/\zeta) \vee \mathcal{R}) \, v\zeta$ so that $\rho^s/\zeta \subseteq (\rho/\zeta) \vee \mathcal{R}$ and therefore $\rho^s/\zeta \subseteq ((\rho/\zeta) \vee \mathcal{R})^0 = \lambda/\zeta$. It follows that $\rho^s \subseteq \lambda$ and thus $\lambda = \rho^s$. $\qquad \square$

Corollary 1.5.7 *Let $\lambda, \rho \in \Delta$ be such that $\lambda, \rho \subseteq \zeta_{\mathcal{LRB}}$. Then $\lambda \, T_r \, \rho \Leftrightarrow \lambda^s = \rho^s$.*

Proof This follows immediately from Theorem 1.5.6. $\qquad \square$

In Theorem 1.4.11 we have shown that mapping to the lower end of each K-class yields a complete endomorphism. A parallel result holds for the upper ends of each T_r-class.

Theorem 1.5.8 *The mapping $\chi : \rho \mapsto \rho^{T_r}$ ($\rho \in \Gamma(\mathrm{CR}_X)$) is a complete endomorphism of $\Gamma(\mathrm{CR}_X)$.*

Proof By Theorem 1.3.4, T_r is a complete congruence on $\Gamma(\mathrm{CR}_X)$). Hence, by the dual of [PR1, Lemma I.2.2], χ respects arbitrary intersections and, in particular, χ is order preserving.

Now we consider arbitrary joins. In the light of Lemma 1.2.3, it suffices to establish the corresponding result for Δ. So let $\rho_\alpha \in \Delta$ for all $\alpha \in A$. Since χ is order preserving and for all $\alpha \in A$, we have $\rho_\alpha \subseteq \bigvee_{\alpha \in A} \rho_\alpha$, it follows that $\rho_\alpha^{T_r} \subseteq \left(\bigvee_{\alpha \in A} \rho_\alpha \right)^{T_r}$ for all $\alpha \in A$, and therefore that

$$\bigvee_{\alpha \in A} \rho_\alpha^{T_r} \subseteq \left(\bigvee_{\alpha \in A} \rho_\alpha \right)^{T_r}.$$

It remains to establish the reverse inclusion. Once again we break the argument down to several cases, reflecting the possibilities in the statement of Theorem 1.5.6.

Case: There exists $\rho_\beta \in [\zeta_{\mathcal{RG}}, \omega]$. *Then* $\rho_\beta^{T_r} = \omega$ *so that*

$$\bigvee_{\alpha \in A} \rho_\alpha^{T_r} = \omega \supseteq \left(\bigvee_{\alpha \in A} \rho_\alpha \right)^{T_r}$$

and equality follows.

Case: There exists $\rho_\beta \in [\zeta_{\mathcal{CS}}, \zeta_{\mathcal{LZ}}]$ *but no* ρ_α *lies in* $[\zeta_{\mathcal{RG}}, \omega] \cup [\zeta_{\mathcal{RRO}}, \zeta_{\mathcal{S}}]$. *Then* $\rho_\alpha \subseteq \zeta_{\mathcal{LZ}}$ *for all* $\alpha \in A$ *so that*

$$\left(\bigvee_{\alpha \in A} \rho_\alpha \right)^{T_r} \subseteq (\zeta_{\mathcal{LZ}})^{T_r} = \zeta_{\mathcal{LZ}} = \bigvee_{\alpha \in A} \rho_\alpha^{T_r}$$

and equality follows.

Case: There exists $\rho_\beta \in [\zeta_{\mathcal{RRO}}, \zeta_{\mathcal{S}}]$ *but there is no* ρ_α *in* $[\zeta_{\mathcal{CS}}, \omega]$. *Then* $\rho_\alpha \subseteq \zeta_{\mathcal{S}}$, *for all* $\alpha \in A$, *so that* $\rho_\alpha^{T_r} \subseteq \zeta_{\mathcal{S}}$ *while* $\rho_\beta^{T_r} = \zeta_{\mathcal{S}}$. *Hence*

$$\bigvee_{\alpha \in A} \rho_\alpha^{T_r} = \zeta_{\mathcal{S}} = (\zeta_{\mathcal{S}})^{T_r} \supseteq \left(\bigvee_{\alpha \in A} \rho_\alpha \right)^{T_r}$$

and equality follows.

Case: There exists $\rho_\beta \in [\zeta_{\mathcal{LRRO}}, \zeta_{\mathcal{LNB}}]$ *but no* ρ_α *lies in* $[\zeta_{\mathcal{CS}}, \omega] \cup [\zeta_{\mathcal{RRO}}, \zeta_{\mathcal{S}}]$. *Then* $\rho_\alpha \subseteq \zeta_{\mathcal{LNB}}$, *for all* $\alpha \in A$, *so that*

$$\bigvee_{\alpha \in A} \rho_\alpha^{T_r} = \zeta_{\mathcal{LNB}} \supseteq \left(\bigvee_{\alpha \in A} \rho_\alpha \right)^{T_r}$$

and equality follows.

Case: There exists $\rho_\beta \in [\zeta_{\mathcal{RRO}}, \zeta_{\mathcal{S}}]$ *and a* $\rho_\gamma \in [\zeta_{\mathcal{CS}}, \zeta_{\mathcal{LZ}}]$ *but no* ρ_α *lies in* $[\zeta_{\mathcal{RG}}, \omega]$. *Then*

$$\bigvee_{\alpha \in A} \rho_\alpha^{T_r} \supseteq \rho_\beta^{T_r} \vee \rho_\gamma^{T_r} = \zeta_{\mathcal{S}} \vee \zeta_{\mathcal{LZ}} = \omega \supseteq \left(\bigvee_{\alpha \in A} \rho_\alpha \right)^{T_r}$$

and equality follows.

Case: For all $\alpha \in A$, $\rho_\alpha \in [\zeta, \zeta_{\mathcal{LRB}}]$. Let $u, v \in U$ be such that $u \left(\bigvee_{\alpha \in A} \rho_\alpha \right)^s v$. Then there exist $w, t \in U$ with $u = s(w)$, $v = s(t)$, $w \bigvee_{\alpha \in A} \rho_\alpha t$, and therefore there exist $p_0, \ldots, p_n$ and $1, 2, \ldots, n \in A$ such that $w = p_0 \rho_1 p_1 \rho_2 \ldots \rho_n p_n = t$. Then

$$u = s(w) \rho_1^s s(p_1) \rho_2^s \ldots \rho_n^s s(t) = v$$

so that $u \bigvee_{\alpha \in A} \rho_\alpha^s v$. Thus $\left(\bigvee_{\alpha \in A} \rho_\alpha \right)^s \subseteq \bigvee_{\alpha \in A} \rho_\alpha^s$. Since $\rho_\alpha \subseteq \zeta_{\mathcal{LRB}}$ for all $\alpha \in A$, it follows from Theorem 1.5.6 that

$$\left(\bigvee_{\alpha\in A} \rho_\alpha\right)^{T_r} \subseteq \bigvee_{\alpha\in A} \rho_\alpha^{T_r}$$

and equality follows.

Thus χ respects arbitrary joins and meets and the proof is complete.   □

We know, by [PR1, Lemma VI.3.1], that a congruence on a completely regular semigroup is determined by its kernel and its trace. The next two results essentially interpret this fact in the context of the constructions of this chapter.

Lemma 1.5.9 *Let* $\lambda, \rho \in \Delta$, $\lambda_K \subseteq \rho \subseteq \lambda \cap \eta$ *and* $u, v \in U$. *Then*

$$u \, \rho \, v \iff u \, \lambda \, v, \quad c(u) = c(v), \quad s(u) \, \rho^s \, s(v), \quad e(u) \, \rho^e \, e(v).$$

Proof The direct part follows immediately from Corollary 1.5.4 and the hypothesis. Conversely, we have

$$\ker(\lambda/\zeta) = \ker(\lambda_K/\zeta) \subseteq \ker(\rho/\zeta) \subseteq \ker((\lambda \cap \eta)/\zeta) \subseteq \ker(\lambda/\zeta)$$

which implies that $\ker(\rho/\zeta) = \ker(\lambda/\zeta)$. Hence $(uv^{-1})\zeta \in \ker(\lambda/\zeta) = \ker(\rho/\zeta)$ while, from Corollary 1.5.4, $u^0\zeta \ (\rho/\zeta) \ v^0\zeta$. By [PR1, Lemma VI.3.1], $u\zeta \ (\rho/\zeta) \ v\zeta$ and the converse is established.   □

Note that ρ^K is always a suitable choice for λ in Lemma 1.5.9. The corollary to follow is very helpful in the study of word problems since it reduces the word problem for $\zeta_\mathcal{V}$ to the word problems for $\zeta_\mathcal{V}^K$, $\zeta_\mathcal{V}^s$ and $\zeta_\mathcal{V}^e$.

Corollary 1.5.10 *Let* $\mathcal{V} \in [\mathcal{S}, \mathcal{CR}]$ *and* $u, v \in U$. *Then*

$$u \, \zeta_\mathcal{V} \, v \iff u \, \zeta_\mathcal{V}^K \, v, \quad c(u) = c(v), \quad s(u) \, \zeta_\mathcal{V}^s \, s(v), \quad e(u) \, \zeta_\mathcal{V}^e \, e(v).$$

Basic properties of ρ^s can be found in Polák [1] and Corollary 1.5.4 can be found in Polák [2]. For Theorem 1.5.5, see Pastijn [1, 2] and Polák [2]. Pastijn [2] established the last part of Theorem 1.5.6 and showed that the mapping $\rho \mapsto \rho^s$ extends to a complete endomorphism of $\Gamma(\mathrm{CR}_X)$, which implies Theorem 1.5.8.

1.6   Congruence ρ_{K_r}

In this section we consider certain combinations of the operators introduced earlier in this chapter. Keeping in mind that $K_r = K \cap T_r$, we characterize the congruence ρ_{K_r} in the same general spirit as we did ρ_K and ρ^{T_r}, thereby obtaining some unexpected results. In particular $\rho_{K_r} = \rho_K \vee \rho_{T_r}$ where we know next to nothing about ρ_{T_r}. But direct consideration of ρ_{K_r} in this section will lead to some of its fundamental properties, properties that are important when studying the detailed structure of K-classes.

By [PR1, Lemma VII.4.13], for any $\rho \in \Gamma(CR_X)$ we have $\rho_{K_r} = \rho_K \vee \rho_{T_r}$. In general, the join of two congruences does not lend itself to a simple direct description. In this case, however, it is possible to do so, provided we work with the corresponding congruences on U.

Notation 1.6.1 For any $\rho \in \Delta$, define a relation ρ_e on U as follows: for $u, v \in U$,

$$u \, \rho_e \, v \quad \text{if} \quad e^k(u) \, (\rho \cap \eta) \, e^k(v), \quad \text{for all} \ k \ge 0.$$

The relation ρ_s is defined dually. For any $t_1, \ldots, t_n \in \{s, e\}$ with $t_i \ne t_j$, we define $\rho_{t_1 t_2 \cdots t_n}$ inductively by $\rho_{t_1 t_2 \cdots t_n} = (\rho_{t_1 t_2 \cdots t_{n-1}})_{t_n}$.

In part (ii) of the next lemma, we will take advantage of the fact that when calculating $e^k(u)$ for $u \in U$ we do not need to eliminate unmatched parentheses at each step for $e(u)$, then $e^2(u)$ and so on but can leave the elimination of matched parentheses to the final stage, as in

$$e^2(((x)^{-1}(yz)^{-1})^{-1}) = e(\overline{()^{-1}(yz)^{-1})^{-1}}) = \overline{z)^{-1})^{-1}} = z.$$

Lemma 1.6.2 *Let* $\rho \in \Delta$.

(i) $\zeta \subseteq \rho_e = (\rho \cap \eta)_e \subseteq \rho \cap \zeta_{RRB}$.   (iii) $(\rho_e)_K = \rho_K = (\rho_K)_e$.

(ii) $\rho_e \in \Delta$.   (iv) $\overline{\rho} = \bigcap_{w \in \{s,e\}^*} \rho_w$.

Proof (i) That $\rho_e \subseteq \zeta_{RRB}$ follows immediately from the definition of ρ_e and the dual of [PR1, Lemma V.1.4]. The remaining inclusions in part (i) are immediate consequences of the definitions of ζ and ρ_e.

(ii) The first step is to show that ρ_e is a congruence. By part (i) we may assume that $\eta \subseteq \rho$. Clearly ρ_e is an equivalence relation. Now let $p, u, v \in U$ be such that $u \, \rho_e \, v$. If $c(p) \subseteq c(u) = c(v)$, then

$$e^k(pu) = e^k(u) \, (\rho \cap \eta) \, e^k(v) = e^k(pv) \qquad \text{for all } k \ge 0.$$

Hence $pu \, \rho_e \, pv$. Alternatively, $c(p) \setminus c(u) \ne \emptyset$ say $c(p) \setminus c(u) = \{x_1, \ldots, x_m\}$. Then

$$e^i(pu) = e^i(p)u \, (\rho \cap \eta) \, e^i(p)v = e^i(pv) \qquad \text{for } 0 \le i \le m,$$
$$e^{m+i}(pu) = e^i(u) \, (\rho \cap \eta) \, e^i(v) = e^{m+i}(pv) \qquad \text{for } 1 \le i \le \#(u) = \#(v).$$

Thus $e^k(pu) \, (\rho \cap \eta) \, e^k(pv)$ for all $k \ge 0$ and $pu \, \rho_e \, pv$.

With regard to right compatibility, if $c(u) = c(v) \subseteq c(p)$, then

$$e^0(up) = up \ (\rho \cap \eta) \ vp = e^0(vp),$$
$$e^k(up) = e^k(p) = e^k(vp), \quad \text{for } k \geq 1.$$

Thus $up \ \rho_e \ vp$. Now suppose that $c(u) \setminus c(p) = c(v) \setminus c(p) = \{x_1, \ldots, x_m\}$. By the hypothesis that $u \ \rho_e \ v$, we have

$$e^i(up) = e^i(u)p \ (\rho \cap \eta) \ e^i(v)p \qquad \text{for } 0 \leq i \leq m,$$
$$e^{m+i}(up) = e^i(p) = e^{m+i}(vp) \qquad \text{for } 1 \leq i \leq \#(u) - m.$$

In other words, $e^k(up) \ (\rho \cap \eta) \ e^k(vp)$ for $k \geq 0$ and $up \ \rho_e \ vp$. Therefore $up \ \rho_e \ vp$ and ρ_e is a congruence.

Let φ be an endomorphism of U. Recall that this means that φ is a *unary* endomorphism of the unary semigroup U.

Define an endomorphism φ^* on P^+ by stipulating that φ^* agrees with φ on X and leaves the parentheses " (" and ")$^{-1}$" fixed.

Claim: $\varphi^*|_U = \varphi$. This is clear.

Note that, for any $x \in X$, $x\varphi \in U$ so that $x\varphi$ has no unmatched parentheses. Therefore, when φ^* is applied to a word $w \in P^+$, it introduces no new unmatched parentheses. Consequently we have that $\overline{w\varphi^*} = (\overline{w})\varphi^*$.

We now address the claim that ρ_e is fully invariant. Let φ be an endomorphism of U and let φ^* be the endomorphism of P^+ defined as above.

Let $u, v \in U$ and $u \ \rho_e \ v$. In order to establish that $u\varphi \ \rho_e \ v\varphi$, we must show that $e^k(u\varphi) \ (\rho \cap \eta) \ e^k(v\varphi)$ for all $k \geq 0$.

We have seen that $\rho_e \subseteq \zeta_{\mathcal{RRB}}$ so that, by the dual of [PR1, Lemma V.1.4], u and v have the same final parts, say $f(u) = f(v) = x_1 x_2 \cdots x_m$. This enables us to write

$$u = \ u_1 x_1 u_2 \cdots u_m x_m)^{-1} \ldots)^{-1}, \quad v = \ v_1 x_1 v_2 \cdots v_m x_m)^{-1} \ldots)^{-1}$$

for some $u_i, v_i \in P^*$, where there may be no final closing parentheses ")$^{-1}$" and where

$$x_i \notin c \ (u_{i+1} x_{i+1} \cdots x_m) \cup c \ (v_{i+1} x_{i+1} \cdots x_m) \,.$$

It is possible that $c(x_i\varphi) \cap c(x_j\varphi) \neq \emptyset$ for some distinct i, j. So we let

$$c(x_i\varphi) \setminus c((u_{i+1}x_{i+1} \cdots x_m)\varphi^*) = \{x_{i1}, \ldots, x_{in_i}\}, \ i \neq m$$
$$c(x_m\varphi) = \{x_{m1}, \ldots, x_{mn_m}\}.$$

where some of these sets may be empty, that is, for some i we may have $n_i = 0$. Let $x_{i1}, \ldots, x_{in_i}$ be the order in which these variables appear in $x_i\varphi$ (and therefore $u\varphi$) for the last time. Since $c(u) = c(v)$, we must have $c(u\varphi) = c(v\varphi)$. Note that every variable in

$c(u\varphi) = c(v\varphi)$ appears as one of the variables x_{ij} and, importantly, that the final parts of $u\varphi$ and $v\varphi$ are both equal to

$$x_{11}x_{12}\cdots x_{1n_1}x_{21}\cdots x_{2n_2}\cdots x_{m1}x_{m2}\cdots x_{mn_m}.$$

Thus $e^k(u\varphi)\ \eta\ e^k(v\varphi)$ for all $k \geq 0$. Therefore it remains to show that

$$e^k(u\varphi)\ \rho\ e^k(v\varphi) \qquad \text{for all } k \geq 0. \tag{1.6.1}$$

The assumption that $u\ \rho_e\ v$ implies that $u\ \rho\ v$. Since ρ is fully invariant we then have $u\varphi\ \rho\ v\varphi$ and (1.6.1) holds for $k = 0$.

Now assume that $e^k(u\varphi)\ \rho\ e^k(v\varphi)$ for $0 \leq k \leq n$ and consider $e^{n+1}(u\varphi)$, $e^{n+1}(v\varphi)$. Since $c(u\varphi) = c(v\varphi)$, if $e^{n+1}(u\varphi) = \emptyset$ then so also does $e^{n+1}(v\varphi) = \emptyset$ and vice versa. So we may assume that $e^{n+1}(u\varphi) \neq \emptyset \neq e^{n+1}(v\varphi)$.

Let ℓ be the least integer such that $n_1 + \cdots + n_\ell \geq n + 1$ and let d be such that $n + 1 = n_1 + \cdots + n_{\ell-1} + d$. From the definition of ℓ, we must have $d \geq 1$. Note that this includes the possibility that $d = n_\ell$. Let

$$(x_\ell)\varphi^* = (x_\ell)\varphi = a_{\ell 1}x_{\ell 1}a_{\ell 2}x_{\ell 2}\cdots a_{\ell n_\ell}x_{\ell n_\ell}a_{\ell n_\ell + 1}$$

where $a_{\ell j} \in P^*$, $x_{\ell j} \in X$ and $x_{\ell j} \notin c(a_{\ell j+1}x_{\ell j+1}\cdots a_{\ell n_\ell + 1})$ for $j = 1, \ldots, n_\ell$. Since $x\varphi \in U$, for all $x \in X$, there are no unmatched parentheses in $(x_\ell)\varphi$ and therefore there are no unmatched parentheses of the form "(" in any final segment of $\varphi(x_\ell)$ of the form $a_{\ell j}x_{\ell j}\cdots a_{\ell n_\ell + 1}$. Similarly, there are no unmatched parentheses in $x_\ell\varphi^*$. We have (where there may or may not be a final series of parentheses of the form ")$^{-1}$")

$$u\varphi = u_1x_1u_2x_2u_3\cdots u_mx_m)^{-1}\ldots)^{-1}\varphi$$
$$= u_1\varphi^*(x_1\varphi)u_2\varphi^*(x_2\varphi)\cdot u_{\ell-1}\varphi^*(x_{\ell-1}\varphi)u_\ell\varphi^*(x_\ell\varphi)u_{\ell+1}\varphi^*\cdot(x_m\varphi))^{-1}\cdot)^{-1}.$$

From this we find that

$$e^{n_1}(u\varphi) = a_{1n_1+1}(e(u)\varphi)$$
$$= a_{1n_1+1}u_2\varphi^*(x_2\varphi)\cdots(x_m\varphi))^{-1}\ldots)^{-1}.$$
$$e^{n_1+n_2\cdots n_{\ell-1}}(u\varphi) = a_{\ell-1n_{\ell-1}}(e^{\ell-1}(u))\varphi$$
$$= a_{\ell-1n_{\ell-1}}u_\ell\varphi^*(x_\ell\varphi)u_{\ell+1}\varphi^*\cdots(x_m\varphi))^{-1}\cdots)^{-1}$$
$$= a_{\ell-1n_{\ell-1}}u_\ell\varphi^*a_{l1}x_{\ell 1}a_{\ell 2}x_{\ell 2}\cdots a_{\ell d}x_{\ell d}a_{\ell d+1}x_{\ell d+1}\cdots$$
$$a_{\ell n_\ell}x_{\ell n_\ell}a_{\ell n_\ell + 1}(e^\ell(u))\varphi))^{-1}\ldots)^{-1}.$$

Hence

$$e^{n+1}(u\varphi) = \overline{a_{\ell d+1} x_{\ell d+1} \cdots a_{\ell n_\ell} x_{\ell n_\ell} a_{\ell n_\ell+1} (e^\ell(u))\varphi))^{-1} \ldots)^{-1}},$$

$$e^{n+1}(v\varphi) = \overline{a_{\ell d+1} x_{\ell d+1} \cdots a_{\ell n_\ell} x_{\ell n_\ell} a_{\ell n_\ell+1} (e^\ell(v))\varphi))^{-1} \ldots)^{-1}}.$$

By the induction hypothesis, we have $e^\ell(u) \rho \, e^\ell(v)$. In [PR1, Section I.10], we saw how to extend the congruence ρ (on U) to a congruence ρ^P on P^+. It then follows that

$$a_{\ell d+1} x_{\ell d+1} \cdots a_{\ell n_\ell} x_{\ell n_\ell} a_{\ell n_\ell+1} (e^\ell(u))\varphi))^{-1} \ldots)^{-1}$$
$$\rho^P \, a_{\ell d+1} x_{\ell d+1} \cdots a_{\ell n_\ell} x_{\ell n_\ell} a_{\ell n_\ell+1} (e^\ell(v))\varphi))^{-1} \ldots)^{-1}.$$

Since $e^\ell(u), e^\ell(v) \in U$, the deletion of unmatched parentheses in $e^{n+1}(u\varphi), e^{n+1}(v\varphi)$ introduces no new differences between $e^{n+1}(u\varphi)$ and $e^{n+1}(v\varphi)$. Hence we can deduce that

$$e^{n+1}(u\varphi) = \overline{a_{\ell d+1} x_{\ell d+1} \cdots a_{\ell n_\ell} x_{\ell n_\ell} a_{\ell n_\ell+1} (e^\ell(u))\varphi))^{-1} \ldots)^{-1}}$$
$$\rho^P \, \overline{a_{\ell d+1} x_{\ell d+1} \cdots a_{\ell n_\ell} x_{\ell n_\ell} a_{\ell n_\ell+1} (e^\ell(v))\varphi))^{-1} \ldots)^{-1}} = e^{n+1}(v\varphi).$$

By [PR1, Lemma I.10.6], we have $\rho^P|_U = \rho$ from which we get $e^{n+1}(u\varphi) \, \rho \, e^{n+1}(v\varphi)$.

By induction, we have now shown that $e^k(u\varphi) \rho \, e^k(v\varphi)$ for all $k \geq 0$ so that $u\varphi \, \rho_e \, v\varphi$ and therefore ρ_e is a fully invariant congruence, that is, $\rho_e \in \Delta$.

(iii) Recall that by Theorem 1.4.9, $\rho_K = \overline{\rho}$ for all $\rho \in \Delta$. By part (i) we have $\rho_e \subseteq \rho$ and, by Lemma 1.4.4(ii), it follows that $(\rho_e)_K \subseteq \rho_K$. On the other hand, by Lemma 1.4.2 and Theorem 1.4.9, for any $u, v \in U, u \, \rho_K \, v \Rightarrow u \, \rho_e \, v$. Thus $\rho_K \subseteq \rho_e$ so that $(\rho_e)_K \subseteq \rho_K \subseteq \rho_e$. We now have $(\rho_e)_K \subseteq \rho_K = (\rho_K)_K \subseteq (\rho_e)_K$ whence $(\rho_e)_K = \rho_K$. Additionally $\rho_K = (\rho_K)_K = ((\rho_K)_e)_K \subseteq (\rho_K)_e \subseteq \rho_K$ which implies that $\rho_K = (\rho_K)_e$.

(iv) This follows immediately from the definitions of $\overline{\rho}, \rho_s$ and ρ_e. $\square$

We will now establish the connection between $(\rho/\zeta)_{K_r}$ and ρ_e.

Theorem 1.6.3 *Let $\rho \in \Delta$. Then*

$$(\rho/\zeta)_{K_r} = \begin{cases} (\overline{\rho} \vee \zeta_{\mathcal{RG}})/\zeta & \text{if } \zeta_{\mathcal{RG}} \subseteq \rho \subseteq \omega, \\ (\overline{\rho} \vee \zeta_{\mathcal{CS}})/\zeta & \text{if } \zeta_{\mathcal{CS}} \subseteq \rho \subseteq \zeta_{\mathcal{LZ}}, \\ \rho_e/\zeta & \text{if } \rho \subseteq \eta. \end{cases}$$

Moreover, for $\rho \subseteq \eta$, $(\rho/\zeta)_{K_r K} = (\rho/\zeta)_{K K_r} = (\rho/\zeta)_K$.

Proof Let $\zeta_{\mathcal{RG}} \subseteq \rho \subseteq \omega$. By Theorem 1.4.9, we have $\rho_K = \overline{\rho}$ and by Theorem 1.5.5, we have $\rho_{T_r} = \zeta_{\mathcal{RG}}$ so that $\rho_{K_r} = \rho_K \vee \rho_{T_r} = \overline{\rho} \vee \zeta_{\mathcal{RG}}$. Likewise, if $\zeta_{\mathcal{CS}} \subseteq \rho \subseteq \zeta_{\mathcal{LZ}}$ then by the same references, we obtain $\rho_{K_r} = \rho_K \vee \rho_{T_r} = \overline{\rho} \vee \zeta_{\mathcal{CS}}$. This covers the first two cases, so now let us consider the case $\rho \subseteq \eta$. By Lemma 1.6.2(iii), we have $\rho_K = (\rho_e)_K \subseteq \rho_e$. In order to show that $\rho_{K_r} \subseteq \rho_e$, we consider three separate cases.

Case: $\rho \subseteq \zeta_{\mathcal{LRB}}$. Then $\rho_e \subseteq \rho \subseteq \zeta_{\mathcal{LRB}}$. Let $u, v \in U$ be such that $u \, \rho^s \, v$. There exist $w, t \in U$ such that $u = s(w)$, $v = s(t)$ and $w \, \rho \, t$. Since $c(w) = c(t) = \{x_1, \ldots, x_n\}$, say, it is straightforward to verify that

$$w x_1 \ldots x_n \, \rho_e \, t x_1 \ldots x_n, \quad u = s(w x_1 \ldots x_n), \quad v = s(t x_1 \ldots x_n)$$

so that $u \, (\rho_e)^s \, v$. Thus $\rho^s \subseteq (\rho_e)^s$. Since the reverse inclusion follows immediately from the fact that $\rho_e \subseteq \rho$, we have established the identity: $\rho^s = (\rho_e)^s$. By Corollary 1.5.7, we have $\rho \, T_r \, \rho_e$ and $\rho_{T_r} = (\rho_e)_{T_r} \subseteq \rho_e$. Consequently, by Lemma 1.6.2(iii)(iii), $\rho_{K_r} = \rho_K \vee \rho_{T_r} \subseteq \rho_e$.

Case: $\rho \in [\zeta_{\mathcal{LRRO}}, \zeta_{\mathcal{LNB}}]$. By [PR1, Theorem V.4.3], for any $a, x, y \in X$ we have $(yxa, (ya)^0 yxa) \in \zeta_{\mathcal{LRRO}} \subseteq \rho$. In addition, $e^k(yxa) = e^k((ya)^0 yxa)$ for all $k \geq 1$ and so $(yxa, (ya)^0 yxa) \in \rho_e$. Since the identity $yxa = (ya)^0 yxa$ defines the variety $\mathcal{LRRO}$, it follows that $\zeta_{\mathcal{LRRO}} \subseteq \rho_e$ so that $\rho_e \in [\zeta_{\mathcal{LRRO}}, \rho] \subseteq [\zeta_{\mathcal{LRRO}}, \zeta_{\mathcal{LNB}}]$ and therefore, by Theorem 1.5.5(i), we obtain $\rho_{T_r} = \zeta_{\mathcal{LRRO}} = (\rho_e)_{T_r} \subseteq \rho_e$. Hence $\rho_{K_r} = \rho_K \vee \rho_{T_r} \subseteq \rho_e$.

Case: $\rho \in [\zeta_{\mathcal{RRO}}, \zeta_S]$. For any $a, x \in X$ we have $(xa, a^0 xa) \in \zeta_{\mathcal{RRO}} \subseteq \rho$, by [PR1, Lemma V.3.1]. Clearly $e^k(xa) = e^k(a^0 xa)$ for all $k \geq 1$ and so $(xa, a^0 xa) \in \rho_e$. Since the identity $xa = a^0 xa$ defines the variety $\mathcal{RRO}$, it follows that $\zeta_{\mathcal{RRO}} \subseteq \rho_e$ so that $\rho_e \in [\zeta_{\mathcal{RRO}}, \rho] \subseteq [\zeta_{\mathcal{RRO}}, \eta]$ and therefore, by Theorem 1.5.5(i), we get $\rho_{T_r} = \zeta_{\mathcal{RRO}} = (\rho_e)_{T_r} \subseteq \rho_e$. Hence $\rho_K \vee \rho_{T_r} \subseteq \rho_e$. Since $\rho \subseteq \eta$, it follows from [PR1, Corollary IX.5.3(i)] that we have considered all the possible cases and that in all three cases $\rho_{K_r} = \rho_K \vee \rho_{T_r} \subseteq \rho_e$.

Now we consider the reverse inclusion.

Let $u \, \rho_e \, v$ and $\lambda = \rho_{K_r} = \rho_K \vee \rho_{T_r}$. By hypothesis, $\rho \subseteq \eta$. We will show by induction on $\#(u) = \#(v)$ that $u \, \lambda \, v$. Since $\rho_K \subseteq \lambda \subseteq \rho_e \subseteq \rho$, we have that $\lambda \, K \, \rho_e$. If $c(u) = c(v) = \{x\}$, where $x \in X$, then $u \zeta \, \mathcal{H} \, v \zeta$. By [PR1, Corollary VII.1.8], λ / ζ and ρ_e / ζ agree on $H_{u\zeta}$. Therefore λ and ρ_e agree on the set of elements with content of size one and the claim holds for $\#(u) = 1$.

So now suppose that $\#(u) = \#(v) = n + 1 > 1$ and that the claim holds for elements of smaller content, that is, $\rho_e|_{U_n^*} = \lambda|_{U_n^*}$. The assumption that $\rho \subseteq \eta$, by the dual of [PR1, Theorem I.5.5], implies that

$$\rho_{T_r} \subseteq \eta_{T_r} = \zeta_{\mathcal{RRO}} \subseteq \zeta_{\mathcal{RRB}}. \tag{1.6.2}$$

Clearly $\eta \, K \, \zeta_{\mathcal{B}}$ so that $\eta_K \subseteq \zeta_{\mathcal{B}}$. The fact that $\rho \subseteq \eta$ implies, by Lemma 1.4.4(ii) and Theorem 1.4.9, that

$$\rho_K \subseteq \eta_K \subseteq \zeta_{\mathcal{B}} \subseteq \zeta_{\mathcal{RRB}}. \tag{1.6.3}$$

Combining (1.6.2) and (1.6.3), we obtain

$$\lambda = \rho_K \vee \rho_{T_r} \subseteq \zeta_{\mathcal{RRB}}. \tag{1.6.4}$$

It follows immediately from $u\ \rho_e\ v$ and the definition of ρ_e that $e(u)\ \rho_e\ e(v)$ and so, by the induction hypothesis, $e(u)\ \lambda\ e(v)$. Since, by the dual of Lemma 1.5.2(i), $\lambda \subseteq \lambda^e$, this means that $e(u)\ \lambda^e\ e(v)$ and therefore, by (1.6.4) and the dual of Lemma 1.5.2, that $u\lambda\ \mathcal{L}\ v\lambda$. Now $\rho_{T_r} \subseteq \lambda \subseteq \rho_e \subseteq \rho$. By the dual of [PR1, Corollary VII.4.9(iv)], we have $\rho/\rho_{T_r} \subseteq \mathcal{R}$ which by the dual of [PR1, Lemma VI.4.9] implies that ρ/ρ_{T_r} is over $\mathcal{RG}$. Hence ρ_e/λ is also over $\mathcal{RG}$ and, since $u\ \rho_e\ v$, we have $u\lambda\ \mathcal{R}\ v\lambda$. Therefore $u\lambda\ \mathcal{H}\ v\lambda$. But we also have that $\lambda\ K\ \rho_e$ which implies that the restriction of ρ_e/λ to any $\mathcal{H}$-class is the identity relation. Since $u\rho_e = v\rho_e$ we must have $u\lambda = v\lambda$ and $\rho_e \subseteq \lambda$. Thus $\rho_e|_{U^*_{n+1}} = \lambda_{U^*_{n+1}}$. Hence, by induction, $\rho_e = \lambda$ and $\rho_e = \rho_{K_r}$.

The final claim follows from Lemma 1.6.2(iii). $\square$

We have arrived at the main result of this section.

Theorem 1.6.4 *The mapping* $\theta : \rho \mapsto \rho_{K_r}\ (\zeta \subseteq \rho \subseteq \eta)$ *is a complete endomorphism of* $[\zeta, \eta]$.

Proof In order to show that this mapping respects arbitrary intersections, it suffices by Theorem 1.6.3 to show that the mapping $\chi : \rho \mapsto \rho_e$ respects arbitrary intersections in the stated interval. Towards that end, for any $\rho_\alpha \in \Delta, \alpha \in A$, and any $u, v \in U$, we have

$$u \left(\bigcap_{\alpha \in A} \rho_\alpha \right)_e v \iff e^k(u) \left(\left(\bigcap_{\alpha \in A} \rho_\alpha \right) \cap \eta \right) e^k(v) \quad \text{for all} \quad k \geq 0$$

$$\iff e^k(u)\ (\rho_\alpha \cap \eta)\ e^k(v) \quad \text{for all} \quad k \geq 0,\ \alpha \in A$$

$$\iff u \bigcap_{\alpha \in A} (\rho_\alpha)_e\ v.$$

Thus, χ respects arbitrary intersections, as required. By [PR1, Lemma I.2.2] and the fact that K_r is a complete congruence on Δ, we know that the mapping $\rho \mapsto \rho_{K_r}$ respects arbitrary joins and so the proof is complete. $\square$

Most of this section is an amplification of certain results due to Polák [3].

1.7   Arguesian Property

This section differs from the three preceding ones: instead of studying a single congruence, we study the totality of congruences of the stipulated kind. It may come as a surprise that this imposing edifice, the lattice $\mathcal{L}(\mathcal{CR})$ of varieties of completely regular semigroups, which seems so complicated, almost forbidding, and of which, so far, we were able to comprehend only small chips, should turn out to be arguesian. This is deduced from the remarkable fact that the fully invariant congruences on CR_X above $\rho_{\mathcal{ES}}$ and those below $\rho_{\mathcal{S}}$ commute among themselves. The proof of the latter again demonstrates the power of the construction of ρ^s. Indeed, the latter makes use of the invariant s which, in its turn, takes advantage of the "layered" structure of U induced by the cardinality of the contents of words. This may be the clue to the undoubted success of this approach of which the arguesian property seems only one, albeit very important, instance.

It would be appropriate for the reader to review [PR1, Sections I.2 and I.3] for the definitions and basic results concerning the arguesian property.

We begin with $\Gamma(CS_X)$ where we can take advantage of the characterization of $\Gamma(CS_X)$ given in [PR1, Theorem VIII.3.5].

Proposition 1.7.1 $\Gamma(CS_X)$ *is a lattice of commuting congruences.*

Proof Let $\lambda = \rho_{(r,N,\pi)}$, $\rho = \rho_{(t,M,\nu)} \in \Gamma(CS_X)$. By [PR1, Theorem VIII.3.5], the following cases exhaust all the possibilities.

Case: $r = \pi = \varepsilon$ *or* $t = \nu = \varepsilon$. Then either $\lambda \subseteq \mathcal{H}$ or $\rho \subseteq \mathcal{H}$ and, since $\mathcal{D} = \mathcal{J} = \omega$ on CS_X, the claim follows from [PR1, Lemma IX.2.4].

Case: $r = \nu = \varepsilon$, $\pi = t = \omega$. Then $\lambda \subseteq \mathcal{R}$ and $\rho \subseteq \mathcal{L}$ and the claim follows from [PR1, Lemma I.7.2].

Case: $\pi = t = \varepsilon$, $r = \nu = \omega$. This case is the dual of the preceding case.

Case: $r = t = \varepsilon$, $\pi = \nu = \omega$. Then $\lambda, \rho \subseteq \mathcal{R}$. Let $a, b, c \in CS_X$ be such that $a \lambda b \rho c$. Since $\nu = \omega$ and $a \mathcal{R} b$, it follows that $a\rho \mathcal{H} b\rho$. Hence $a^0\rho b^0$ and since $b \rho c$ implies that $b \mathcal{R} c$,

$$a = aa^0\rho \, ab^0 = ab^{-1}b \, \rho \, ab^{-1}c \, \lambda \, bb^{-1}c = c.$$

Thus $a \rho\lambda c$ so that $\lambda\rho \subseteq \rho\lambda$ and equality follows by symmetry.

Case: $\pi = \nu = \varepsilon$, $r = t = \omega$. This is the dual of the preceding case.

Case: At least three of r, π, t, ν *equal* ω. As these cases are all similar, it suffices to consider just one of them, say the case where $r = \pi = t = \omega$. Let a, b, c be as above. Since $t = \omega$, we have $a \rho (ca)^0$ while, since $r = \pi = \omega$, we have $b^{-1}b \lambda c^{-1}c$. Hence

$$a = a^0 a \, \rho \, (ca)^0 a = c^0(ca)^0 a \, \rho \, c^0 a^0 a = c^0 a$$
$$= cc^{-1}a \, \rho \, cb^{-1}a \, \lambda \, cb^{-1}b \, \lambda \, cc^{-1}c = c,$$

so that $a \rho\lambda c$. Thus $\lambda\rho \subseteq \rho\lambda$ and equality follows by symmetry. □

Before considering the interval $[\mathcal{S}, C\mathcal{R}]$ we require some preparation.

Corollary 1.7.1 *Let $u, v \in U$.*

(i) *If $\mathcal{V} \in [\mathcal{J}, \mathcal{RG}]$, then $u\zeta_\mathcal{V} \mathcal{R} v\zeta_\mathcal{V}$.*
(ii) *If $\mathcal{V} \in [\mathcal{LZ}, \mathcal{CS}]$, then $u\zeta_\mathcal{V} \mathcal{R} v\zeta_\mathcal{V}$ if and only if $h(u) = h(v)$.*
(iii) *If $\mathcal{V} \in [\mathcal{S}, \mathcal{RRO}]$, then $u\zeta_\mathcal{V} \mathcal{R} v\zeta_\mathcal{V}$ if and only if $c(u) = c(v)$.*
(iv) *If $\mathcal{V} \in [\mathcal{LNB}, \mathcal{LRRO}]$, then $u\zeta_\mathcal{V} \mathcal{R} v\zeta_\mathcal{V}$ if and only if $h(u) = h(v)$ and $c(u) = c(v)$.*

Proof We consider only part (iv). The arguments for parts (i)–(iii) are similar but simpler.

Let $\mathcal{V} \in [\mathcal{LNB}, \mathcal{LRRO}]$ and $F = F\mathcal{V}_X$. By [PR1, Theorem IX.5.1(iv)] there exists an $\mathcal{LNB}$-congruence ρ on F over $\mathcal{RG}$. By the dual of [PR1, Lemma VI.4.9], we have $\rho \subseteq \mathcal{R}$. Consequently

$$u\zeta_\mathcal{V} \mathcal{R} v\zeta_\mathcal{V} \iff u\zeta_{\mathcal{LNB}} \mathcal{R} v\zeta_{\mathcal{LNB}}.$$

But by [PR1, Lemma V.1.2], $\mathcal{R}$ is trivial on $U/\zeta_{\mathcal{LNB}}$. Therefore, by [PR1, Theorem V.1.9(v)], we obtain

$$u\zeta_{\mathcal{LNB}} \mathcal{R} v\zeta_{\mathcal{LNB}} \iff u\zeta_{\mathcal{LNB}} = v\zeta_{\mathcal{LNB}} \iff h(u) = h(v), \quad c(u) = c(v)$$

and the claim follows.   □

We are now able to prove the central result of this section.

Theorem 1.7.2 *The interval $[\zeta, \eta]$ in Δ is a lattice of commuting congruences.*

Proof Let $\lambda, \rho \in [\zeta, \eta]$. Then any λ-related, ρ-related, $\lambda\rho$-related or $\rho\lambda$-related elements must have the same content. We will prove by induction on n that

$$(\lambda\rho)\big|_{U_n} = (\rho\lambda)\big|_{U_n}. \tag{1.7.1}$$

Let u, v and w be such that

$$u \lambda v \rho w. \tag{1.7.2}$$

Then $c(u) = c(v) = c(w)$ and $u\zeta \mathcal{D} v\zeta \mathcal{D} w\zeta$ by Theorem 1.1.9(i).

First let $u, v, w \in U_1$ so that $c(u) = \{x\}$, say. Then $u\zeta, v\zeta, w\zeta \in H_{x\zeta}$ and we have $u\zeta (\lambda/\zeta) v\zeta (\rho/\zeta) w\zeta$. But $H_{x\zeta}$ is a group and so the restrictions of λ/ζ and ρ/ζ to $H_{x\zeta}$ commute. Hence there exists an element $v^* \in U_1$ such that

$$u\zeta (\rho/\zeta) v^*\zeta (\lambda/\zeta) w\zeta$$

whence $u \rho v^* \lambda w$ so that $u \rho\lambda w$. Thus $\lambda\rho \subseteq \rho\lambda$ on U_1 and the claim holds for $n = 1$ by symmetry.

Now suppose that (1.7.1) holds for n, for all fully invariant congruences $\lambda, \rho \in [\zeta, \eta]$ and consider the restrictions of $\lambda, \rho, \lambda\rho$ and $\rho\lambda$ to U_{n+1}. Let $u, v, w \in U_{n+1}$ be such that (1.7.2) holds. Our first goal is to establish the following claim:

$$\text{there exists } b \in U \text{ with } u\rho \; \mathcal{R} \; b\rho \text{ and } b\lambda \; \mathcal{R} \; w\lambda. \tag{1.7.3}$$

We consider five cases, all but one of which are straightforward. By [PR1, Corollary IX.5.3(i)] these cases exhaust all the possibilities.

Case: $\zeta_{\mathcal{RRO}} \subseteq \lambda \subseteq \zeta_{\mathcal{S}}$. Then $c(u) = c(w)$ and Corollary 1.7.1(iii) implies that we have $u\rho \; \mathcal{R} \; u\rho$ and $u\lambda \; \mathcal{R} \; w\lambda$.

Case: $\zeta_{\mathcal{RRO}} \subseteq \rho \subseteq \zeta_{\mathcal{S}}$. Then $c(u) = c(w)$ and Corollary 1.7.1(iii) implies that $u\rho \; \mathcal{R} \; w\rho$ and $w\lambda \; \mathcal{R} \; w\lambda$.

Case: $\zeta_{\mathcal{LRRO}} \subseteq \lambda \subseteq \zeta_{\mathcal{LNB}}$ and $\rho \subseteq \zeta_{\mathcal{LNB}}$. Then $h(u) = h(v) = h(w)$ and $c(u) = c(w)$. By Corollary 1.7.1(iv), we have $u\rho \; \mathcal{R} \; u\rho$ and $u\lambda \; \mathcal{R} \; w\lambda$.

Case: $\zeta_{\mathcal{LRRO}} \subseteq \rho \subseteq \zeta_{\mathcal{LNB}}$ and $\lambda \subseteq \zeta_{\mathcal{LNB}}$. Then $h(u) = h(v) = h(w)$ and $c(u) = c(v)$. By Corollary 1.7.1(iv), we have $u\rho \; \mathcal{R} \; w\rho$ and $w\lambda \; \mathcal{R} \; w\lambda$.

Case: $\lambda, \rho \subseteq \zeta_{\mathcal{LRB}}$. From (1.7.2), we have $s(u) \; \lambda^s \; s(v) \; \rho^s \; s(w)$. From Lemma 1.5.2(v), it follows that $\lambda^s, \rho^s \in [\zeta, \zeta_{\mathcal{LRB}}]$ and therefore that $c(s(u)) = c(s(v)) = c(s(w))$. But $\#(s(u)) = n$ so that by the induction hypothesis, we can assert that there exists an element $a \in U_n$ with $s(u) \; \rho^s \; a \; \lambda^s \; s(w)$. Note that $c(a) = c(s(u)) = c(s(w))$. Let $x \in X$ be such that $c(p) \setminus c(s(p)) = \{x\}$ for any $p \in \{u, v, w\}$, and let $b = ax$. Then $s(b) = a \; \rho^s \; s(u)$, $s(b) = a \; \lambda^s \; s(w)$ and $c(b) = c(u) = c(w)$. By Lemma 1.5.3, we get $u\rho \; \mathcal{R} \; b\rho$ and $b\lambda \; \mathcal{R} \; w\lambda$.

Therefore we can conclude that (1.7.3) holds for $\#(u) = n + 1$.

By duality, there exists an element $d \in U$ with $c(d) = c(u) = c(w)$ and

$$u\rho \; \mathcal{L} \; d\rho, \quad d\lambda \; \mathcal{L} \; w\lambda. \tag{1.7.4}$$

By (1.7.3) and (1.7.4), we obtain $u^0\rho \; b^0 u^0 \rho \; b^0 u^0 d^0$. Similarly, we have $w^0 \lambda \; b^0 w^0 d^0$. Note that

$$c(u) = c(w) = c(b) = c(d) \implies u\zeta \; \mathcal{D} \; w\zeta \; \mathcal{D} \; b\zeta \; \mathcal{D} \; d\zeta.$$

Hence

$$u^0 \; \rho \; (b^0 u^0 d^0)^0 \; \zeta \; (b^0 d^0)^0 \; \zeta \; (b^0 w^0 d^0)^0 \; \lambda \; w^0.$$

Consequently, with $e = u^0, f = v^0, g = w^0$ and $h = (b^0 u^0 d^0)^0$, we have $e \; \rho \; h \; \lambda \; g$. By Theorem 1.1.9(ii)(iii), $(ef)^0 \zeta \; \mathcal{R} \; u\zeta \; \mathcal{H} \; e\zeta$ and $(gf)^0 \zeta \; \mathcal{R} \; w\zeta \; \mathcal{H} \; g\zeta$.

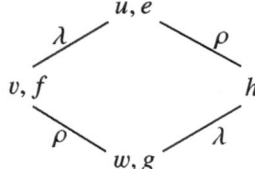

Therefore

$$u \, \zeta \, (ef)^0 ue \, \rho \, (hf)^0 uh \, \lambda \, (hf)^0 vh, \tag{1.7.5}$$

$$w \, \zeta \, (gf)^0 wg \, \lambda \, (hf)^0 wh \, \rho \, (hf)^0 vh. \tag{1.7.6}$$

Now, by Theorem 1.1.9(ii)(iii), the elements

$$p = \big((hf)^0 uh\big)\zeta, \quad q = \big((hf)^0 vh\big)\zeta, \quad r = \big((hf)^0 wh\big)\zeta \tag{1.7.7}$$

all lie in the $\mathcal{H}$-class $H_{h\zeta}$ of CR_X and, by (1.7.5) and (1.7.6), we get $p \, (\lambda/\zeta) \, q \, (\rho/\zeta) \, r$. Since congruences on a group commute, the restrictions of λ/ζ and ρ/ζ to $H_{h\zeta}$ commute so that there must be an element $y \in U$, with $s = y\zeta \in H_{h\zeta}$, such that

$$p \, (\rho/\zeta) \, s \, (\lambda/\zeta) \, r. \tag{1.7.8}$$

From (1.7.5), (1.7.6), (1.7.7) and (1.7.8), we get $u \, \rho \, (hf)^0 uh \, \rho \, y$, and $w \, \lambda \, (hf)^0 wh \, \lambda \, y$ so that $u \, \rho\lambda \, w$. Thus $\lambda\rho \subseteq \rho\lambda$ on U_{n+1} and, by symmetry, we must have $\lambda\rho = \rho\lambda$ on U_{n+1}. By induction, the proof of the theorem is now complete. $\qquad\square$

The main consequence of Theorem 1.7.2 follows.

Theorem 1.7.3 *The lattice $\mathcal{L}(\mathcal{CR})$ is arguesian.*

Proof By [PR1, Lemma I.7.3], the interval $[\zeta, \zeta_S]$ is an arguesian lattice. By the standard antiisomorphism, $[S, \mathcal{CR}]$ is arguesian. On the other hand, $\mathcal{L}(S)$ is just a two-element chain and is also arguesian. By [PR1, Corollary IX.9.3], $\mathcal{L}(\mathcal{CR})$ is a sublattice of the direct product $\mathcal{L}(S) \times [S, \mathcal{CR}]$ and the claim follows. $\qquad\square$

Pastijn [4] proved that fully invariant congruences on an arbitrary completely simple semigroup commute. This implies Proposition 1.7.1. Theorem 1.7.2 was established by Pastijn [2, 4]. The treatment here is modeled on that in Petrich and Reilly [5]. Earlier, Hall and Jones [1] showed that $\mathcal{L}(\mathcal{BG})$ is modular and Reilly [1] that each interval of the form $[\mathcal{V}, \mathcal{V}^T]$ is modular. Completely regular semigroups for which any two congruences commute were determined by Bonzini and Cherubini [1]. Varieties of completely regular semigroups whose free objects have weakly permutable fully invariant congruences were described by Vernikov [1]. Vernikov [2] characterizes those varieties of completely regular semigroups that are lower-modular.

1.8   Review

After the preliminary material on Green's relations and fully invariant relations in the first two sections, the kernel and trace relations are transferred in the third section from fully invariant congruences to varieties. The next two sections have a similar character: we introduce and study certain properties of three important relations on a free unary semigroup in terms of some concrete operations on (unary) words. These are then transferred to the corresponding congruences on a free completely regular semigroup on the same set. In this way, to every fully invariant congruence ρ under the least semilattice congruence one can associate three fully invariant congruences which will play a fundamental role when they are converted to varieties in the next chapter. Somewhat unexpectedly, the first of these turns out to be the lower end of the K-class of ρ, while the second is the upper end of the T_r-class of ρ and the third is again a lower end. We thus arrive at a surprising amalgamation of the kernel-trace approach to congruences and the construction of certain relations on the free unary semigroup by means of an inductive process. This enriches the former and gives a revealing interpretation to the latter. A good concept is worth a hundred theorems. On the one hand, we have relations defined on words, and on the other hand, relations defined on congruence lattices. The latter induce relations on the lattice of varieties as well as some lower and upper ends of the intervals which form their classes. The happy marriage of the first and the last of these concepts corroborates the impression that they are both good concepts, in the sense that they are natural for the objects studied as well as their creators showing remarkable insight and fertile imagination. We may recapitulate, in a nutshell, the main points in this chapter by the following implications

$$\rho/\zeta = \theta \in \Gamma(\mathrm{CR}_X) \implies \rho \in \Delta \implies \overline{\rho} \in \Delta \implies \overline{\rho}/\zeta = \theta_K \in \Gamma(\mathrm{CR}_X),$$

with similar diagrams for ρ^s and ρ_e. The chapter ends by establishing that the lattice $\mathcal{L}(\mathcal{CR})$ is arguesian and therefore also modular.

Kernel and Trace Relations

With this chapter, we initiate a systematic study of the structure of the lattice $\mathcal{L}(\mathcal{CR})$. We start by designing and studying the properties of various decompositions of $\mathcal{L}(\mathcal{CR})$. These are effected by defining various relations which turn out to be complete congruences, so that their classes are intervals, at least some of which may have strong properties.

Kernel and (left, right) trace relations on a free completely regular semigroup of infinite rank restricted to the set of fully invariant congruences can be transferred to $\mathcal{L}(\mathcal{CR})$ via the standard antiisomorphism. Considering this notation in the new roles, we arrive at kernel and (left, right) trace relations on $\mathcal{L}(\mathcal{CR})$, which admit also an intrinsic characterization within $\mathcal{L}(\mathcal{CR})$.

This chapter is dedicated to a detailed study of these relations, some of their variants and certain of their relationships.

The most important objects that we study here are the kernel K- and the trace T-relations. We start with the former, which needs the concept of E-disjunctive semigroups. Related to the T-relation are its one-sided variants T_ℓ- and T_r-relations. In conjunction with the K-relation, they often play an important role. We also study their relatives, namely K_ℓ- and K_r-relations. The role of the last two is secondary, but quite instructive.

This describes in a nutshell, the program of this chapter as the first instalment in the attack on the lattice $\mathcal{L}(\mathcal{CR})$.

The kernel and the trace of a congruence ρ on a completely regular semigroup S determine ρ uniquely. In their turn, they induce the kernel and trace relations on $\mathcal{C}(S)$. These concepts proved of seminal importance for the study of congruences on completely regular semigroups. It was a pleasant surprise when it was revealed that the kernel and trace relations can be transferred from congruences on completely regular semigroups to their varieties. This result quickly led to intrinsic characterizations for these relations independent of their origins, as well as a great variety of results. As a consequence, this produced a great body of

© The Author(s), under exclusive license to Springer Nature Switzerland AG 2024   45
M. Petrich and N. R. Reilly, *Completely Regular Semigroup Varieties*, Synthesis Lectures on Mathematics & Statistics, https://doi.org/10.1007/978-3-031-42891-3_2

knowledge about $\mathcal{L}(\mathcal{CR})$ which otherwise may not have come to the surface. The resulting decompositions of and the induced operators on $\mathcal{L}(\mathcal{CR})$ became a standard arsenal in the study of $\mathcal{L}(\mathcal{CR})$.

These new concepts contributed to the success story, particularly in regard to the emergence of the left and right traces, as well as the left and right kernel relations. As distant relatives, there appeared the local and core relations. Among these relations, the kernel and trace relations retained their basic importance and a suitable tool in the study of the global structure of $\mathcal{L}(\mathcal{CR})$.

Deep results and difficult problems abound in this area of research. Among these a prominent place is held by the kernel relation which still presents formidable difficulties creating lacunae waiting to be filled. The trace relation is an easier customer, while left and right traces slowly replace the trace relation. The introduction of one-sided trace relations gave a strong impulse to this theory.

In this chapter, X will denote a countably infinite set, whose elements are sometimes referred to as variables. We continue to denote by $FV(X)$ or, more simply by FV, the free completely regular semigroup in the variety V on the set X. In addition, we write CR_X for $F\mathcal{CR}(X)$.

2.1   Kernel Relation

We collect here most of the information we have on the varieties V_K and V^K. This includes an expression for V^K in terms of the Mal'cev product, splitting of the values of V_K according to whether $V \in \mathcal{L}(\mathcal{G})$, $V \in \mathcal{L}(L\mathcal{O}) \setminus \mathcal{L}(\mathcal{G})$ or otherwise, the homomorphism properties of the mappings $V \to V_K$ and $V \to V^K$ and statements concerning the solution of the word problem for the varieties V_K and V^K. We shall take advantage of certain results in Sects. 1.3 and 1.7 concerning fully invariant congruences.

Notation 2.1.1 We write

$\mathcal{E}$ for the class of all E-disjunctive completely regular semigroups,

$$\mathcal{PD} = (\mathcal{CS} \cup (\mathcal{CR} \setminus L\mathcal{O})) \cap \mathcal{E},$$
$$\mathcal{PE} = \mathcal{G} \cup (\mathcal{CS} \setminus \mathcal{ReG}) \cup ((\mathcal{CR} \setminus L\mathcal{O}) \cap \mathcal{E}).$$

Clearly $\mathcal{PD}$ consists entirely of E-disjunctive completely regular semigroups and $\mathcal{PD} \subseteq \mathcal{PE}$. However $\mathcal{PD} \neq \mathcal{PE}$ since not all elements of $\mathcal{CS} \setminus \mathcal{ReG}$ are E-disjunctive. Similarly, $\mathcal{PD} \neq \mathcal{E}$ since G^0 is E-disjunctive for any nontrivial group G (so that $G^0 \in \mathcal{E}$) but $G^0 \notin \mathcal{PD}$.

Lemma 2.1.2 (i) *Let* $\mathcal{U}, V \in \mathcal{L}(\mathcal{CR})$. *Then* $\mathcal{U} \ K \ V \Leftrightarrow \zeta_{\mathcal{U}} \ K \ \zeta_V$.
(ii) *For any* $\mathcal{U}, V, W \in \mathcal{L}(\mathcal{CR})$, *we have* $\langle \mathcal{U} \circ \langle V \circ W \rangle \rangle \subseteq \langle \langle \mathcal{U} \circ V \rangle \circ W \rangle$.

Proof (i) This follows from the fact that for $V \in \mathcal{L}(\mathcal{CR})$, we have $\rho_V = \zeta_V/\zeta$.

(ii) It is sufficient to show that $\mathcal{U} \circ \langle V \circ W \rangle \subseteq \langle \langle \mathcal{U} \circ V \rangle \circ W \rangle$. Let $S \in \mathcal{U} \circ \langle V \circ W \rangle$. Then there exists a congruence ρ on S such that $S/\rho \in \langle V \circ W \rangle$ and $e\rho \in \mathcal{U}$ for all $e \in E(S)$. It follows from [PR1, Theorem IX.6.12] that there exists $T \in W$ and $\tau \in \mathcal{RM}(S/\rho, T)$ such that τ is surjective and $f\tau^{-1} \in V$ for all $f \in E(T)$. Let $\rho^{\#} : S \to S/\rho$ be the natural surjective homomorphism defined by $s\rho^{\#} = s\rho$. By [PR1, Lemmas IX.6.3 and IX.6.9], $\rho^{\#}\tau \in \mathcal{RM}(S, T)$ and $\rho^{\#}\tau$ is surjective.

Let $f \in E(T)$. Define $\theta_f : f(\rho^{\#}\tau)^{-1} \to f\tau^{-1}$ by $s\theta = s\rho$. Clearly,

$$\theta_f \in \mathcal{RM}(f(\rho^{\#}\tau)^{-1}, f\tau^{-1})$$

and θ_f is surjective. By [PR1, Lemma II.3.3(iii)], for any $h \in E(f\tau^{-1})$, we have $h = e\rho$ for some $e \in E(S)$, so that $h\theta_f^{-1} = e\rho \in \mathcal{U}$. By [PR1, Theorem IX.6.12], $f(\rho^{\#}\tau)^{-1} \in \langle \mathcal{U} \circ V \rangle$, whence $S \in \langle \langle \mathcal{U} \circ V \rangle \circ W \rangle$, as required.   □

Henceforth, the characterization of the K-relation on $\mathcal{L}(\mathcal{CR})$ given in Lemma 2.1.2(i) is the one that we will rely upon mostly. We refer to K as the *kernel relation* or, more briefly, the *K-relation*. This concept is central in this chapter and plays a critical role in the remainder of the book.

To see that the opposite inclusion in Lemma 2.1.2(ii) does not always hold, let $\mathcal{U} = \mathcal{G}$, $V = \mathcal{RB}$ and $W = \mathcal{S}$. Then $\langle \mathcal{U} \circ \langle V \circ W \rangle \rangle = \mathcal{G} \circ \mathcal{B} = \mathcal{BG}$ while $\langle \langle \mathcal{U} \circ V \rangle \circ W \rangle = \mathcal{CS} \circ \mathcal{S} = \mathcal{CR}$ so that $\langle \mathcal{U} \circ \langle V \circ W \rangle \rangle \neq \langle \langle \mathcal{U} \circ V \rangle \circ W \rangle$. In the next result, consistent with earlier notational conventions, we will write $\mathcal{O} \cap H V = \mathcal{O} H V$ and $(L\mathcal{O}) \cap D V = L\mathcal{O} D V$.

The solution to the word problem for free groups is well known, while that for free completely simple semigroups can be found in [PR1, Theorem VIII.2.6]. Recall that the word problem for a variety V of unary semigroups is *solvable* if there exists an algorithm such that, applied to $u, v \in U$, it will determine in a finite number of steps whether $u \zeta_V v$.

Theorem 2.1.3 *The following statements hold.*

(i) K *is a complete congruence on* $\mathcal{L}(\mathcal{CR})$.
(ii) *For any* $\mathcal{U}, V \in \mathcal{L}(\mathcal{CR})$ *we have*

$$\mathcal{U} \, K \, V \Longleftrightarrow \mathcal{U} \cap \mathcal{PD} = V \cap \mathcal{PD}$$
$$\Longleftrightarrow \mathcal{U} \cap \mathcal{PE} = V \cap \mathcal{PE}$$
$$\Longleftrightarrow (\mathcal{U} \vee \mathcal{S}) \cap \mathcal{E} = (V \vee \mathcal{S}) \cap \mathcal{E}$$
$$\Longleftrightarrow \langle (\mathcal{U} \vee \mathcal{S}) \cap \mathcal{E} \rangle = \langle (V \vee \mathcal{S}) \cap \mathcal{E} \rangle$$
$$\Longleftrightarrow \rho_{\mathcal{U}} \cap \mathcal{H} = \rho_V \cap \mathcal{H}.$$

(iii) *For any* $V \in \mathcal{L}(\mathcal{CR})$, *we have*

$$\mathcal{V}_K = \langle \mathcal{V} \cap \mathcal{PD} \rangle = \langle \mathcal{V} \cap \mathcal{PE} \rangle = \begin{cases} \mathcal{V} \cap \mathcal{G} & \text{if } \mathcal{V} \subseteq \mathcal{O} \\ \mathcal{V} \cap \mathcal{CS} & \text{if } \mathcal{V} \subseteq \mathcal{LO}, \ \mathcal{V} \not\subseteq \mathcal{O} \\ \langle \mathcal{V} \cap \mathcal{E} \rangle & \text{otherwise.} \end{cases}$$

For $\mathcal{V} \in \mathcal{L}(\mathcal{CS})$, we have $\mathcal{V} = \mathcal{V}_K$ if and only if $\mathcal{V} \subseteq \mathcal{G}$ or $\mathcal{V} \not\subseteq \mathcal{ReG}$.

(iv) For any $\mathcal{V} = [u_\alpha = v_\alpha]_{\alpha \in A} \in \mathcal{L}(\mathcal{CR})$, we have

$$\begin{aligned}
\mathcal{V}^K &= \mathcal{B} \circ (\mathcal{V} \vee \mathcal{S}) = \mathcal{RB} \circ (\mathcal{V} \vee \mathcal{S}) = \langle \mathcal{B} \circ \mathcal{V} \rangle \\
&= \{ S \in \mathcal{CR} \mid S/\tau \in \mathcal{V} \vee \mathcal{S} \} \\
&= \{ S \in \mathcal{CR} \mid S/(\tau \cap \mathcal{D}) \in \mathcal{V} \vee \mathcal{S} \} \\
&= \begin{cases} [xu_\alpha y (x v_\alpha y)^{-1} \in E]_{\alpha \in A} & \text{if } \mathcal{S} \subseteq \mathcal{V} \\ [xu_\alpha v_\alpha^0 y (x v_\alpha u_\alpha^{0y})^{-1} \in E]_{\alpha \in A} \cap \mathcal{LO} & \text{if } \mathcal{V} \subseteq \mathcal{CS} \end{cases} \\
&= [u = u^2 \mid u = u^2 \text{ holds in } \mathcal{V}].
\end{aligned}$$

(v) The following intervals constitute the complete set of K-classes of varieties in $\mathcal{L}(\mathcal{LO})$:

$$\{ [\mathcal{V}, \mathcal{OHV}] \mid \mathcal{V} \in \mathcal{L}(\mathcal{G}) \}, \quad \{ [\mathcal{V}, \mathcal{LODV}] \mid \mathcal{V} \in \mathcal{L}(\mathcal{CS}) \setminus \mathcal{L}(\mathcal{ReG}) \}.$$

In particular, the following intervals are K-classes: $[\mathcal{T}, \mathcal{B}]$, $[\mathcal{G}, \mathcal{O}]$, $[\mathcal{CS}, \mathcal{LO}]$.

(vi) The mapping $\mathcal{V} \mapsto \mathcal{V}_K$ ($\mathcal{V} \in \mathcal{L}(\mathcal{CR})$) is a complete $\vee$-endomorphism of $\mathcal{L}(\mathcal{CR})$. Its restriction to $\mathcal{L}(\mathcal{CSHA})$ is not an intersection homomorphism.

(vii) The mapping $\mathcal{V} \mapsto \mathcal{V}^K$ ($\mathcal{V} \in \mathcal{L}(\mathcal{CR})$) is a complete endomorphism of $\mathcal{L}(\mathcal{CR})$.

(viii) If $\mathcal{V} \in \mathcal{L}(\mathcal{CR})$ has a solvable word problem, then so do $\mathcal{V}_K$ and $\mathcal{V}^K$.

Proof Recall the convention introduced in the preamble to this book. The proofs of the different parts will proceed in the order (i), (vii), (iv), (v), (iii), (ii), (vi) and (viii).

(i) See Theorem 1.3.4.

(vii) This follows from Theorem 1.4.11 and the standard antiisomorphism between $\Gamma(CR_X)$ and $\mathcal{L}(\mathcal{CR})$.

(iv) $A = B$. By [PR1, Theorem IX.2.9], we have $\mathcal{B} \circ (\mathcal{V} \vee \mathcal{S}) \in \mathcal{L}(\mathcal{CR})$. Let $\rho = \zeta_\mathcal{V}$. Then ρ/ρ_K is a congruence on $F\mathcal{V}^K$ and

$$F\mathcal{V}^K / (\rho/\rho_K) = (U/\rho_K)/(\rho/\rho_K) \cong U/\rho = F\mathcal{V} \in \mathcal{V}.$$

But, since $\ker \rho_K = \ker \rho$, we must have $\ker(\rho/\rho_K) = E(U/\rho_K)$. Thus ρ/ρ_K is a congruence whose idempotent classes contain only idempotents and which are, therefore, bands. Consequently, $F\mathcal{V}^K \in \mathcal{B} \circ \mathcal{V} \subseteq \mathcal{B} \circ (\mathcal{V} \vee \mathcal{S})$. Therefore $\mathcal{V}^K \subseteq \mathcal{B} \circ (\mathcal{V} \vee \mathcal{S})$.

In order to establish the reverse containment, let $\mathcal{W} = \mathcal{B} \circ (\mathcal{V} \vee \mathcal{S})$. Then there must be a congruence λ on $F\mathcal{W}$ with $\ker \lambda = E(F\mathcal{W})$ and $F\mathcal{W}/\lambda \in \mathcal{V} \vee \mathcal{S}$. However, $\zeta_{\mathcal{V} \vee \mathcal{S}}/\zeta_\mathcal{W}$ is the least congruence on $F\mathcal{W}$ for which the quotient lies in $\mathcal{V} \vee \mathcal{S}$. Hence $\zeta_{\mathcal{V} \vee \mathcal{S}}/\zeta_\mathcal{W} \subseteq \lambda$ so that ker $\zeta_{\mathcal{V}\vee\mathcal{S}}/\zeta_{\mathcal{W}} = E(F\mathcal{W})$ and ker $\zeta_{\mathcal{V}\vee\mathcal{S}}/\zeta = $ ker $\zeta_{\mathcal{W}}/\zeta$. It follows that

$$\ker \rho_{\mathcal{W}} = \ker(\zeta_{\mathcal{W}}/\zeta) = \ker(\zeta_{\mathcal{V}\vee\mathcal{S}}/\zeta) = \ker \rho_{\mathcal{V}\vee\mathcal{S}}$$
$$= \ker(\rho_{\mathcal{V}} \cap \rho_{\mathcal{S}}) = \ker \rho_{\mathcal{V}} \cap \ker \rho_{\mathcal{S}} \qquad \text{by [PR1, Theorem VII.1.2]}$$
$$= \ker \rho_{\mathcal{V}} \cap U = \ker \rho_{\mathcal{V}}$$

so that $(\rho_{\mathcal{V}})_K = (\rho_{\mathcal{W}})_K \subseteq \rho_{\mathcal{W}}$ and, by the usual antiisomorphism between fully invariant congruences and varieties, we have $\mathcal{W} \subseteq \mathcal{V}^K$ and equality follows.

$B = C$. This follows immediately from [PR1, Theorem IX.4.3(i)].

$A = D$. We have

$$\mathcal{V}^K = \mathcal{RB} \circ (\mathcal{V} \vee \mathcal{S})$$
$$\subseteq \mathcal{RB} \circ \langle \mathcal{S} \circ \mathcal{V} \rangle$$
$$\subseteq \langle\langle \mathcal{RB} \circ \mathcal{S} \rangle \circ \mathcal{V} \rangle \quad \text{by Lemma 2.1.2(ii)}$$
$$= \langle \mathcal{B} \circ \mathcal{V} \rangle$$
$$\subseteq \langle \mathcal{B} \circ (\mathcal{V} \vee \mathcal{S}) \rangle$$
$$= \mathcal{B} \circ (\mathcal{V} \vee \mathcal{S}) = \mathcal{V}^K.$$

Therefore $\mathcal{V}^K = \langle \mathcal{B} \circ \mathcal{V} \rangle$.

$B = E = F$. This follows from [PR1, Theorem IX.4.3(i)].

$A = G$. If $\mathcal{S} \subseteq \mathcal{V}$, then the equality follows from [PR1, Theorem IX.4.3(i)]. So let $\mathcal{V} \subseteq \mathcal{CS}$. Since we already know that $A = D$, it follows that

$$\mathcal{NBG}^K = \{S \in \mathcal{CR} \mid S/\tau \in \mathcal{NBG}\} = \mathcal{LO} \quad \text{by [PR1, Theorem IV.4.2(ii)]}.$$

Then, from the fact that $A = B$, we can deduce that

$$\mathcal{V}^K = (\mathcal{S} \vee \mathcal{V})^K$$
$$= \left([u_\alpha v_\alpha^0 = v_\alpha u_\alpha^0]_{\alpha \in A} \cap \mathcal{NBG}\right)^K \quad \text{by [PR1, Proposition IX.9.2]}$$
$$= \left([u_\alpha v_\alpha^0 = v_\alpha u_\alpha^0]_{\alpha \in A}\right)^K \cap \mathcal{NBG}^K \quad \text{by part (vii)}$$
$$= \left(\mathcal{B} \circ [u_\alpha v_\alpha^0 = v_\alpha u_\alpha^0]_{\alpha \in A}\right) \cap \mathcal{NBG}^K \quad \text{since } A = B$$
$$= \left[x u_\alpha v_\alpha^0 y (x v_\alpha u_\alpha^0 y)^{-1} \in E\right]_{\alpha \in A} \cap \mathcal{NBG}^K \quad \text{by [PR1, Theorem IX.4.3(i)]}$$
$$= \left[x u_\alpha v_\alpha^0 y (x v_\alpha u_\alpha^0 y)^{-1} \in E\right]_{\alpha \in A} \cap \mathcal{LO}.$$

where the equality $\mathcal{NBG}^K = \mathcal{LO}$ was proved above.

$A = H$. Let $\mathcal{W} = [u = u^2 \mid u = u^2$ holds in $\mathcal{V}]$ and $\rho_{\mathcal{V}}$ (respectively, $\rho_{\mathcal{W}}$) denote the fully invariant congruence on $\mathbf{CR}_X$ corresponding to $\mathcal{V}$ (respectively, $\mathcal{W}$). Since $\mathcal{W}$ is defined by a subset of the identities satisfied by $\mathcal{V}$, it is clear that $\rho_{\mathcal{W}} \subseteq \rho_{\mathcal{V}}$ whence ker $\rho_{\mathcal{W}} \subseteq$ ker $\rho_{\mathcal{V}}$. On the other hand, by the definition of $\mathcal{W}$, $u \in \ker \rho_{\mathcal{V}} \Rightarrow u \in \ker \rho_{\mathcal{W}}$ so that we get ker $\rho_V \subseteq$ ker $\rho_W \subseteq$ ker ρ_V and ker $\rho_V =$ ker ρ_W. Thus $W \mathcal{K} V$ and therefore $W \subseteq V^K$. However, V^K satisfies all the identities defining W so that $V^K \subseteq W$ and equality prevails.

(v) *Case*: $V \in \mathcal{L}(\mathcal{G})$. Since ker $\rho_V =$ ker ρ_{V_K}, it follows that ρ_{V_K}/ρ_V is an idempotent pure congruence on CR_X/ρ_V. But CR_X/ρ_V is a group and so has no nontrivial idempotent pure congruences. Therefore $\rho_V = \rho_{V_K}$ and $V = V_K$. Hence, by part (iv),

$$V^K = \{S \in \mathcal{CR} \mid S/\tau \in V \vee \mathcal{S}\} \subseteq \mathcal{O} \quad \text{by [PR1, Theorem IV.4.2(i)]}.$$

Also, for $G \in V^K \cap \mathcal{G}$, we have $G = G/\tau \in V \vee \mathcal{S}$ by part (iv) so that

$$G \in (V \vee \mathcal{S}) \cap \mathcal{G} = (V \cap \mathcal{G}) \vee (\mathcal{S} \cap \mathcal{G}) \quad \text{by [PR1, Theorem IX.9.1]}$$
$$= (V \cap \mathcal{G}) \vee \mathcal{T} = V \cap \mathcal{G}.$$

Therefore $V^K \cap \mathcal{G} \subseteq V \cap \mathcal{G}$ and equality prevails. Consequently $V^K \subseteq \mathcal{O}HV$.
    Conversely, since $\mathcal{G} \cap HV = V$, we have

$$S \in \mathcal{O}HV \Longrightarrow S/\tau \in (\mathcal{G} \vee \mathcal{S}) \cap \mathcal{O}HV \quad \text{by [PR1, Theorem IV.4.2(ii)]}$$
$$\Longrightarrow S/\tau \in (\mathcal{G} \vee \mathcal{S}) \cap HV \subseteq V \vee \mathcal{S}$$
$$\Longrightarrow S \in V^K \quad \text{by part (iv)}.$$

Therefore $V^K = \mathcal{O}HV$.

Case: $V \in \mathcal{L}(\mathcal{CS}) \setminus \mathcal{L}(\mathcal{ReG})$. Let λ and ρ be the fully invariant congruences on CS_X corresponding to V_K and V, respectively. Then ker $\lambda =$ ker ρ and $\rho \subseteq \lambda$. Hence λ/ρ is an idempotent pure congruence on CS_X/ρ. By [PR1, Proposition VIII.3.9], $\lambda = \rho$ whence $V = V_K$. By part (iv),

$$V^K = \{S \in \mathcal{CR} \mid S/\tau \in V \vee \mathcal{S}\} \subseteq L\mathcal{O} \quad \text{by [PR1, Theorem IV.4.2(ii)]}.$$

Also, we have

$$S \in V^K \cap \mathcal{CS} \Longrightarrow S/\tau \in (V \vee \mathcal{S}) \cap \mathcal{CS} \quad \text{by part (iv)}$$
$$\Longrightarrow S/\tau \in V \cap \mathcal{CS} = V \quad \text{by [PR1, Theorem IX.9.1]}.$$

On the other hand, since $V \notin \mathcal{L}(\mathcal{ReG})$ we have $\mathcal{RB} \subseteq V$ whence $S/\mu \in V$. But $\tau \cap \mu = \varepsilon$. Therefore S is a subdirect product of S/τ and S/μ so that $S \in V$. Thus $V^K \cap \mathcal{CS} \subseteq V$ and, since $V \in \mathcal{L}(\mathcal{CS})$, equality prevails. Hence $V^K \subseteq DV$. We have already seen that $V^K \subseteq L\mathcal{O}$. Combining these two observations, we obtain $V^K \subseteq L\mathcal{O} \cap DV = L\mathcal{O}DV$.
    Conversely, by [PR1, Theorem IV.4.2(ii)], we obtain

$$S \in L\mathcal{O}DV \Longrightarrow S/\tau \in (\mathcal{CS} \vee \mathcal{S}) \cap DV = V \vee \mathcal{S} \Longrightarrow S \in V^K.$$

Therefore $V^K = L\mathcal{O}DV$.

(iii) $A = D$. By part (v), we have

$$\mathcal{V} \in \mathcal{L}(\mathcal{O}) \Longrightarrow \mathcal{V}_K = \mathcal{V} \cap \mathcal{G},$$

$$\mathcal{V} \in \mathcal{L}(L\mathcal{O}) \setminus \mathcal{L}(\mathcal{O}) \Longrightarrow \mathcal{V}_K = \mathcal{V} \cap \mathcal{CS},$$

and these two cases exhaust all $\mathcal{V} \in \mathcal{L}(\mathcal{CR})$ such that $\mathcal{V}_K \subseteq \mathcal{CS}$.

It remains to consider the case $\mathcal{V}_K \nsubseteq \mathcal{L}(\mathcal{CS})$ so that $\mathcal{V}_K \supseteq \mathcal{S}$. Let $\mathcal{V}$ have this property. Then $\mathcal{S} \subseteq \mathcal{V}$ and hence $\mathcal{V}^K = \mathcal{B} \circ \mathcal{V}$ by part (iv). Let $S \in \mathcal{B} \circ \mathcal{V}$ so that $S/\tau \in \mathcal{V}$. By [PR1, Proposition VI.1.16], S/τ is E-disjunctive. Hence $S/\tau \in \mathcal{V} \cap \mathcal{E} \subseteq \langle \mathcal{V} \cap \mathcal{E} \rangle$ which implies that $S \in \mathcal{B} \circ \langle \mathcal{V} \cap \mathcal{E} \rangle$. Therefore

$$\mathcal{B} \circ \mathcal{V} \subseteq \mathcal{B} \circ \langle \mathcal{V} \cap \mathcal{E} \rangle \subseteq \mathcal{B} \circ (\langle \mathcal{V} \cap \mathcal{E} \rangle \vee \mathcal{S}) \subseteq \mathcal{B} \circ \mathcal{V}$$

and thus $\mathcal{V}^K = \mathcal{B} \circ \mathcal{V} = \mathcal{B} \circ \langle \mathcal{V} \cap \mathcal{E} \rangle$.

Since $\mathcal{V}_K \nsubseteq \mathcal{L}(\mathcal{CS})$, we also have $\mathcal{V} \notin \mathcal{L}(\mathcal{O})$ so that $\mathcal{V} \cap \mathcal{G} \neq \mathcal{T}$; say $G \in (\mathcal{V} \cap \mathcal{G}) \setminus \mathcal{T}$. Then $G \times Y_2 \in \mathcal{V} \vee \mathcal{S} = \mathcal{V}$. With $I = G \times \{0\}$ we then obtain $G^0 \cong (G \times Y_2)/I \in \mathcal{V}$. Clearly $G^0 \in \mathcal{E}$ so that $Y_2 \cong E(G^0) \in \langle \mathcal{V} \cap \mathcal{E} \rangle$. [PR1, Proposition I.5.10(iii)] implies that $\mathcal{S} \subseteq \langle \mathcal{V} \cap \mathcal{E} \rangle$. Hence $\mathcal{V}^K = \mathcal{B} \circ \langle \mathcal{V} \cap \mathcal{E} \rangle = \langle \mathcal{V} \cap \mathcal{E} \rangle^K$ and therefore $\mathcal{V} \, K \, \langle \mathcal{V} \cap \mathcal{E} \rangle$.

Now let $\mathcal{U} \in \mathcal{L}(\mathcal{CR})$ be such that $\mathcal{U} \, K \, \mathcal{V}$. Then $\mathcal{S} \subseteq \mathcal{V}_K = \mathcal{U}_K \subseteq \mathcal{U}$. For any $S \in \mathcal{V} \cap \mathcal{E}$, we have $S \in \mathcal{V} \subseteq \mathcal{V}^K = \mathcal{U}^K = \mathcal{B} \circ \mathcal{U}$, whence $S = S/\tau \in \mathcal{U}$. Consequently $\mathcal{V} \cap \mathcal{E} \subseteq \mathcal{U}$ so that $\langle \mathcal{V} \cap \mathcal{E} \rangle \subseteq \mathcal{U}$. Hence $\mathcal{V}_K = \langle \mathcal{V} \cap \mathcal{E} \rangle$.

That there is no overlap among the three cases follows from part (v) and the fact established above that for $\mathcal{V} \notin \mathcal{L}(L\mathcal{O})$, we have $\mathcal{S} \subseteq \langle \mathcal{V} \cap \mathcal{E} \rangle$.

$A = C$. We consider three cases.

Case: $\mathcal{V} \in \mathcal{L}(\mathcal{O})$. By part (v), we have $\mathcal{V}_K = \mathcal{V} \cap \mathcal{G}$. On the other hand,

$$\mathcal{V} \cap (\mathcal{CS} \setminus \mathcal{Re}\mathcal{G}) = \emptyset = \mathcal{V} \cap (\mathcal{CR} \setminus L\mathcal{O}).$$

Therefore $\mathcal{V}_K = \mathcal{V} \cap \mathcal{G} = \mathcal{V} \cap \mathcal{PE} = \langle \mathcal{V} \cap \mathcal{PE} \rangle$.

Case: $\mathcal{V} \in \mathcal{L}(L\mathcal{O}) \setminus \mathcal{L}(\mathcal{O})$. Then $\mathcal{V} \cap \mathcal{CS} \nsubseteq \mathcal{Re}\mathcal{G}$. By [PR1, Proposition VIII.3.9],

$$F(\mathcal{V} \cap \mathcal{CS})(X) \in \mathcal{V} \cap \mathcal{CS} \cap \mathcal{E} \subseteq (\mathcal{V} \cap \mathcal{G}) \cup (\mathcal{V} \cap (\mathcal{CS} \setminus \mathcal{Re}\mathcal{G})) \subseteq \mathcal{V} \cap \mathcal{PE}$$

so that $\mathcal{V} \cap \mathcal{CS} \subseteq \langle \mathcal{V} \cap \mathcal{PE} \rangle$. But by part (v), $\mathcal{V}_K = \mathcal{V} \cap \mathcal{CS}$. Hence $\mathcal{V}_K \subseteq \langle \mathcal{V} \cap \mathcal{PE} \rangle$. On the other hand,

$$\mathcal{V} \cap \mathcal{PE} = (\mathcal{V} \cap \mathcal{G}) \cup (\mathcal{V} \cap (\mathcal{CS} \setminus \mathcal{Re}\mathcal{G})) \subseteq \mathcal{V} \cap \mathcal{CS}$$

so that $\langle \mathcal{V} \cap \mathcal{PE} \rangle \subseteq \mathcal{V} \cap \mathcal{CS}$ and equality prevails. Thus $\langle \mathcal{V} \cap \mathcal{PE} \rangle = \mathcal{V} \cap \mathcal{CS} = \mathcal{V}_K$.

Case: $\mathcal{V} \notin \mathcal{L}(L\mathcal{O})$. Then, there exists $S \in \mathcal{V}$ and $e \in E(S)$ such that $eSe \notin \mathcal{O}$. By [PR1, Theorem II.5.3], there exists a $\mathcal{D}$-class D in eSe that is not orthodox. Then $D \cup \{e\}$ is a completely regular subsemigroup of eSe with identity e that is also not orthodox so that $(D \cup \{e\})/\tau \in (\mathcal{CR} \setminus L\mathcal{O}) \cap \mathcal{E}$. Since $(D \cup \{e\})/\tau$ contains a subsemilattice isomorphic to Y_2, we have $\mathcal{S} \subseteq \langle (\mathcal{CR} \setminus L\mathcal{O}) \cap \mathcal{E} \cap \mathcal{V} \rangle \subseteq \langle \mathcal{E} \cap \mathcal{PE} \cap \mathcal{V} \rangle$ and since D/τ is not orthodox

$(D/\tau)/\mu$ is a rectangular band with nontrivial $\mathcal{L}$- and $\mathcal{R}$-classes which implies that

$$\mathcal{RB} \subseteq \langle (\mathcal{CR} \setminus L\mathcal{O}) \cap \mathcal{E} \cap \mathcal{V} \rangle \subseteq \langle \mathcal{E} \cap \mathcal{PE} \cap \mathcal{V} \rangle.$$

Thus $\mathcal{RB} \vee \mathcal{S} \subseteq \langle \mathcal{V} \cap \mathcal{PE} \rangle$.

To see that $\langle \mathcal{V} \cap \mathcal{PE} \rangle \subseteq \langle \mathcal{V} \cap \mathcal{E} \rangle$ we first note that $\mathcal{V} \cap \mathcal{G} \subseteq \mathcal{V} \cap \mathcal{E}$. Next we have

$$S \in \mathcal{V} \cap \mathcal{CS} \Longrightarrow S/\tau \in \mathcal{V} \cap \mathcal{E}, \quad S/\mu \in \mathcal{RB} \subseteq \langle \mathcal{V} \cap \mathcal{E} \rangle$$
$$\Longrightarrow S \in \langle \mathcal{V} \cap \mathcal{E} \rangle \qquad \text{since } \mu \cap \tau = \epsilon.$$

Finally

$$S \in ((\mathcal{CR} \setminus L\mathcal{O}) \cap \mathcal{E}) \cap \mathcal{V} \Longrightarrow S \in \mathcal{V} \cap \mathcal{E}.$$

Thus $\mathcal{V} \cap \mathcal{PE} \subseteq \langle \mathcal{V} \cap \mathcal{E} \rangle$ so that $\langle \mathcal{V} \cap \mathcal{PE} \rangle \subseteq \langle \mathcal{V} \cap \mathcal{E} \rangle$.

Conversely, let $S \in \mathcal{V} \cap \mathcal{E}$. Then by [PR1, Theorem IV.4.2(i)],

$$S \in \mathcal{O} \Longrightarrow S/\tau \in (\mathcal{G} \vee \mathcal{S}) \cap \mathcal{V} = (\mathcal{G} \vee \mathcal{S}) \cap (\mathcal{V} \vee \mathcal{S})$$
$$\Longrightarrow S/\tau \in (\mathcal{G} \cap \mathcal{V}) \vee \mathcal{S} \qquad \text{by [PR1, Corollary IX.9.3]}$$
$$\Longrightarrow S \cong S/\tau \in \langle \mathcal{V} \cap \mathcal{PE} \rangle.$$

Similarly $S \in L\mathcal{O} \setminus \mathcal{O}$ implies $S \in \langle \mathcal{V} \cap \mathcal{E} \rangle$. Finally

$$S \in \mathcal{CR} \setminus L\mathcal{O} \Longrightarrow S \in (\mathcal{CR} \setminus L\mathcal{O}) \cap \mathcal{E} \cap \mathcal{V} \subseteq \mathcal{V} \cap \mathcal{PE}.$$

Thus, in all cases, $\mathcal{V} \cap \mathcal{E} \subseteq \langle \mathcal{V} \cap \mathcal{PE} \rangle$ so that $\langle \mathcal{V} \cap \mathcal{E} \rangle \subseteq \langle \mathcal{V} \cap \mathcal{PE} \rangle$. Therefore equality follows and $\mathcal{V}_K = \langle \mathcal{V} \cap \mathcal{E} \rangle = \langle \mathcal{V} \cap \mathcal{PE} \rangle$.

$B = C$. First we have $\mathcal{V} \cap \mathcal{PD} \subseteq \mathcal{V} \cap \mathcal{PE} \Rightarrow \langle \mathcal{V} \cap \mathcal{PD} \rangle \subseteq \langle \mathcal{V} \cap \mathcal{PE} \rangle$. For the reverse containment, consider any $S \in \mathcal{V} \cap \mathcal{PE}$. Then

$$S \in \mathcal{V} \cap \mathcal{G} \Longrightarrow S \in \mathcal{V} \cap (\mathcal{CS} \cap \mathcal{E}) \subseteq \mathcal{V} \cap \mathcal{PD}.$$

Now let $S \in \mathcal{V} \cap (\mathcal{CS} \setminus \mathcal{ReG})$ and let F denote the free object in $\mathcal{V} \cap \mathcal{CS}$. By [PR1, Theorem VIII.3.9], F is E-disjunctive. Hence $F \in \mathcal{V} \cap (\mathcal{CS} \cap \mathcal{E}) \subseteq \mathcal{V} \cap \mathcal{PD}$. Consequently $S \in \langle \mathcal{V} \cap \mathcal{PD} \rangle$.

Finally, let $S \in (\mathcal{CR} \setminus L\mathcal{O}) \cap \mathcal{E} \cap \mathcal{V}$. Then immediately we have $S \in \mathcal{PD}$. Hence $\langle \mathcal{V} \cap \mathcal{PE} \rangle \subseteq \langle \mathcal{V} \cap \mathcal{PD} \rangle$. Therefore equality prevails.

The final claim now follows from the above and the observations that $\mathcal{CS} \in \mathcal{L}(L\mathcal{O})$ and $\mathcal{CS} \cap \mathcal{O} = \mathcal{ReG}$.

(ii) $A \Leftrightarrow B$. That $B \Rightarrow A$ follows immediately from part (iii). So suppose that $\mathcal{U} \ K \ \mathcal{V}$. Then, by part (iii), $S \in \mathcal{U} \cap \mathcal{PD} \Rightarrow S \in \mathcal{U}_K = \mathcal{V}_K \subseteq \mathcal{V}$ so that $\mathcal{U} \cap \mathcal{PD} \subseteq \mathcal{V} \cap \mathcal{PD}$. Switching the roles of $\mathcal{U}$ and $\mathcal{V}$, we see that $\mathcal{U} \cap \mathcal{PD} = \mathcal{V} \cap \mathcal{PD}$ so that $A \Rightarrow B$.

$A \Leftrightarrow C$. The proof here is identical to the proof of $A \Leftrightarrow B$, substituting $\mathcal{PE}$ for $\mathcal{PD}$ throughout.

$A \Leftrightarrow D$. Since $\mathcal{U} \, K \, \mathcal{V}$ if and only if $\mathcal{U}^K = \mathcal{V}^K$, the equivalence follows immediately from part (iv).

$D \Leftrightarrow E$. We have

$$S \in \langle (\mathcal{U} \vee \mathcal{S}) \cap \mathcal{E} \rangle \cap \mathcal{E} \Longrightarrow S \cong S/\tau \in \mathcal{U}^K$$
$$\Longrightarrow S \cong (S/\tau)/\tau \in \mathcal{U} \vee \mathcal{S}$$
$$\Longrightarrow S \in (\mathcal{U} \vee \mathcal{S}) \cap \mathcal{E}.$$

Thus $\langle (\mathcal{U} \vee \mathcal{S}) \cap \mathcal{E} \rangle \cap \mathcal{E} \subseteq (\mathcal{U} \vee \mathcal{S}) \cap \mathcal{E}$ and the reverse inclusion being obvious, equality prevails. The asserted equivalence then follows trivially.

$A \Leftrightarrow F$. This has a different character from the other equivalences in this part as it concerns the fully invariant congruence corresponding to the varieties $\mathcal{U}$, $\mathcal{V}$. It follows immediately from [PR1, Corollary VII.1.8].

(vi) The positive assertion follows from [PR1, Theorem I.3.4 and Lemma I.2.2]. By [PR1, Theorem VIII.5.5(i)], we have

$$\mathcal{C}\mathcal{S}\mathcal{H}\mathcal{A}_2 \cap \mathcal{C}\mathcal{S}\mathcal{H}\mathcal{A}_3 = \mathcal{C}\mathcal{S}\mathcal{H}(\mathcal{A}_2 \cap \mathcal{A}_3) = \mathcal{C}\mathcal{S}\mathcal{H}\mathcal{J} = \mathcal{R}\mathcal{B}$$

so that we have $(\mathcal{C}\mathcal{S}\mathcal{H}\mathcal{A}_2 \cap \mathcal{C}\mathcal{S}\mathcal{H}\mathcal{A}_3)_K = \mathcal{J}$, by part (v). On the other hand, by part (iii), we have

$$(\mathcal{C}\mathcal{S}\mathcal{H}\mathcal{A}_2)_K \cap (\mathcal{C}\mathcal{S}\mathcal{H}\mathcal{A}_3)_K = \mathcal{C}\mathcal{S}\mathcal{H}\mathcal{A}_2 \cap \mathcal{C}\mathcal{S}\mathcal{H}\mathcal{A}_3 = \mathcal{R}\mathcal{B}.$$

This establishes the negative assertion.

(viii) The first assertion follows immediately from Theorem 1.2.12(iv) while the second is a consequence of Theorem 1.4.9 and the definition of $\overline{\rho}$ (in Notation 1.4.1). $\qquad\square$

Corollary 2.1.4 *Let* $\mathcal{U}, \mathcal{V} \in \mathcal{L}(\mathcal{C}\mathcal{R})$.

(i) $\mathcal{V}^K \cap \mathcal{G} = \mathcal{V} \cap \mathcal{G}$, $\mathcal{V}^K \subseteq H\mathcal{V} = H\mathcal{V}^K$.
(ii) $\mathcal{V}^K \cap \mathcal{C}\mathcal{S} = (\mathcal{V} \cap \mathcal{C}\mathcal{S}) \vee \mathcal{R}\mathcal{B}$, $\mathcal{V}^K \subseteq D(\mathcal{V} \vee \mathcal{R}\mathcal{B}) = D\mathcal{V}^K$.
(iii) $\mathcal{U} \, K \, \mathcal{V} \Rightarrow \mathcal{U} \cap \mathcal{G} = \mathcal{V} \cap \mathcal{G}$ *and* $\mathcal{U} \cap (\mathcal{C}\mathcal{S} \setminus \mathcal{R}e\mathcal{G}) = \mathcal{V} \cap (\mathcal{C}\mathcal{S} \setminus \mathcal{R}e\mathcal{G})$.

Proof (i) The first equality follows from Theorem 2.1.3(ii) and the fact that $\mathcal{G} \subseteq \mathcal{P}\mathcal{E}$. The second equality follows from the first. The inclusion follows from the first equality.

(ii) First assume that $\mathcal{R}\mathcal{B} \subseteq \mathcal{V}$. Then we have

$$\mathcal{V} \, K \, \mathcal{V}^K \Longrightarrow \mathcal{V} \cap \mathcal{P}\mathcal{E} = \mathcal{V}^K \cap \mathcal{P}\mathcal{E} \qquad \text{by Theorem 2.1.3(ii)}$$
$$\Longrightarrow \mathcal{V} \cap \mathcal{G} = \mathcal{V}^K \cap \mathcal{G}, \quad \mathcal{V} \cap (\mathcal{C}\mathcal{S} \setminus \mathcal{R}e\mathcal{G}) = \mathcal{V}^K \cap (\mathcal{C}\mathcal{S} \setminus \mathcal{R}e\mathcal{G})$$
$$\Longrightarrow \mathcal{V} \cap \mathcal{R}e\mathcal{G} = \mathcal{V}^K \cap \mathcal{R}e\mathcal{G}, \quad \mathcal{V} \cap (\mathcal{C}\mathcal{S} \setminus \mathcal{R}e\mathcal{G}) = \mathcal{V}^K \cap (\mathcal{C}\mathcal{S} \setminus \mathcal{R}e\mathcal{G})$$
$$\Longrightarrow \mathcal{V} \cap \mathcal{C}\mathcal{S} = \mathcal{V}^K \cap \mathcal{C}\mathcal{S}.$$

Then for arbitrary $\mathcal{V} \in \mathcal{L}(\mathcal{CR})$, we have

$$
\begin{aligned}
\mathcal{V}^K \cap \mathcal{CS} &= (\mathcal{V} \vee \mathcal{RB})^K \cap \mathcal{CS} \\
&= (\mathcal{V} \vee \mathcal{RB}) \cap \mathcal{CS} \quad \text{by the above} \\
&= (\mathcal{V} \cap \mathcal{CS}) \vee (\mathcal{RB} \cap \mathcal{CS}) \quad \text{by [PR1, Theorem IX.9.1]} \\
&= (\mathcal{V} \cap \mathcal{CS}) \vee \mathcal{RB}.
\end{aligned}
$$

In passing, we have demonstrated that $\mathcal{V}^K \cap \mathcal{CS} = (\mathcal{V} \vee \mathcal{RB}) \cap \mathcal{CS}$ and the containment then follows from the maximality of $D\mathcal{V}^K$.

(iii) This follows from parts (i) and (ii). $\qquad\qquad\qquad\qquad\qquad\qquad\qquad\square$

As a further simple consequence of Theorem 2.1.3, we present some observations concerning relatively simple word problems.

Corollary 2.1.5 *Let $u, v \in U$ and η be the least semilattice congruence on U. Then*

$$
\begin{aligned}
&u\, \eta\, v \iff c(u) = c(v), \\
&u\, \zeta_{\mathcal{B}}\, v \iff w(u)\, \eta\, w(v) && \textit{for all } w \in \{s, e\}^*, \\
&u\, \zeta_{\mathcal{O}}\, v \iff w(u)\, (\zeta_{\mathcal{G}} \cap \eta)\, w(v) && \textit{for all } w \in \{s, e\}^*, \\
&u\, \zeta_{L\mathcal{O}}\, v \iff w(u)\, (\zeta_{\mathcal{CS}} \cap \eta)\, w(v) && \textit{for all } w \in \{s, e\}^*.
\end{aligned}
$$

Proof The characterization of η was given in [PR1, Theorem V.1.9(iv)]. The remaining characterizations follow from Lemma 1.4.2, Theorems 1.4.9 and 2.1.3(v) as well as the definition of $\overline{\rho}$ since we have:

$$
\zeta_{\mathcal{B}} = \eta^K = \overline{\eta}, \quad \zeta_{\mathcal{O}} = (\zeta_{\mathcal{G} \vee \mathcal{S}})^K = \overline{\zeta_{\mathcal{G} \vee \mathcal{S}}}, \quad \zeta_{L\mathcal{O}} = (\zeta_{\mathcal{CS} \vee \mathcal{S}})^K.
$$

$\qquad\qquad\qquad\qquad\qquad\qquad\qquad\qquad\qquad\qquad\qquad\qquad\qquad\qquad\qquad\square$

More concrete descriptions than those in Theorem 2.1.3 are possible in some cases such as $\mathcal{L}(\mathcal{O})$ and $\mathcal{L}(L\mathcal{O})$.

Theorem 2.1.6 *The following statements hold.*

(i) *For every $\mathcal{V} \in \mathcal{L}(\mathcal{O})$, we have $\mathcal{V}_K = \mathcal{V} \cap \mathcal{G}$ and $\mathcal{V}^K$ is the greatest orthogroup variety $\mathcal{U}$ such that $\mathcal{U} \cap \mathcal{G} = \mathcal{V} \cap \mathcal{G}$.*

(ii) *The mapping $\mathcal{V} \to \mathcal{V}_K$ ($\mathcal{V} \in \mathcal{L}(\mathcal{O})$) is a complete retraction of $\mathcal{L}(\mathcal{O})$ onto $\mathcal{L}(\mathcal{G})$.*

(iii) *The mapping $\chi : \mathcal{V} \to \mathcal{V}_K \vee \mathcal{RB}$ ($\mathcal{V} \in \mathcal{L}(L\mathcal{O})$) is a complete retraction of $\mathcal{L}(L\mathcal{O})$ onto the interval $[\mathcal{RB}, \mathcal{CS}]$ which induces K on $\mathcal{L}(L\mathcal{O})$.*

Proof (i) By Theorem 2.1.3(v), we have $\mathcal{V}^K \in [\mathcal{V} \cap \mathcal{G}, \mathcal{O}H(\mathcal{V} \cap \mathcal{G})]$ so that $\mathcal{V}_K = \mathcal{V} \cap \mathcal{G}$ and $\mathcal{V}^K = \mathcal{O}H(\mathcal{V} \cap \mathcal{G})$, by Theorem 2.1.3(iv) . This establishes the first claim. In addition, $\mathcal{V}^K$ is an orthogroup variety with $\mathcal{V}^K \cap \mathcal{G} = \mathcal{O}H(\mathcal{V} \cap \mathcal{G}) \cap \mathcal{G} = \mathcal{V} \cap \mathcal{G}$ where the latter equality follows from [PR1, Lemma VIII.1.2]. If $\mathcal{U}$ is any orthogroup variety with $\mathcal{U} \cap \mathcal{G} = \mathcal{V} \cap \mathcal{G}$, then, as above for $\mathcal{V}$,

$$\mathcal{U} \subseteq \mathcal{U}^K = \mathcal{O}H(\mathcal{U} \cap \mathcal{G}) = \mathcal{O}H(\mathcal{V} \cap \mathcal{G}) = \mathcal{V}^K$$

and the claim for $\mathcal{V}^K$ holds.

(ii) By Theorem 2.1.3(iii), for any $\mathcal{V} \in \mathcal{L}(\mathcal{O})$, we have $\mathcal{V}_K = \mathcal{V} \cap \mathcal{G}$. Therefore, within $\mathcal{L}(\mathcal{O})$ the mapping $\mathcal{V} \to \mathcal{V}_K$ is simply the mapping $\mathcal{V} \to \mathcal{V} \cap \mathcal{G}$ which, by [PR1, Theorem IX.9.1] is a complete homomorphism and it is then clearly a complete retraction of $\mathcal{L}(\mathcal{O})$ onto $\mathcal{L}(\mathcal{G})$.

(iii) It follows easily from [PR1, Theorem IX.9.1] that the mapping $\mathcal{V} \to \mathcal{V} \cap \mathcal{CS}$ is a complete retraction of $\mathcal{L}(L\mathcal{O})$ onto $\mathcal{L}(\mathcal{CS})$. It is a simple exercise, in the style of [PR1, Theorem IX.9.2], to show that the mapping $\mathcal{V} \to \mathcal{V} \vee \mathcal{RB}$ is a complete retraction of $\mathcal{L}(\mathcal{CS})$ onto $[\mathcal{RB}, \mathcal{CS}]$. It follows that combining those mappings we obtain the mapping

$$\mathcal{V} \mapsto (\mathcal{V} \cap \mathcal{CS}) \vee \mathcal{RB} \quad (\mathcal{V} \in \mathcal{L}(L\mathcal{O})) \tag{2.1.1}$$

which yields a complete retraction of $\mathcal{L}(L\mathcal{O})$ onto $[\mathcal{RB}, \mathcal{CS}]$. If $\mathcal{V} \in \mathcal{L}(\mathcal{O})$, then

$$\begin{aligned}
(\mathcal{V} \cap \mathcal{CS}) \vee \mathcal{RB} &= (\mathcal{V} \cap \mathcal{ReG}) \vee \mathcal{RB} \\
&= ((\mathcal{V} \cap \mathcal{G}) \vee \mathcal{RB}) \vee \mathcal{RB} \quad \text{by [PR1, Corollary III.5.11(ii)]} \\
&= \mathcal{V}_K \vee \mathcal{RB} \quad \text{by Theorem 2.1.3(iii).}
\end{aligned}$$

If $\mathcal{V} \in \mathcal{L}(L\mathcal{O}) \setminus \mathcal{L}(\mathcal{O})$, then by Theorem 2.1.3(iii), $(\mathcal{V} \cap \mathcal{CS}) \vee \mathcal{RB} = \mathcal{V}_K \vee \mathcal{RB}$. Thus χ and the mapping in (2.1.1) are one and the same so that χ is a complete retraction, as claimed.

Since K is a complete congruence on $\mathcal{L}(\mathcal{CR})$ by Theorem 1.3.4, while $\mathcal{V}_K \, K \, \mathcal{V}$ and $\mathcal{RB} \, K \, \mathcal{T}$ by Theorem 2.1.3(v), it follows that $(\mathcal{V}_K \vee \mathcal{RB}) \, K \, \mathcal{V}$ so that χ induces K on $\mathcal{L}(L\mathcal{O})$. □

We conclude this section with some basic facts concerning congruences of the form τ_S which, recall, can also be described as $\pi_{E(S)}$.

Proposition 2.1.7 *The following statements hold.*

(i) *For any completely regular semigroup S, we have $\tau_{S/\tau} = \varepsilon$.*
(ii) *For any $S_\alpha \in \mathcal{CR}$ where $\alpha \in A$, we have $\tau_{\prod S_\alpha} = \prod \tau_{S_\alpha}$.*
(iii) *The classes $\mathcal{E}$ and $\mathcal{PD}$ are closed under the formation of direct products, but $\mathcal{PE}$ is not.*
(iv) *The classes $\mathcal{E}$, $\mathcal{PD}$ and $\mathcal{PE}$ are closed with respect to the operation $S \to S/\tau$.*
(v) *$\mathcal{E} \cap \mathcal{ReG} = \mathcal{G}$.*

Proof (i) Straightforward.

(ii) Since $\prod_{\alpha \in A} \tau_{S_\alpha}$ is clearly an idempotent pure congruence on S it follows immediately that $\prod_{\alpha \in A} \tau_{S_\alpha} \subseteq \tau_{\prod_{\alpha \in A} S_\alpha}$ since the latter is the largest idempotent pure congruence. Let $\tau = \tau_{\prod_{\alpha \in A} S_\alpha}$ and $g = (g_\alpha)$, $h = (h_\alpha) \in \prod_{\alpha \in A} S_\alpha$ and $g \tau h$. Let $\beta \in A$ and $a, b \in S_\beta^1$ be such that $ag_\beta b \in E(S_\beta)$. Then there exist $x = (x_\alpha)$, $y = (y_\alpha)$ with $x_\beta = a$ and $y_\beta = b$ and $xgy \in E(\prod_{\alpha \in A} S_\alpha)$. Since $g \tau h$ we must also have $xhy \in E(\prod_{\alpha \in A} S_\alpha)$ and therefore $ah_\beta b = x_\beta h_\beta y_\beta \in E(S_\beta)$. Reversing the roles of g and h we see that $g_\beta \tau_{S_\beta} h_\beta$. Thus $\tau \subseteq \prod_{\alpha \in A} \tau_{S_\alpha}$ and equality prevails.

(iii) That the classes $\mathcal{E}$, $\mathcal{PD}$ are closed under products follows easily from part (ii). To see that $\mathcal{PE}$ is not closed under direct products, simply take $S \in \mathcal{CS} \setminus \mathcal{ReG}$ which is not E-disjunctive and $T \in (\mathcal{CR} \setminus \mathcal{LO}) \cap \mathcal{E}$. Then $S \times T \notin \mathcal{PE}$.

(iv) The claim concerning S/τ follows easily from the fact that τ is the largest idempotent pure congruence on S and the fact that S/τ is E-disjunctive.

(v) This follows immediately from [PR1, Corollary III.5.3(v)]. $\square$

In Theorem 2.1.3, (i) is due to Pastijn [2] and Polák [2]; (iii) is due to Polák [1]; the first and fourth equalities in (iv) are due to Jones [1], see also Polák [1]; the second equality in (iv) is due to Pastijn [2]; (v) is due to Polák [2], see also Pastijn [2] and Petrich and Reilly [4]; (vi) is due to Petrich and Reilly [4]; (vii) is due to Polák [2]. Proposition 2.1.7 (ii) and (iv) are due to Petrich and Reilly [4]. For more information on kernel classes, see Sect. 4.6 and [PR3, Section I].

2.2   *E*-Disjunctive Relation

We have encountered the E-disjunctive relation in the preceding section in conjunction with the K-relation. These two relations are closely interwoven. While the K-relation stems from its namesake in congruences on regular semigroups, this new relation is determined by the E-disjunctive membership in varieties. We start with some general properties of this relation which will be supplemented by its relationship with the K-relation and completed with properties of the congruence τ.

We call E defined by $\mathcal{U} E \mathcal{V} \Leftrightarrow \mathcal{U} \cap \mathcal{E} = \mathcal{V} \cap \mathcal{E}$ $(\mathcal{U}, \mathcal{V} \in L(\mathcal{CR}))$ the *E-disjunctive relation* . It is evident that E is a convex relation, in the sense that if $\mathcal{U} \subseteq \mathcal{V} \subseteq \mathcal{W}$ and $\mathcal{U} E \mathcal{W}$, then $\mathcal{U} E \mathcal{V}$.

Theorem 2.2.1 *The following statements hold.*

(i) *The relation E is a complete $\cap$-congruence on $L(\mathcal{CR})$. Its restriction to $L(\mathcal{S} \vee \mathcal{A}_2)$ is not a $\vee$-congruence.*

(ii) *For any $\mathcal{V} \in L(\mathcal{CR})$, $\mathcal{V}E$ has a least element $\mathcal{V}_E = \langle \mathcal{V} \cap \mathcal{E} \rangle$.*

(iii) *Let* $\mathcal{V} = [u_\alpha = v_\alpha]_{\alpha \in A} \in \mathcal{L}(\mathcal{CR})$, *then* $\mathcal{V}E$ *has greatest element*

$$\mathcal{V}^E = \begin{cases} \mathcal{RB} \circ \mathcal{V} \text{ if } \mathcal{V} \in \mathcal{L}(\mathcal{CS}) \setminus \mathcal{L}(\mathcal{RB}), \\ \mathcal{B} \circ \mathcal{V} \quad \text{otherwise.} \end{cases}$$

$$= \begin{cases} [xu_\alpha x = xv_\alpha x]_{\alpha \in A} & \text{if } \mathcal{V} \in \mathcal{L}(\mathcal{CS}), \\ [x^2 = x] & \text{if } \mathcal{V} \in \mathcal{L}(\mathcal{B}), \\ [xu_\alpha y(xv_\alpha y)^{-1} \in E]_{\alpha \in A} & \text{if } \mathcal{V} \in \mathcal{L}(\mathcal{CR}) \setminus (\mathcal{L}(\mathcal{CS}) \cup \mathcal{L}(\mathcal{B})), \end{cases}$$

where $x, y \notin \bigcup_{\alpha \in A} c(u_\alpha v_\alpha)$.
(iv) *Let* $\mathcal{V} \in \mathcal{L}(\mathcal{CR})$. *Then*

$$\mathcal{V}E = \begin{cases} [\mathcal{T}, \mathcal{B}] & \text{if } \mathcal{V} \in \mathcal{L}(\mathcal{B}), \\ [\mathcal{V} \cap \mathcal{G}, \mathcal{RB} \circ \mathcal{V}] & \text{if } \mathcal{V} \in \mathcal{L}(\mathcal{ReG}) \setminus \mathcal{L}(\mathcal{RB}), \\ \{\mathcal{V}\} & \text{if } \mathcal{V} \in \mathcal{L}(\mathcal{CS}) \setminus \mathcal{L}(\mathcal{ReG}), \\ [(\mathcal{V} \cap \mathcal{E}), \mathcal{B} \circ \mathcal{V}] & \text{if } \mathcal{V} \in \mathcal{L}(\mathcal{CR}) \setminus (\mathcal{L}(\mathcal{CS}) \cup \mathcal{L}(\mathcal{B})). \end{cases}$$

This gives all E-classes on $\mathcal{L}(\mathcal{CR})$.
  In particular, for $\mathcal{V} \in (\mathcal{L}(\mathcal{G}) \setminus \mathcal{T}) \cup (\mathcal{L}(\mathcal{CS}) \setminus \mathcal{L}(\mathcal{ReG}))$, *we have*

$$(\mathcal{V} \vee \mathcal{S})E = [\mathcal{V} \vee \mathcal{S}, \mathcal{B} \circ (\mathcal{V} \vee \mathcal{S})] = [\mathcal{V} \vee \mathcal{S}, \mathcal{V}^K].$$

The following intervals are E-classes:

$$[\mathcal{T}, \mathcal{B}], \quad [\mathcal{G}, \mathcal{ReG}], \quad \{\mathcal{CS}\}, \quad [\mathcal{SG}, \mathcal{O}], \quad [\mathcal{NBG}, \mathcal{LO}].$$

Proof Recall the convention introduced in the preamble to this chapter.
(i) Clearly for $\mathcal{U}_\alpha, \mathcal{V}_\alpha \in \mathcal{L}(\mathcal{CR})$, $\alpha \in A$, we have

$$\mathcal{U}_\alpha \ E \ \mathcal{V}_\alpha \text{ for all } \alpha \in A \implies \left(\bigcap_{\alpha \in A} \mathcal{U}_\alpha \right) E \left(\bigcap_{\alpha \in A} \mathcal{V}_\alpha \right)$$

and E is a complete $\cap$-congruence. For the final claim, it is easy to see that $\mathcal{T} E \mathcal{S}$ yet $(\mathcal{T} \vee \mathcal{A}_2) \cap \mathcal{E} = \mathcal{A}_2 \cap \mathcal{E} = \mathcal{A}_2$ while $\mathbb{Z}_2^0 \in \mathcal{S} \vee \mathcal{A}_2$. Therefore $(\mathcal{T} \vee \mathcal{A}_2, \mathcal{S} \vee \mathcal{A}_2) \notin E$.
(ii) For any $\mathcal{V} \in \mathcal{L}(\mathcal{CR})$, we have $\mathcal{V} \cap \mathcal{E} \subseteq (\mathcal{V} \cap \mathcal{E}) \cap \mathcal{E} \subseteq \mathcal{V} \cap \mathcal{E}$. Thus $\mathcal{V} \cap \mathcal{E} = (\mathcal{V} \cap \mathcal{E}) \cap \mathcal{E}$ and $\mathcal{V} E \langle \mathcal{V} \cap \mathcal{E} \rangle$. If $\mathcal{U} \in \mathcal{L}(\mathcal{CR})$ is such that $\mathcal{U} E \mathcal{V}$, then we have $\langle \mathcal{V} \cap \mathcal{E} \rangle = \langle \mathcal{U} \cap \mathcal{E} \rangle \subseteq \mathcal{U}$ from which the claim follows.
(iii) $A = B$.
Case: $\mathcal{V} \in \mathcal{L}(\mathcal{CS}) \setminus \mathcal{L}(\mathcal{RB})$. By [PR1, Theorem IX.3.3], we know that $\mathcal{RB} \circ \mathcal{V}$ is a variety. In this case, $\mathcal{V} \cap \mathcal{E} \supseteq \mathcal{V} \cap \mathcal{G} \neq \mathcal{T}$ and so, for any $\mathcal{U} \in \mathcal{V}E$, there exists a nontrivial group $G \in \mathcal{U}$. If $\mathcal{S} \subseteq \mathcal{U}$ we then have $G^0 \in \mathcal{U} \cap \mathcal{E} = \mathcal{V} \cap \mathcal{E} \subseteq \mathcal{V} \subseteq \mathcal{CS}$, which is a contradiction. Hence $\mathcal{U} \subseteq \mathcal{CS}$ and therefore $\mathcal{V}E \subseteq \mathcal{L}(\mathcal{CS})$.

Let $S \in \mathcal{RB} \circ \mathcal{V}$. By definition, there exists a congruence ρ over $\mathcal{RB}$ such that $S/\rho \in \mathcal{V}$. Then $\rho \subseteq \tau$ so that $S/\tau \in \mathcal{V}$. Consequently

$$S \in (\mathcal{RB} \circ \mathcal{V}) \cap \mathcal{E} \Longrightarrow S \cong S/\tau \in \mathcal{V} \cap \mathcal{E} \Longrightarrow (\mathcal{RB} \circ \mathcal{V}) \cap \mathcal{E} \subseteq \mathcal{V} \cap \mathcal{E}$$

and equality follows, implying that $\mathcal{RB} \circ \mathcal{V} \in \mathcal{V}E$.

On the other hand, if $\mathcal{U} \in \mathcal{L}(\mathcal{CS}), \mathcal{U} E \mathcal{V}$ and $S \in \mathcal{U}$, then $S \cong S/\tau \in \mathcal{U} \cap \mathcal{E} = \mathcal{V} \cap \mathcal{E} \subseteq \mathcal{V}$ so that $S \in \mathcal{RB} \circ \mathcal{V}$. Therefore $\mathcal{RB} \circ \mathcal{V}$ is the largest element of $\mathcal{V}E$.

Case: $\mathcal{V} \notin \mathcal{L}(\mathcal{CS}) \setminus \mathcal{L}(\mathcal{RB})$. In this case $\mathcal{V} \in \mathcal{L}(\mathcal{RB}) \cup [\mathcal{S}, \mathcal{CR}]$. For $\mathcal{V} \in \mathcal{L}(\mathcal{RB})$, we have $\mathcal{B} \circ \mathcal{V} = \mathcal{B}$ while, for $\mathcal{V} \in [\mathcal{S}, \mathcal{CR}]$, $\mathcal{B} \circ \mathcal{V}$ is a variety by [PR1, Theorem IX.4.3(i)]. Thus $\mathcal{B} \circ \mathcal{V}$ is a variety in this case. For any $S \in (\mathcal{B} \circ \mathcal{V}) \cap \mathcal{E}$, we have $S \cong S/\tau \in \mathcal{V} \cap \mathcal{E}$ and thus $(\mathcal{B} \circ \mathcal{V}) \cap \mathcal{E} \subseteq \mathcal{V} \cap \mathcal{E}$. The reverse containment is clear. Hence we have equality and $\mathcal{B} \circ \mathcal{V} \in \mathcal{V}E$. On the other hand, for any $S \in \mathcal{U} \in \mathcal{V}E$, we have $S/\tau \in \mathcal{U} \cap \mathcal{E} = \mathcal{V} \cap \mathcal{E} \subseteq \mathcal{V}$ so that $S \in \mathcal{B} \circ \mathcal{V}$. Thus $\mathcal{U} \subseteq \mathcal{B} \circ \mathcal{V}$ and $\mathcal{B} \circ \mathcal{V}$ is the largest element of $\mathcal{V}E$.

$B = C$. We consider three cases.

Case: $\mathcal{V} \in \mathcal{L}(\mathcal{CS})$. Let $S \in \mathcal{RB} \circ \mathcal{V}$. By [PR1, Theorem IX.3.3], $S \in \mathcal{CS}$ and by the definition of $\mathcal{RB} \circ \mathcal{V}$, there must exist a congruence ρ on S such that ρ is over $\mathcal{RB}$ and $S/\rho \in \mathcal{V}$. Let $\alpha \in A$ and $c(u_\alpha v_\alpha) = \{x_1, \ldots, x_n\}$. Consider any choice $x, x_1, \ldots, x_n \in S$. Then $u_\alpha(x_1, \ldots, x_n) \, \rho \, v_\alpha(x_1, \ldots, x_n)$. Let us write u_α for $u_\alpha(x_1, \ldots, x_n)$ and v_α for $v_\alpha(x_1, \ldots, x_n)$. Then $xu_\alpha x \, \rho \, xv_\alpha x$. Also $B = (xu_\alpha x)\rho = (xv_\alpha x)\rho$ must be a rectangular band subsemigroup of S. Since S is completely simple, $xu_\alpha x \, \mathcal{H} \, xv_\alpha x$ in S, whence we also have $xu_\alpha x \, \mathcal{H} \, xv_\alpha x$ in B. But B is a rectangular band. Therefore $xu_\alpha x = xv_\alpha x$. Since this is true for all $x, x_1, \ldots, x_n \in S$ and $\alpha \in A$ it follows that $S \in [xu_\alpha x = xv_\alpha x]_{\alpha \in A}$ and so $\mathcal{RB} \circ \mathcal{V} \subseteq [xu_\alpha x = xv_\alpha x]_{\alpha \in A}$.

For the converse, since $\mathcal{V} \in \mathcal{L}(\mathcal{CS})$, there must be $\alpha \in A$ such that the identity $u_\alpha = v_\alpha$ is not homotypical. Then the identity $xu_\alpha x = xv_\alpha x$ is also not homotypical. Hence $\mathcal{W} = [xu_\alpha x = xv_\alpha x]_{\alpha \in A} \in \mathcal{L}(\mathcal{CS})$. Let $S \in \mathcal{W}$ and consider any substitution of the variables into S and write u for u_α and v for v_α. We have

$$xuy \in E_S \Longleftrightarrow uyx \in E_S \Longleftrightarrow uyx(yx)^{-1}yx \in E_S$$
$$\Longleftrightarrow (yx)u(yx)(yx)^{-1} \in E_S \Longleftrightarrow (yx)v(yx)(yx)^{-1} \in E_S$$
$$\Longleftrightarrow \cdots \Longleftrightarrow xvy \in E_S.$$

Therefore $u \, \tau_S \, v$ and $S/\tau_S \in \mathcal{V}$. Clearly τ_S is over $\mathcal{RB}$. Hence $S \in \mathcal{RB} \circ \mathcal{V}$ and $\mathcal{W} \subseteq \mathcal{RB} \circ \mathcal{V}$ and equality follows.

Case: $\mathcal{V} \in \mathcal{L}(\mathcal{B})$. Then $\mathcal{V} \in \mathcal{L}(\mathcal{RB}) \cup [\mathcal{S}, \mathcal{B}]$. Either way, $\mathcal{B} \circ \mathcal{V} = \mathcal{B} = [x^2 = x]$.

Case: $\mathcal{V} \in \mathcal{L}(\mathcal{CR}) \setminus (\mathcal{L}(\mathcal{CS}) \cup \mathcal{L}(\mathcal{B}))$. Then $\mathcal{B} \circ \mathcal{V} = [xu_\alpha y(xv_\alpha y)^{-1} \in E]_{\alpha \in A}$ by [PR1, Theorem IX.4.3(i)].

(iv) *Case*: $\mathcal{V} \in \mathcal{L}(\mathcal{B})$. This follows trivially from parts (ii) and (iii).

Case: $\mathcal{V} \in \mathcal{L}(\mathcal{Re}\mathcal{G}) \setminus \mathcal{L}(\mathcal{RB})$. Then $\mathcal{V} \cap \mathcal{E} \subseteq \mathcal{Re}\mathcal{G} \cap \mathcal{E} \subseteq \mathcal{G}$ so that $\mathcal{V} \cap \mathcal{E} \subseteq \mathcal{V} \cap \mathcal{G}$. Since the reverse inclusion is clear, we have $\mathcal{V} \cap \mathcal{G} = \mathcal{V} \cap \mathcal{E} = \langle \mathcal{V} \cap \mathcal{E} \rangle = \mathcal{V}_E$.

Case: $V \in \mathcal{L}(\mathcal{CS}) \setminus \mathcal{L}(\mathcal{ReG})$. By [PR1, Proposition VIII.3.9], the free object FV_X in V is E-disjunctive. Therefore $FV_X \in V \cap \mathcal{E}$ so that $V_E = \langle V \cap \mathcal{E} \rangle = V$. That $V^E = \mathcal{RB} \circ V$ follows from part (iii).

Case: $V \in \mathcal{L}(\mathcal{CR}) \setminus (\mathcal{L}(\mathcal{CS}) \cup \mathcal{L}(\mathcal{B}))$. The claim follows immediately from parts (ii) and (iii).

That completes the general description of E-classes.

Now let $V \in (\mathcal{L}(\mathcal{CS}) \setminus \mathcal{L}(\mathcal{ReG}))$. By [PR1, Proposition VIII.3.9], the free object FV_X in V is E-disjunctive. Hence so also is FV_X^0. Therefore $FV_X^0 \in (V \vee \mathcal{S}) \cap \mathcal{E}$. But FV_X^0 generates $V \vee \mathcal{S}$. Hence $(V \vee \mathcal{S})_E = \langle (V \vee \mathcal{S}) \cap \mathcal{E} \rangle = V \vee \mathcal{S}$. Exactly the same argument applies to any nontrivial $V \in \mathcal{L}(\mathcal{G})$.

For the upper end of the interval containing $V \vee \mathcal{S}$, where V belongs to $\mathcal{L}(\mathcal{CS}) \setminus \mathcal{L}(\mathcal{ReG})$ or V is a nontrivial variety of groups, by the main description of E-classes in this part, we obtain $(V \vee \mathcal{S})^E = \mathcal{B} \circ (V \vee \mathcal{S})$ which, by Theorem 2.1.3(iv), is just V^K. Thus

$$(V \vee \mathcal{S})E = [V \vee \mathcal{S}, \mathcal{B} \circ (V \vee \mathcal{S})] = [V \vee \mathcal{S}, V^K].$$

The specific examples now follow trivially. □

Notation 2.2.2 We define two equivalence relations by stating their classes:

$$\Lambda : \mathcal{L}(\mathcal{CS}) \cup \mathcal{L}(\mathcal{B}), \quad \text{its complement in } \mathcal{L}(\mathcal{CR}),$$
$$\Xi : \mathcal{L}(\mathcal{CS}) \setminus \mathcal{L}(\mathcal{RB}), \quad \text{its complement in } \mathcal{L}(\mathcal{CR})$$

and define a third as the intersection of these two: $\Pi = \Lambda \cap \Xi$. Its classes are:

$$\mathcal{L}(\mathcal{B}), \quad \mathcal{L}(\mathcal{CS}) \setminus \mathcal{L}(\mathcal{RB}), \quad \mathcal{L}(\mathcal{CR}) \setminus (\mathcal{L}(\mathcal{B}) \cup \mathcal{L}(\mathcal{CS})).$$

We examine the relationship between E and K in detail in the next theorem.

Theorem 2.2.3 *It holds that* $E = K \cap \Lambda = K \cap \Xi = K \cap \Pi$. *In particular,* $VE = VK$ *for* $V \in \mathcal{L}(\mathcal{CR}) \setminus \mathcal{L}(\mathcal{LO})$.

Proof Recall the convention introduced in the preamble to this chapter. We also take advantage of Theorem 2.1.3(iv) without further comment.

$A = B$. *Case*: $V \in \mathcal{L}(\mathcal{B})$. We have $V(K \cap \Lambda) = \mathcal{L}(\mathcal{B}) = VE$.

Case: $V \in \mathcal{L}(\mathcal{ReG}) \setminus \mathcal{L}(\mathcal{RB})$. We have

$$V(K \cap \Lambda) = VK \cap V\Lambda = [V \cap \mathcal{G}, \mathcal{OH}(V \cap \mathcal{G})] \cap V\Lambda$$
$$= [V \cap \mathcal{G}, (\mathcal{RB} \circ (V \cap \mathcal{G})) \cap \mathcal{CS}]$$
$$= [V \cap \mathcal{G}, \mathcal{RB} \circ V]$$
$$= VE \quad \text{by Theorem 2.1.3(iii).}$$

Case: $\mathcal{V} \in \mathcal{L}(\mathcal{CS}) \setminus \mathcal{L}(\mathcal{ReG})$. In this case we must have $\mathcal{RB} \subseteq \mathcal{V}$. Let $S \in \mathcal{RB} \circ \mathcal{V}$. By [PR1, Theorem IX.3.3], $\mathcal{RB} \circ \mathcal{V} \in \mathcal{L}(\mathcal{CS})$ so that, in particular, S is completely simple. Also, there must exist a congruence ρ on S such that ρ is over $\mathcal{RB}$ and $S/\rho \in \mathcal{V}$. In addition, since S is completely simple, $S/\mu \in \mathcal{RB} \subseteq \mathcal{V}$. Hence $S/(\rho \cap \mu) \in \mathcal{V}$. But ρ is a congruence over $\mathcal{RB}$ and $\mu \subseteq \mathcal{H}$. Therefore $\rho \cap \mu = \varepsilon$ and $S \in \mathcal{V}$. The reverse containment is trivial and so we have $\mathcal{RB} \circ \mathcal{V} = \mathcal{V}$. Consequently,

$$\mathcal{V}(K \cap \Lambda) = [\mathcal{V}, L\mathcal{O}D\mathcal{V}] \cap \mathcal{V}\Lambda = [\mathcal{V}, L\mathcal{O}D\mathcal{V} \cap \mathcal{CS}]$$
$$= [\mathcal{V}, D\mathcal{V} \cap \mathcal{CS}] = \{\mathcal{V}\} = \{\mathcal{RB} \circ \mathcal{V}\} = \mathcal{V}E.$$

Case: $\mathcal{V} \in \mathcal{L}(\mathcal{CR}) \setminus (\mathcal{L}(\mathcal{CS}) \cup \mathcal{L}(\mathcal{B}))$. In this case $\mathcal{S} \subseteq \mathcal{V}$ so that $\mathcal{V}K = \mathcal{V}E \subseteq \mathcal{V}\Lambda$. Hence $\mathcal{V}(K \cap \Lambda) = \mathcal{V}K = \mathcal{V}E$.

$A = C$. We consider four cases, as in Theorem 2.2.1(iv).

Case: $\mathcal{V} \in \mathcal{L}(\mathcal{B})$. We have $\mathcal{V}(K \cap \Xi) = \mathcal{L}(\mathcal{B}) = \mathcal{V}E$.

Case: $\mathcal{V} \in \mathcal{L}(\mathcal{ReG}) \setminus \mathcal{L}(\mathcal{RB})$. By Theorem 2.1.3(ii), we have $\mathcal{V}_K = \mathcal{V} \cap \mathcal{G}$ and clearly $\mathcal{V} \cap \mathcal{G} \equiv \mathcal{V}$. Hence $\mathcal{V} \cap \mathcal{G}$ must be the smallest member of $\mathcal{V}(K \cap \Xi)$. On the other hand, by Theorem 2.1.3(iii), for any $S \in \mathcal{U} \in \mathcal{V}(K \cap \Xi)$, we have

$$\mathcal{U} \in \mathcal{V}(K \cap \Xi) \Longrightarrow \mathcal{U} \in \mathcal{L}(\mathcal{CS}), \quad \mathcal{U} \subseteq \mathcal{V}^K$$
$$\Longrightarrow S \in \mathcal{CS}, \quad S/\tau_S \in \mathcal{V} \vee \mathcal{S}$$
$$\Longrightarrow S \in \mathcal{CS}, \quad S/\tau_S \in (\mathcal{V} \vee \mathcal{S}) \cap \mathcal{CS}$$
$$\Longrightarrow S/\tau_S \in \mathcal{V} \quad \text{by [PR1, Theorem IX.9.1]}$$
$$\Longrightarrow S \in \mathcal{RB} \circ \mathcal{V}$$
$$\Longrightarrow \mathcal{U} \subseteq \mathcal{RB} \circ \mathcal{V}.$$

By Theorem 2.1.3(iii) and the definition of Ξ, it is clear that $\mathcal{RB} \circ \mathcal{V} \in \mathcal{V}(K \cap \Xi)$ and so $\mathcal{RB} \circ \mathcal{V}$ must be the greatest element in $\mathcal{V}(K \cap \Xi)$ and $\mathcal{V}(K \cap \Xi) = [\mathcal{V} \cap \mathcal{G}, \mathcal{RB} \circ \mathcal{V}] = \mathcal{V}E$, by Theorem 2.2.1(iv).

Case: $\mathcal{V} \in \mathcal{L}(\mathcal{CS}) \setminus \mathcal{L}(\mathcal{ReG})$. The argument in this case is entirely analogous to that of the previous case (but with $\mathcal{V} \cap \mathcal{CS}$ replacing $\mathcal{V} \cap \mathcal{G}$).

Case: $\mathcal{V} \in \mathcal{L}(\mathcal{CR}) \setminus (\mathcal{L}(\mathcal{CS}) \cup \mathcal{L}(\mathcal{B}))$. This divides into three subcases corresponding to the division in Theorem 2.1.3(ii).

Subcase: $\mathcal{V} \in \mathcal{L}(\mathcal{O})$. Then $\mathcal{V} \in \mathcal{L}(\mathcal{O}) \setminus (\mathcal{L}(\mathcal{CS}) \cup \mathcal{L}(\mathcal{B}))$, which implies that $\mathcal{S} \subseteq \mathcal{V}$ and $\mathcal{V} \cap \mathcal{G} \neq \mathcal{J}$ while, by Theorem 2.1.3(ii), $\mathcal{V}_K = \mathcal{V} \cap \mathcal{G}$. It follows that $(\mathcal{V} \cap \mathcal{G}) \vee \mathcal{S} \in \mathcal{V}(K \cap \Xi)$. On the other hand,

$$\mathcal{U} \in \mathcal{V}(K \cap \Xi) \Longrightarrow \mathcal{U} \cap \mathcal{G} = \mathcal{U}_K = \mathcal{V}_K = \mathcal{V} \cap \mathcal{G} \quad \text{and} \quad \mathcal{S} \subseteq \mathcal{U}$$
$$\Longrightarrow (\mathcal{V} \cap \mathcal{G}) \vee \mathcal{S} \subseteq \mathcal{U}.$$

Therefore $(\mathcal{V} \cap \mathcal{G}) \vee \mathcal{S}$ must be the smallest member of $\mathcal{V}(K \cap \Xi)$. Now, for any $\mathcal{S} \in \mathcal{V} \cap \mathcal{E}$, we have $\mathcal{S} \cong \mathcal{S}/\tau \in (\mathcal{V} \cap \mathcal{G}) \vee \mathcal{S}$. Consequently by Theorem 2.2.1(ii)

$$\mathcal{V}_E = \langle \mathcal{V} \cap \mathcal{E} \rangle \subseteq (\mathcal{V} \cap \mathcal{G}) \vee \mathcal{S}.$$

On the other hand, $(\mathcal{V} \cap \mathcal{G}) \vee \mathcal{S}$ is a variety of Clifford semigroups and so is generated by groups and groups with a zero adjoined, all of which belong to $\mathcal{V} \cap \mathcal{E}$. Therefore $(\mathcal{V} \cap \mathcal{G}) \vee \mathcal{S} \subseteq \langle \mathcal{V} \cap \mathcal{E} \rangle = \mathcal{V}_E$. Hence $\mathcal{V}_E$ is the smallest member of $\mathcal{V}(K \cap \Xi)$.

To determine the greatest member of $\mathcal{V}(K \cap \Xi)$, note that $\mathcal{S} \subseteq \mathcal{V}$ so that, by Theorem 2.1.3(iii), $\mathcal{V}^K = \mathcal{B} \circ (\mathcal{V} \vee \mathcal{S}) = \mathcal{B} \circ \mathcal{V}$. Since $\mathcal{V} \, \Xi \, \mathcal{B} \circ \mathcal{V}$, it follows that $\mathcal{B} \circ \mathcal{V}$ is the greatest member of $\mathcal{V}(K \cap \Xi)$. But by Theorem 2.2.1(iii), we have $\mathcal{B} \circ \mathcal{V} = \mathcal{V}^E$. Hence $\mathcal{V}(K \cap \Xi) = [\mathcal{V}_E, \mathcal{V}^E] = \mathcal{V}E$, as desired.

Subcase: $\mathcal{V} \in \mathcal{L}(L\mathcal{O}) \setminus \mathcal{L}(\mathcal{O})$. The argument is entirely analogous to that in the preceding subcase, with $\mathcal{V} \cap \mathcal{C}\mathcal{S}$ replacing $\mathcal{V} \cap \mathcal{G}$.

Subcase: $\mathcal{V} \in \mathcal{L}(\mathcal{C}\mathcal{R}) \setminus \mathcal{L}(L\mathcal{O})$. By Theorem 2.1.3(ii), $\mathcal{V}_K = \langle \mathcal{V} \cap \mathcal{E} \rangle$ and clearly $\langle \mathcal{V} \cap \mathcal{E} \rangle \in \mathcal{V}\Xi$. By Theorem 2.1.3(iii), we have $\mathcal{V}^K = \mathcal{B} \circ (\mathcal{V} \vee \mathcal{S}) = \mathcal{B} \circ \mathcal{V}$ which also belongs to $\mathcal{V}\Xi$. Hence $\mathcal{V}(K \cap \Xi) = \mathcal{V}K = [\langle \mathcal{V} \cap \mathcal{E} \rangle, \mathcal{B} \circ \mathcal{V}] = \mathcal{V}E$, where the last equality follows from Theorem 2.2.1(ii)(iii).

Thus in all cases $\mathcal{V}(K \cap \Xi) = \mathcal{V}E$ so that $E = K \cap \Xi$.

$A = D$. This follows from $A = B = C$ and the definition of Π.

For the final claim, let $\mathcal{V} \in \mathcal{L}(\mathcal{C}\mathcal{R}) \setminus \mathcal{L}(L\mathcal{O})$. Let $\mathcal{U} \in \mathcal{L}(\mathcal{C}\mathcal{R})$ be such that $\mathcal{U} \, K \, \mathcal{V}$. Then $\mathcal{U}^K \subseteq (L\mathcal{O})^K = L\mathcal{O}$, by Theorem 2.1.3(v). But then $\mathcal{V} \not\subseteq \mathcal{U}^K$ which would contradict the assumption that $\mathcal{U} \, K \, \mathcal{V}$.

If $\mathcal{U} \in L\mathcal{O}$, then $\mathcal{U}^K = \{\mathcal{S} \mid \mathcal{S}/\tau \in \mathcal{U} \vee \mathcal{S}\} \in L\mathcal{O}$. But then $\mathcal{V} \notin \mathcal{L}(\mathcal{U}^K)$, which would contradict the assumption that $\mathcal{U} \, K \, \mathcal{V}$. Hence $\mathcal{U} \notin L\mathcal{O}$ and we have, in particular, that $\mathcal{S} \subseteq \mathcal{U} \cap \mathcal{V}$. Then

$$S \in \mathcal{U} \cap \mathcal{E} \Longrightarrow S = S/\tau \in \mathcal{U}^K = \mathcal{V}^K$$
$$\Longrightarrow S = S/\tau \in (\mathcal{V} \vee \mathcal{S}) \cap \mathcal{E} = \mathcal{V} \cap \mathcal{E} \quad \text{by Theorem 2.1.3(iv)}.$$

Thus $\mathcal{U} \cap \mathcal{E} \subseteq \mathcal{V} \cap \mathcal{E}$ and, reversing the roles, we have equality: $\mathcal{U} \cap \mathcal{E} = \mathcal{V} \cap \mathcal{E}$.

Conversely, let $\mathcal{U} \, E \, \mathcal{V}$, that is, $\mathcal{U} \cap \mathcal{E} = \mathcal{V} \cap \mathcal{E}$. Then

$$S \in \mathcal{U}^K \Longrightarrow S/\tau \in (\mathcal{U} \vee \mathcal{S}) \cap \mathcal{E} = \mathcal{U} \cap \mathcal{E} \quad \text{by Theorem 2.1.3(iv)}$$
$$\Longrightarrow S/\tau \in \mathcal{V} \cap \mathcal{E} \subseteq \mathcal{V} = \mathcal{V} \vee \mathcal{S}$$
$$\Longrightarrow S \in \mathcal{V}^K \qquad \text{by Theorem 2.1.3(iv)}.$$

Therefore $\mathcal{U}^K \subseteq \mathcal{V}^K$. Reversing the roles of $\mathcal{U}$ and $\mathcal{V}$ we obtain equality: $\mathcal{U}^K = \mathcal{V}^K$. Thus $\mathcal{U} \, K \, \mathcal{V}$. $\qquad \square$

The next proposition will clarify the relationship between the relations E and K.

Proposition 2.2.4 *The congruence on $\mathcal{L}(\mathcal{CR})$ generated by the relation E equals K.*

Proof By Theorem 2.1.3(i), K is a congruence on $\mathcal{L}(\mathcal{CR})$ and by Theorem 2.2.3, we have $E = K \cap \Xi$. Let D be any congruence on $\mathcal{L}(\mathcal{CR})$ which contains E. Since $E \subseteq K$ and K is a congruence, in order to prove the contention, it suffices to show that $K \subseteq D$. Let

$$A = \mathcal{L}(\mathcal{B}), \quad B = \mathcal{L}(\mathcal{CS}) \setminus \mathcal{L}(\mathcal{RB}), \quad C = \mathcal{L}(\mathcal{CR}) \setminus (\mathcal{L}(\mathcal{B}) \cup \mathcal{L}(\mathcal{CS})).$$

First let $\mathcal{U}, \mathcal{V} \in \mathcal{L}(\mathcal{CR})$ be varieties that are K- but not E-related. Since A is both an E- and a K-class, we have $\mathcal{U}, \mathcal{V} \notin A$. Hence $\mathcal{U}, \mathcal{V} \in B \cup C$ and thus either $\mathcal{U} \in B, \mathcal{V} \in C$ or $\mathcal{U} \in C, \mathcal{V} \in B$, so we may assume the former. Next $\mathcal{T} K \mathcal{S}$ implies that $\mathcal{U} \vee \mathcal{T} K \mathcal{U} \vee \mathcal{S}$ and thus $\mathcal{U} K \mathcal{U} \vee \mathcal{S}$. This together with $\mathcal{U} K \mathcal{V}$ gives $\mathcal{U} \vee \mathcal{S} K \mathcal{V}$. Since $\mathcal{U} \vee \mathcal{S}, \mathcal{V} \in C$, we also have $\mathcal{U} \vee \mathcal{S} \Xi \mathcal{V}$ which gives $\mathcal{U} \vee \mathcal{S} E \mathcal{V}$. But then the hypothesis yields $\mathcal{U} \vee \mathcal{S} D \mathcal{V}$. On the other hand, $\mathcal{T} E \mathcal{S}$ implies that $\mathcal{T} D \mathcal{S}$ whence $\mathcal{U} \vee \mathcal{T} D \mathcal{U} \vee \mathcal{S}$ and hence $\mathcal{U} D \mathcal{U} \vee \mathcal{S}$ which, in view of the above displayed relation, yields $\mathcal{U} D \mathcal{V}$.

We have proved that $\mathcal{U} K \mathcal{V}$ and $(\mathcal{U}, \mathcal{V}) \notin E$ imply $\mathcal{U} D \mathcal{V}$. If $\mathcal{U} E \mathcal{V}$, then $\mathcal{U} D \mathcal{V}$. Therefore $K \subseteq D$ and the assertion of the proposition follows.   □

The following detail should help clear up the picture at the very bottom of the lattice $\mathcal{L}(\mathcal{CR})$.

Proposition 2.2.5 *The restrictions of the relations E and K to $\mathcal{CS} \cup \mathcal{L}(\mathcal{B})$ coincide. In particular, for $\mathcal{U}, \mathcal{V} \in \mathcal{CS} \cup \mathcal{L}(\mathcal{B})$, we have*

$$\mathcal{U} K \mathcal{V} \Longleftrightarrow \begin{cases} \mathcal{U} = \mathcal{V} & \text{if } \mathcal{U}, \mathcal{V} \in \mathcal{L}(\mathcal{CS}) \setminus \mathcal{L}(\mathcal{ReG}), \\ \mathcal{U} \vee \mathcal{RB} = \mathcal{V} \vee \mathcal{RB} & \text{if } \mathcal{U}, \mathcal{V} \in \mathcal{L}(\mathcal{ReG}) \setminus \mathcal{L}(\mathcal{RB}), \\ \text{no condition} & \text{if } \mathcal{U}, \mathcal{V} \in \mathcal{L}(\mathcal{B}). \end{cases}$$

Proof Straightforward verification.   □

It is a natural question to ask for the structure of E-disjunctive completely regular semigroups belonging to various familiar varieties. This problem is solved in [PR1, Lemma III.4.13] for completely simple semigroups. Next we characterize the class of E-disjunctive Clifford semigroups. Toward this goal the following lemma will be helpful.

Lemma 2.2.6 *Let $S = [Y; S_\alpha, \chi_{\alpha,\beta}]$ where $S_\alpha \in \mathcal{G}$ and e_α is its identity element for all $\alpha \in Y$. Assume that $\alpha > \beta$. Then $e_\alpha \tau e_\beta$ if and only if for every $\gamma \in Y$, $\chi_{\alpha\gamma,\beta\gamma}$ is injective.*

Proof We will take advantage of the one-sided characterization of $\tau = \pi_K$ found in [PR1, Lemma VI.1.6]. Let $e_\alpha \tau e_\beta$ and $a, b \in S_{\alpha\gamma}$ be such that $a\chi_{\alpha\gamma,\beta\gamma} = b\chi_{\alpha\gamma,\beta\gamma}$. Then

$$ab^{-1}\chi_{\alpha\gamma,\beta\gamma} = bb^{-1}\chi_{\alpha\gamma,\beta\gamma} = e_{\beta\gamma}$$

implying that

$$ab^{-1} = ab^{-1}e_\alpha \ \tau \ ab^{-1}e_\beta = ab^{-1}\chi_{\alpha\gamma,\beta\gamma} = e_{\beta\gamma}.$$

Therefore $ab^{-1} = e_{\alpha\gamma}$ so that $a = b$ and it ensues that $\chi_{\alpha\gamma,\beta\gamma}$ is injective.

Conversely, suppose that $\chi_{\alpha\gamma,\beta\gamma}$ is injective for all $\gamma \in Y$, and let $a \in S_\gamma$. If $ae_\alpha \in E(S)$, then

$$ae_\beta = (a\chi_{\gamma,\beta\gamma})(e_\beta\chi_{\beta,\beta\gamma}) = a\chi_{\gamma,\beta\gamma} = (ae_\alpha)\chi_{\alpha\gamma,\beta\gamma} \in E(S).$$

On the other hand, if $ae_\beta \in E(S)$ where $a \in S_\gamma$, then

$$(ae_\alpha)\chi_{\alpha\gamma,\beta\gamma} = (ae_\alpha)e_\beta = ae_\beta \in E(S).$$

But $\chi_{\alpha\gamma,\beta\gamma}$ is injective and therefore $ae_\alpha \in E(S)$. Hence $e_\alpha \ \tau \ e_\beta$.   □

Theorem 2.2.7 Let $S = [Y; S_\alpha, \chi_{\alpha,\beta}]$ where $S_\alpha \in \mathcal{G}$ for all $\alpha \in Y$. Then S is E-disjunctive if and only if for any $\alpha > \beta$, there exists $\gamma \in Y$ such that $\chi_{\alpha\gamma,\beta\gamma}$ is not injective.

Proof By contrapositive, we will prove that $\tau \neq \varepsilon$ if and only if there exists $\alpha > \beta$ such that $\chi_{\alpha\gamma,\beta\gamma}$ is injective for all $\gamma \in Y$. Note that $\ker \tau = E(S)$ so that $\tau \neq \varepsilon$ if and only if $\mathrm{tr}\,\tau \neq \varepsilon$. Denote by e_α the identity element of S_α for all $\alpha \in Y$.

Necessity. By hypothesis, there exist $\alpha, \beta \in Y$ such that $\alpha \neq \beta$ and $e_\alpha \ \tau \ e_\beta$. If α and β are not comparable, then either $\alpha > \alpha\beta$ or $\beta > \alpha\beta$. Hence we may suppose that $\alpha > \beta$. By Lemma 2.2.6, $\chi_{\alpha\gamma,\beta\gamma}$ is injective for all $\gamma \in Y$.

Sufficiency. Suppose $\alpha > \beta$ and $\chi_{\alpha\gamma,\beta\gamma}$ is injective for all $\gamma \in Y$. By Lemma 2.2.6, we have $e_\alpha \ \tau \ e_\beta$ which shows that $\tau \neq \varepsilon$.   □

We will also need the following construction.

Lemma 2.2.8 Let $S = [Y; S_\alpha, \chi_{\alpha,\beta}]$ where $S_\alpha \in \mathcal{G}$ for all $\alpha \in Y$. Next let $G = \{e, a\}$ be a nontrivial group with identity element e, and for every $\alpha \in Y$, let $T_\alpha = S_\alpha \times G$. For any $\alpha \in Y$, let $\varphi_{\alpha,\alpha}$ be the identity mapping on T_α, and for any $\alpha > \beta$, define a function $\varphi_{\alpha,\beta}$ by $\varphi_{\alpha,\beta} : (a, x) \mapsto (a\chi_{\alpha,\beta}, e)$ $((a, x) \in T_\alpha)$. Then $T = [Y; T_\alpha, \varphi_{\alpha,\beta}]$ is defined, and the mapping θ defined by $\theta : (a, x) \mapsto a$ $((a, x) \in S)$ is a homomorphism of T onto S.

Proof The conditions for $[Y; T_\alpha, \varphi_{\alpha,\beta}]$ being defined follow at once. For $\alpha, \beta \in Y$ with $\alpha \geq \beta$, define a function $\psi_{\alpha,\beta}$ by

$$\psi_{\alpha,\beta} : x \mapsto \begin{cases} e & \text{if } \alpha > \beta \\ x & \text{if } \alpha = \beta \end{cases} \quad (x \in G).$$

Let $(a, x) \in S_\alpha$ and $(b, y) \in S_\beta$. Then

$$((a, x)(b, y))\theta = ((a, x)\varphi_{\alpha,\alpha\beta}(b, y)\varphi_{\beta,\alpha\beta})\theta$$
$$= ((a\chi_{\alpha,\alpha\beta}, x\psi_{\alpha,\alpha\beta})(b\chi_{\beta,\alpha\beta}, y\psi_{\beta,\alpha\beta}))\theta$$
$$= ((a\chi_{\alpha,\alpha\beta})(b\chi_{\beta,\alpha\beta}), (x\psi_{\alpha,\alpha\beta})(y\psi_{\beta,\alpha\beta}))\theta$$
$$= (ab, (x\psi_{\alpha,\alpha\beta})(y\psi_{\beta,\alpha\beta}))\theta = ab = (a, x)\theta(b, y)\theta$$

and θ is a homomorphism of S onto T.   □

Theorem 2.2.9 *Let $\mathcal{U}$ be a nontrivial variety of groups and $S \in \mathcal{U} \vee \mathcal{S}$. Then S is a homomorphic image of an E-disjunctive semigroup $T \in \mathcal{U} \vee \mathcal{S}$. In particular, every Clifford semigroup is a homomorphic image of an E-disjunctive Clifford semigroup.*

Proof If S is a group, then the assertion follows trivially. So let $S = [Y; S_\alpha, \chi_{\alpha,\beta}]$ where $S_\alpha \in \mathcal{U}$ for all $\alpha \in Y$ and let G be a nontrivial group in $\mathcal{U}$. In the notation of Lemma 2.2.8, we then have $T_\alpha \in \mathcal{U}$, $T \in \mathcal{U} \vee \mathcal{S}$ and, whenever $\alpha > \beta$, the mapping $\varphi_{\alpha,\beta}$ is not injective. Hence, by Theorem 2.2.7, we may set $\gamma = \alpha$ to get that in T, we have $\tau = \varepsilon$. Thus S is a homomorphic image of the E-disjunctive semigroup $T \in \mathcal{U} \vee \mathcal{S}$.   □

Theorem 2.2.1(i)(ii)(iii) are due to Petrich and Reilly [4].

2.3   Congruence Systems and Integral Classes

There are two concepts that play an important role in the first three parts of Theorem 2.1.3. First there is the class of $\mathcal{E}$ of E-disjunctive completely regular semigroups (together with the associated classes $\mathcal{PE}$, and $\mathcal{PD}$) and then there is the class $\{\tau_S \mid S \in \mathcal{CR}\}$ of greatest idempotent pure congruences. These concepts were cast in secondary roles because we had already established in Chap. 1, via the study of fully invariant congruences, that K is a congruence on $\mathcal{L}(\mathcal{CR})$. We will see in this section how congruence systems and their associated integral classes provide an alternative route to similar conclusions in other contexts. This slightly more general treatment will not only highlight the generality of this alternative approach and demonstrate the strong relationship between the apparently disparate concept of congruences on $\mathcal{L}(\mathcal{CR})$, congruence systems and subdirect product systems, but also simplify our work in later sections.

Definition 2.3.1 By a *congruence family* $\mathbb{P}$ we mean a collection of congruences $\{\pi_S \mid S \in \mathcal{CR}\}$, where each π_S is a congruence on S, satisfying the following conditions:

$\mathbb{P}$(i) For $S, T \in \mathcal{CR}$, an isomorphism $\varphi : S \to T$ and $a, b \in S$, we have

$$a \, \pi_S \, b \Longleftrightarrow a\varphi \, \pi_S \, b\varphi;$$

$\mathbb{P}$(ii) For any $S \in \mathcal{CR}$, we have $\pi_{S/\pi_S} = \varepsilon$ and π_S is the least congruence ρ on S such that $\pi_{S/\rho} = \varepsilon$. We refer to the property $\mathbb{P}$(ii) as the *radical property*.

We say that $\mathbb{P}$ is a *congruence system* if, in addition, it satisfies the following conditions:

$\mathbb{P}$(iii) if $S_\alpha \in \mathcal{CR}$ where $\alpha \in A$ and S is a subdirect product of $\prod_{\alpha \in A} S_\alpha$, then

$$\pi_S = \left(\prod_{\alpha \in A} \pi_{S_\alpha} \right)\Big|_S \, ;$$

$\mathbb{P}$(iv) if $S, T \in \mathcal{CR}$ and S is a subsemigroup of T, then $\pi_T|_S \subseteq \pi_S$.

The following simple consequence of $\mathbb{P}$(iii) is useful:

$\mathbb{P}$(iii)* if $S_\alpha \in \mathcal{CR}$ where $\alpha \in A$ and $S = \prod_{\alpha \in A} S_\alpha$, then

$$S/\pi_S \cong \left(\prod_{\alpha \in A} S_\alpha / \pi_{S_\alpha} \right).$$

The next lemma provides some basic information concerning these concepts.

Lemma 2.3.2 *Let* $\mathbb{P} = \{\pi_S \mid S \in \mathcal{CR}\}$ *be a congruence family.*

(i) *For* $S, T \in \mathcal{CR}$, *an epimorphism* $\varphi : S \to T$ *and* $a, b \in S$, *we have*

$$a \, \pi_S \, b \implies a\varphi \, \pi_T \, b\varphi.$$

(ii) *Let* $S, S_\alpha \in \mathcal{CR}$ *where* $\alpha \in A$, *and* S *be a subdirect product of the* S_α. *Then*

$$\pi_S \subseteq \left(\prod_{\alpha \in A} \pi_{S_\alpha} \right)\Big|_S .$$

Proof (i) Let ρ be the relation on S defined by $a \, \rho \, b \Leftrightarrow a\varphi \, \pi_T \, b\varphi$. Then $S/\rho \cong T/\pi_T$ where $\pi_{T/\pi_T} = \varepsilon$ so that, by $\mathbb{P}$(i), we have $\pi_{S/\rho} = \varepsilon$. By $\mathbb{P}$(ii), it follows that $\pi_S \subseteq \rho$. Hence for $a, b \in S$, we have $a \, \pi_S \, b \Rightarrow a\varphi \, \pi_T \, b\varphi$, as required.

(ii) Let S and S_α where $\alpha \in A$ be as in the statement of the lemma. Let $\theta_\alpha : \prod_{\alpha \in A} S_\alpha \to S_\alpha$ be the natural projection mapping $(a_\alpha) \to a_\alpha$. Then θ_α is an epimorphism so that, by part (i), we get

$$(a_\alpha) \, \pi_S \, (b_\alpha) \implies a_\alpha = (a_\alpha)\theta_\alpha \, \pi_{S_\alpha} \, (b_\alpha)\theta_\alpha = b_\alpha.$$

Thus $\pi_S \subseteq \left(\prod_{\alpha \in A} \pi_{S_\alpha} \right)\big|_S$, as required. $\square$

Associated with each congruence family is an important class of semigroups consisting of those for which the corresponding congruence is the equality relation.

Definition 2.3.3 For each congruence family $\mathbb{P} = \{\pi_S \in \mathcal{C}(S) \mid S \in \mathcal{CR}\}$, we define the *integral class* of $\mathbb{P}$ to be $\mathcal{P} = \{S \in \mathcal{CR} \mid \pi_S = \varepsilon\}$.

If $\mathbb{P}$ is a congruence family, then it is clear that, to within isomorphisms, $\mathcal{P} = \{S/\pi_S \mid S \in \mathcal{CR}\}$.

Definition 2.3.4 A class of completely regular semigroups is said to be a *subdirect product class* if it contains the class of all one element semigroups and is closed under the isomorphisms and subdirect products.

The most familiar examples of subdirect product classes are varieties. However, the most interesting examples for us (see subsequent sections) will not be varieties but rather integral classes of certain congruence systems. The next result provides the basic observations required to establish the relationship between congruence systems and subdirect product classes.

Lemma 2.3.5 *Let* $\mathbb{P} = \{\pi_S \mid S \in \mathcal{CR}\}$ *be a congruence family,* $\mathcal{Q}$ *be a subdirect product class and* $S \in \mathcal{CR}$.

(i) *The integral class* $\mathcal{P}$ *of* $\mathbb{P}$ *is a subdirect product class.*
(ii) *There exists a least congruence* $\pi(S : \mathcal{Q})$ *on* S *such that* $S/\pi(S : \mathcal{Q}) \in \mathcal{Q}$.

Proof (i) Clearly $\mathcal{P}$ contains all trivial semigroups. Now let $S \in \mathcal{CR}$, $S_\alpha \in \mathcal{P}$, $\alpha \in A$ and S be a subdirect product of the S_α. By hypothesis, $\pi_{S_\alpha} = \varepsilon_{S_\alpha}$ for all $\alpha \in A$, and, by Lemma 2.3.2(ii), we get

$$\pi_S \subseteq \left(\prod_{\alpha \in A} \pi_{S_\alpha} \right)\Big|_S = \left(\prod_{\alpha \in A} \varepsilon_{S_\alpha} \right)\Big|_S = \varepsilon$$

so that $\pi_S = \varepsilon$. Thus $S \in \mathcal{P}$ and $\mathcal{P}$ is a subdirect product class.
(ii) Let $\{\pi_\alpha \mid \alpha \in A\}$ be the set of all congruences on S such that $S/\pi_\alpha \in \mathcal{P}$. Since $\mathcal{P}$ contains all one element semigroups and $|S/\omega_S| = 1$, it follows that $A \neq \emptyset$. Now $S/\bigcap_{\alpha \in A} \pi_\alpha$ is isomorphic to a subdirect product of the semigroups $S/\pi_\alpha \in \mathcal{P}$ so that, since $\mathcal{P}$ is a subdirect product class, $S/\bigcap_{\alpha \in A} \pi_\alpha \in \mathcal{P}$ and $\bigcap_{\alpha \in A} \pi_\alpha$ must be the least congruence on S for which the quotient lies in $\mathcal{P}$. $\qquad\square$

We can now establish the correspondence between congruence families and subdirect product classes.

Proposition 2.3.6 *The mapping* $\chi : \mathbb{P} \to$ *integral class of* $\mathbb{P}$ *is a bijection between congruence families and subdirect product classes.*

Proof By Lemma 2.3.5(i), for each congruence system $\mathbb{P}$, $\mathbb{P}\chi$ is a subdirect product class.

For each subdirect product class $\mathfrak{Q}$, let $\mathfrak{Q}^* = \{\pi(S : \mathfrak{Q}) \mid S \in \mathcal{CR}\}$. It is evident that $\mathfrak{Q}^*$ satisfies $\mathbb{P}$(i). For any $S \in \mathcal{CR}$, we have $S/\pi(S : \mathfrak{Q}) \in \mathfrak{Q}$ so that $\pi(S/\pi(S : \mathfrak{Q}) : \mathfrak{Q}) = \varepsilon$ and, by Lemma 2.3.5(ii), $\pi(S : \mathfrak{Q})$ is the smallest congruence ρ on S for which $\pi(S/\rho : \mathfrak{Q}) = \varepsilon$. Thus $\mathfrak{Q}^*$ is a congruence family. Clearly the integral class of $\mathfrak{Q}^*$ is $\mathfrak{Q}$ so that χ is surjective.

Now let $\mathbb{P}$ be a congruence system with integral class $\mathcal{P}$. For any $S \in \mathcal{CR}$, we have, by $\mathbb{P}$(ii), that $\pi_S = \pi(S : \mathcal{P})$. Hence $\mathbb{P} = \mathcal{P}^*$. Thus χ is injective and therefore bijective.   □

In subsequent sections, we will encounter several interesting congruence systems. We digress briefly here to consider one family of congruences, which has already figured prominently in this chapter, that falls short of being a congruence system; namely $\mathbb{T} = \{\tau_S \mid S \in \mathcal{CR}\}$. This forced us in Sect. 2.2 to develop singular arguments to deal with it. The reason that $\mathbb{T}$ fails to be a congruence system is because this family fails conditions $\mathbb{P}$(ii) and $\mathbb{P}$(iii). By [PR1, Proposition VI.1.16], τ_S is the unique idempotent pure congruence such that S/τ_S is E-disjunctive. But it need not be the smallest congruence such that the quotient is E-disjunctive.

Let $S = G \times Y_2$ where $G = \{e, a\}$ denotes the cyclic group of order two with identity e and $Y_2 = \{0, 1\}$ denotes the semilattice of order two with identity 1. Then τ_S has classes $\{(e, 1), (e, 0)\}$ and $\{(a, 1), (a, 0)\}$. However, if ρ denotes the Rees congruence determined by the ideal $G \times \{0\}$, then S/ρ is also E-disjunctive, but ρ does not contain τ_S. Thus $\mathbb{T}$ does not satisfy $\mathbb{P}$(ii).

This complication regarding $\mathbb{T}$ appears in another way. By Lemma 2.3.2(ii), a subdirect product of elements from the integral class of a congruence system must lie in the integral class. The class $\mathcal{E}$ of E-disjunctive completely regular semigroups does not have this property. Let $M = \mathcal{M}^0(\{1, 2\}, G, \{1, 2\}; P)$ where G is as above and $P = \begin{bmatrix} e & e \\ e & a \end{bmatrix}$. Let $\varphi_{10} : G \to M$ be the function defined by $e\varphi_{10} = (1, e, 1)$, and $a\varphi_{10} = (1, a, 1)$. Then φ_{10} is an isomorphism of G onto the $\mathcal{H}$-class H_{11} of M. Let S be the strong semilattice of semigroups determined by φ_{10}.

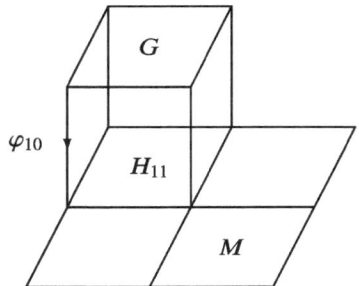

The connecting homomorphism φ_{10} extends to a retraction γ of S onto M and it is easy to see that $\overline{\gamma}$ is a nontrivial idempotent pure congruence on S. Thus S is not E-disjunctive.

Let $\delta : S \rightarrow G^0$ be the natural homomorphism of S onto the Rees quotient S/M. Then $\gamma \times \delta : S \rightarrow (S\gamma, S\delta)$ embeds S as a subdirect product of $G^0 \times M$ where both G^0 and M are E-disjunctive. Thus a subdirect product of E-disjunctive semigroups need not be E-disjunctive. A slight modification of this example (replacing G by T^1 where T is an E-disjunctive completely simple semigroup) will provide an example of a subdirect product of elements of $\mathcal{PD} \cap \mathcal{PE}$ that does not belong to $\mathcal{PD} \cup \mathcal{PE}$. Thus neither $\mathcal{PD}$ nor $\mathcal{PE}$ is a subdirect product class.

Definition 2.3.7 Let $\mathbb{P} = \{\pi_S \mid S \in \mathcal{CR}\}$ be a congruence system with integral class $\mathcal{P}$. A subclass $\mathcal{C}$ of $\mathcal{P}$ is a $\mathbb{P}$-*variety* if it satisfies the following conditions.

$\mathcal{P}(i)$ $\mathcal{C}$ is closed under isomorphisms and direct products,

$\mathcal{P}(ii)$ if $S \in \mathbb{HSC}$, then $S/\pi_S \in \mathcal{C}$.

Let $\mathcal{L}_{\mathbb{P}}(\mathcal{CR})$ denote the class of all $\mathbb{P}$-varieties ordered by inclusion and, for any $\mathcal{V} \in \mathcal{L}(\mathcal{CR})$ let $\Pi_{\mathcal{V}} = \{S \in \mathcal{CR} \mid S/\pi_S \in \mathcal{V}\}$. For example, $\Pi_{(\mathcal{P})} = \mathcal{CR}$.

The next theorem will have multiple applications in later sections. In particular, it establishes the strong link between congruence systems and their integral classes on the one hand and complete homomorphisms and their induced complete congruences on the other.

Theorem 2.3.8 *Let* $\mathbb{P} = \{\pi_S \mid S \in \mathcal{CR}\}$ *be a congruence system with integral class* $\mathcal{P}$.

(i) $\mathcal{L}_{\mathbb{P}}(\mathcal{CR})$ *is a complete lattice with greatest element* $\mathcal{P}$ *and least element* $\mathcal{T}$.
(ii) *The mapping* $\theta_{\mathcal{P}} : \mathcal{V} \mapsto \mathcal{V} \cap \mathcal{P}$ $(\mathcal{V} \in \mathcal{L}(\mathcal{CR}))$ *is a complete homomorphism of* $\mathcal{L}(\mathcal{CR})$ *onto* $\mathcal{L}_{\mathbb{P}}(\mathcal{CR})$.
(iii) *For* $\mathcal{V} \in \mathcal{L}(\mathcal{CR})$, *the class of* $\mathcal{V}$ *in the relation induced by* $\theta_{\mathcal{P}}$ *is an interval:*

$$\mathcal{V}\overline{\theta}_{\mathcal{P}} = [\langle \mathcal{V} \cap \mathcal{P}\rangle, \Pi_{\mathcal{V}}].$$

Proof By $\mathbb{P}(i)$ and $\mathbb{P}(iii)$, $\mathcal{P}$ is closed under isomorphisms and direct products. By $\mathbb{P}(ii)$, for any $S \in \mathcal{CR}$, we have $\pi_{S/\pi_S} = \varepsilon$ so that $S/\pi_S \in \mathcal{P}$. Thus $\mathcal{P}$ satisfies conditions $\mathcal{P}(i)$ and $\mathcal{P}(ii)$, so that $\mathcal{P} \in \mathcal{L}_{\mathbb{P}}(\mathcal{CR})$. It is then clear that $\mathcal{P}$ must be the greatest element of $\mathcal{L}_{\mathbb{P}}(\mathcal{CR})$. Similarly, $\mathcal{T}$ is the least. It now follows that in order to establish that $\mathcal{L}_{\mathbb{P}}(\mathcal{CR})$ is a complete lattice it suffices to show that $\mathcal{L}_{\mathbb{P}}(\mathcal{CR})$ is closed under intersections.

Let $\mathcal{U}_\alpha \in \mathcal{L}_{\mathbb{P}}(\mathcal{CR})$ where $\alpha \in A$ and set $\mathcal{U} = \bigcap_{\alpha \in A} \mathcal{U}_\alpha$. Since each $\mathcal{U}_\alpha$ is closed under direct products, so also is $\mathcal{U}$. Let $S \in \mathbb{HSU}$. Then $S \in \mathbb{HSU}_\alpha$ for each $\alpha \in A$ so that $S/\pi_S \in \mathcal{U}_\alpha$ and thus $S/\pi_S \in \mathcal{U}$. Since $\mathcal{U}$ is clearly closed under isomorphisms, it follows that $\mathcal{U} \in \mathcal{L}_{\mathbb{P}}(\mathcal{CR})$. Hence $\mathcal{L}_{\mathbb{P}}(\mathcal{CR})$ is a complete lattice.

Next we will show that for any $\mathcal{V} \in \mathcal{L}(\mathcal{CR})$, we have $\mathcal{V} \cap \mathcal{P} \in \mathcal{L}_{\mathbb{P}}(\mathcal{CR})$. Clearly $\mathcal{V} \cap \mathcal{P}$ is closed under isomorphisms, since both $\mathcal{V}$ and $\mathcal{P}$ are. For each $\alpha \in A$, let $S_\alpha \in \mathcal{V} \cap \mathcal{P}$ and set $S = \prod_{\alpha \in A} S_\alpha$. Since $\mathcal{V}$ is a variety, we have $S \in \mathcal{V}$ while, by $\mathbb{P}(\text{iii})$,

$$\pi_S = \left(\prod_{\alpha \in A} \pi_{S_\alpha} \right)\Big|_S = \left(\prod_{\alpha \in A} \varepsilon_{S_\alpha} \right)\Big|_S = \varepsilon_S$$

so that $S \in \mathcal{V} \cap \mathcal{P}$ and $\mathcal{V} \cap \mathcal{P}$ is closed under direct products, thereby satisfying $\mathcal{P}(\text{i})$.

Now, for $S \in \mathbb{HS}(\mathcal{V} \cap \mathcal{P})$, we have $S \in \mathcal{V}$ since $\mathcal{V}$ is a variety and $S/\pi_S \in \mathcal{P}$ by $\mathbb{P}(\text{ii})$. Hence $S/\pi_S \in \mathcal{V} \cap \mathcal{P}$ and $\mathcal{V} \cap \mathcal{P}$ satisfies $\mathcal{P}(\text{ii})$. Thus $\mathcal{V} \cap \mathcal{P} \in \mathcal{L}_{\mathbb{P}}(\mathcal{CR})$ and $\theta_{\mathcal{P}}$ maps $\mathcal{L}(\mathcal{CR})$ into $\mathcal{L}_{\mathbb{P}}(\mathcal{CR})$.

Let $\mathcal{A} \in \mathcal{L}_{\mathbb{P}}(\mathcal{CR})$ and set $\mathcal{V} = \langle \mathcal{A} \rangle$. Since $\mathcal{A}$ is closed under direct products, we have $\mathcal{V} = \mathbb{HS}\mathcal{A}$. Hence, if $S \in \mathcal{V} \cap \mathcal{P}$, then $S \in \mathbb{HS}\mathcal{A} \cap \mathcal{P}$ and, by $\mathcal{P}(\text{ii})$, we have $S \cong S/\pi_S \in \mathcal{A} \cap \mathcal{P} = \mathcal{A}$. Therefore $\mathcal{V} \cap \mathcal{P} \subseteq \mathcal{A}$ and, since the reverse inclusion is trivial, we have equality. Consequently $\mathcal{V}\theta_{\mathcal{P}} = \mathcal{V} \cap \mathcal{P} = \mathcal{A}$ and $\theta_{\mathcal{P}}$ maps $\mathcal{L}(\mathcal{CR})$ onto $\mathcal{L}_{\mathbb{P}}(\mathcal{CR})$.

We defer the proof that $\theta_{\mathcal{P}}$ is a complete homomorphism until we prove part (iii).

(iii) Clearly $\mathcal{V} \cap \mathcal{P} \subseteq \langle \mathcal{V} \cap \mathcal{P} \rangle \cap \mathcal{P} \subseteq \mathcal{V} \cap \mathcal{P}$ so that $\langle \mathcal{V} \cap \mathcal{P} \rangle \cap \mathcal{P} = \mathcal{V} \cap \mathcal{P}$ and $\langle \mathcal{V} \cap \mathcal{P} \rangle \in \mathcal{V}\overline{\theta}_{\mathcal{P}}$. Hence $\langle \mathcal{V} \cap \mathcal{P} \rangle$ must be the least element of $\mathcal{V}\overline{\theta}_{\mathcal{P}}$.

For any variety $\mathcal{W} \in \mathcal{V}\overline{\theta}_{\mathcal{P}}$ and any semigroup $S \in \mathcal{W}$ we have $S/\pi_S \in \mathcal{W} \cap \mathcal{P} = \mathcal{V} \cap \mathcal{P}$ which implies that $S \in \Pi_\mathcal{V}$, whence $\mathcal{W} \subseteq \Pi_\mathcal{V}$. Therefore in order to show that $\Pi_\mathcal{V}$ is the greatest element in $\mathcal{V}\overline{\theta}_{\mathcal{P}}$, it suffices to show that $\Pi_\mathcal{V}$ is a variety.

Let $S_\alpha \in \Pi\mathcal{V}$ where $\alpha \in A$, and $S = \prod_{\alpha \in A} S_\alpha$. By $\mathbb{P}(\text{iii})$, $S/\pi_S = \prod_{\alpha \in A} S_\alpha/\pi_{S_\alpha} \in \mathcal{V}$ so we have that $S \in \Pi\mathcal{V}$ and $\Pi_\mathcal{V}$ is closed with respect to direct products. Now let $S \in \Pi\mathcal{V}$ and $\varphi : S \to T$ be an epimorphism. By Lemma 2.3.2(i), T/π_T is a homomorphic image of $S/\pi_S \in \mathcal{V}$. Hence $T/\pi_T \in \mathcal{V}$ and $T \in \Pi_\mathcal{V}$. Thus $\Pi\mathcal{V}$ is closed under homomorphisms. Finally, let $S \in \mathcal{CR}$ be a subsemigroup of $T \in \Pi_\mathcal{V}$. Then $S/(\pi_T|_S)$ is a subsemigroup of T/π_T so that $S/(\pi_T|_S) \in \mathcal{V}$. But, by $\mathbb{P}(\text{iv})$, S/π_S is a homomorphic image of $S/(\pi_T|_S)$. Hence $S/\pi_S \in \mathcal{V}$ and $S \in \Pi_\mathcal{V}$. Therefore $\Pi_\mathcal{V}$ is closed under taking subsemigroups, is a variety and is the greatest element of $\mathcal{V}\overline{\theta}_{\mathcal{P}}$. This completes the proof of part (iii).

Returning to the proof of part (ii), it is clear that the mapping $\theta_{\mathcal{P}}$ as well as the mappings $\mathcal{V} \mapsto \langle \mathcal{V} \cap \mathcal{P} \rangle$, $\mathcal{V} \mapsto \Pi_\mathcal{V}$ are order preserving. Therefore, by [PR1, Lemma I.2.5], we have that $\theta_{\mathcal{P}}$ is a complete lattice homomorphism. $\square$

We conclude this section with an observation that will facilitate the application of the results of this section in the sections to follow. Clearly this result would be unnecessary if we restricted our attention to completely regular semigroups that are either finite or countably infinite.

Recall from [PR1, Definition VI.2.10] that a completely regular semigroup S is said to be *fundamental* (respectively, *left fundamental*) if the equality relation on S is the only congruence contained in $\mathcal{H}$ (respectively, $\mathcal{L}$) that is, if $\mu = \mathcal{H}^0 = \varepsilon$ (respectively, $\mathcal{L}^0 = \varepsilon$. Likewise, recall from [PR1, Definition VI.1.15] that a completely regular semigroup S is

E-disjunctive (respectively, right *E-disjunctive* if the equality relation on S is the only idempotent pure congruence on S respectively, the only idempotent pure congruence on S contained in $\mathcal{R}$.

Lemma 2.3.9 *Let S be a fundamental (respectively, right fundamental or right E-disjunctive) completely regular semigroup and A be a finite or countably infinite subset of S. Then there exists a finite or countably infinite completely regular subsemigroup B of S such that $A \subseteq B$ and B is fundamental (respectively, right fundamental or right E-disjunctive).*

Proof We consider only the case of $S \in \mathcal{RD}$ as the other two cases are similar but simpler. Define a sequence of completely regular subsemigroups $A_1, A_2, \ldots$ of S inductively as follows.

Let $A_1 = A$. For each pair of distinct elements $a, b \in A_n$, since S is right E-disjunctive, $(a, b) \notin \tau_S \cap \mathcal{R}$ and therefore there exist elements $x_{a,b}, y_{a,b} \in S$ such that either $(x_{a,b}ay_{a,b}, x_{a,b}by_{a,b}) \notin \tau_S$ or $(x_{a,b}ay_{a,b}, x_{a,b}by_{a,b}) \notin \mathcal{R}$.

We define A_{n+1} to be the completely regular subsemigroup of S generated by A_n and the set of elements $\{x_{a,b}, y_{a,b} \mid a, b \in A_n\}$. Clearly A_{n+1} is either finite or countably infinite. Also, for each pair of distinct elements $a, b \in A_n$ we have that $(a, b) \notin \tau_{A_{n+1}} \cap \mathcal{R}^0$.

Now let $B = \bigcup_{n \geq 1} A_n$. Then B has the required properties. $\square$

Congruence systems and $\mathbb{P}$-varieties were introduced by Petrich and Reilly [4] where various examples can be found. Theorem 2.3.8(ii)(iii) are due to Reilly and Zhang [1]. The remainder of this section can be found in Petrich and Reilly [4]. The relationship between complete congruences on the lattice of subpseudovarieties of a pseudovariety and congruence families has been extensively studied by Pastijn and Trotter [2]. Auinger et al. [3] and Auinger [1] apply the ideas of this section to complete congruences on lattices of pseudovarieties.

2.4   Trace Relation

We discuss the relation induced on $\mathcal{L}(\mathcal{CR})$ by the trace relation T on $\Gamma(CR_X)$ via the usual antiisomorphism between $\Gamma(\mathrm{CR}_X)$ and $\mathcal{L}(\mathcal{CR})$ (see Notation 1.3.2). We will introduce the congruence system $\{\mu_S \mid S \in \mathcal{CR}\}$ and its integral class $\mathcal{F}$.

Indeed, the properties of this congruence system and T are so strongly related that we can combine their most basic properties in one main theorem. The basic properties of the operators resulting from the characterization of the upper and lower ends of the intervals associated with T are presented.

Recall from Notation 1.3.2 that, for $\mathcal{U}, \mathcal{V} \in \mathcal{L}(\mathcal{CR})$,

$$\mathcal{U} \ T \ \mathcal{V} \Longleftrightarrow \rho_{\mathcal{U}} \ T \ \rho_{\mathcal{V}}.$$

Notation 2.4.1 We denote by $\mathcal{F}$ the class of all fundamental completely regular semigroups and write $\mathbb{M} = \{\mu_S \mid S \in \mathcal{CR}\}$ and $\theta_{\mathcal{F}} : \mathcal{V} \mapsto \mathcal{V} \cap \mathcal{F} \ (\mathcal{V} \in \mathcal{L}(\mathcal{CR}))$.

We will need the following general observation.

Lemma 2.4.2 *Every completely regular semigroup satisfies the identity* $(ab)^{-1}a = a(ba)^{-1}$.

Proof For any $a, b \in S$ where S is completely regular, we have

$$(ab)^{-1}a = (ab)^{-1}b^0a = (ab)^{-1}b^{-1}ba = (ab)^{-1}b^{-1}baba(ba)^{-1}$$
$$= (ab)^{-1}aba(ba)^{-1} = (ab)^0a(ba)^{-1} = (ab)^0ab^0(ba)^{-1}$$
$$= (ab)^0abb^{-1}(ba)^{-1} = abb^{-1}(ba)^{-1} = ab^0(ba)^{-1} = a(ba)^{-1}. \qquad \square$$

Theorem 2.4.3 *The following claims hold.*

  (i) *T is a complete congruence on $\mathcal{L}(\mathcal{CR})$.*
  (ii) *$\mathbb{M}$ is a congruence system and $\mathcal{F}$ is its integral class.*
  (iii) *$T = \bar{\theta}_{\mathcal{F}}$.*
  (iv) *For $\mathcal{U}, \mathcal{V} \in \mathcal{L}(\mathcal{CR})$,*

$$\mathcal{U} \, T \, \mathcal{V} \iff \mathcal{U} \cap \mathcal{F} = \mathcal{V} \cap \mathcal{F} \iff \langle \mathcal{U} \cap \mathcal{F} \rangle = \langle \mathcal{V} \cap \mathcal{F} \rangle \iff \rho_{\mathcal{U}} \vee \mathcal{H} = \rho_{\mathcal{V}} \vee \mathcal{H}.$$

  (v) *For any $\mathcal{V} \in \mathcal{L}(\mathcal{CR})$, we have*

$$\mathcal{V}_T = \langle \mathcal{V} \cap \mathcal{F} \rangle = \begin{cases} \mathcal{V} \cap \mathcal{B} & \text{if } \mathcal{V} \subseteq \mathcal{BG}, \\ \langle \mathcal{V} \cap \mathcal{F} \rangle & \text{otherwise.} \end{cases}$$

  (vi) *Let $\mathcal{V} = [u_\alpha = v_\alpha]_{\alpha \in A} \in \mathcal{L}(\mathcal{CR})$, and $W = \bigcup_{\alpha \in A} c(u_\alpha v_\alpha)$. Then*

$$\mathcal{V}^T = \mathcal{G} \circ \mathcal{V} = \{S \in \mathcal{CR} \mid S/\mu \in \mathcal{V}\}$$
$$= \left[u_\alpha^0 = v_\alpha^0, \ (xu_\alpha y)^0 = (xv_\alpha y)^0 \right]_{\alpha \in A} \quad \text{where } x, y \notin W$$
$$= \left[u_\alpha (xu_\alpha)^{-1} = v_\alpha (xv_\alpha)^{-1} \right]_{\alpha \in A} \quad \text{where } x \notin W$$
$$= \left[u^0 = v^0 \mid u^0 = v^0 \text{ holds in } \mathcal{V} \right]$$

and, if $\mathcal{RB} \subseteq \mathcal{V}$, then also

$$\mathcal{V}^T = \left[(xu_\alpha y)^0 = (xv_\alpha y)^0 \right]_{\alpha \in A} \quad \text{where } x, y \notin W.$$

  (vii) *The mapping $\theta_{\mathcal{F}} : \mathcal{V} \mapsto \mathcal{V} \cap \mathcal{F} \ (\mathcal{V} \in \mathcal{L}(\mathcal{CR}))$ is a complete homomorphism of $\mathcal{L}(\mathcal{CR})$ onto $\mathcal{L}_{\mathbb{M}}(\mathcal{CR})$.*

(viii) *The mapping $V \to V_T$ ($V \in \mathcal{L}(\mathcal{CR})$) is a complete $\vee$-endomorphism of $\mathcal{L}(\mathcal{CR})$. Its restriction to $\mathcal{L}(\mathcal{RO})$ is not an $\cap$-homomorphism.*

(ix) *The mapping $V \mapsto V^T$ ($V \in \mathcal{L}(\mathcal{CR})$) is a complete $\cap$-endomorphism of $\mathcal{L}(\mathcal{CR})$. Its restriction to $\mathcal{L}(\mathcal{CS})$ is not a $\vee$-homomorphism.*

(x) $K \cap T = \varepsilon$.

(xi) *For any $V \in \mathcal{L}(\mathcal{CR})$, we have $V = V_K \vee V_T = V^K \cap V^T$.*

(xii) *For $V \in \mathcal{L}(\mathcal{CR})$, we have $V^T = (V^T)_K$ if and only if $V \in [\mathcal{T}, \mathcal{G}] \cup [\mathcal{RB}, \mathcal{CS}] \cup [\mathcal{ReB}, \mathcal{CR}]$.*

(xiii) $\bigvee_{n \geq 1} \mathcal{T}^{(KT)^n} = \mathcal{CR}$.

(xiv) *The kernel class of $\mathcal{CR}$ is a singleton. In particular, $\mathcal{CR}_K = \mathcal{CR}$. Equivalently, the equality relation is the only idempotent pure congruence on $F\mathcal{CR}(X)$.*

Proof Recall the convention introduced in the preamble to this book regarding sequences of equalities and equivalent statements.

We begin with part (ii) and then proceed with parts (i), (iii), …

(ii) Clearly $\mathbb{M}$ satisfies $\mathbb{P}$(i). That $\mathbb{M}$ satisfies $\mathbb{P}$(ii) follows immediately from [PR1, Proposition VI.2.11(i)]. Thus $\mathbb{M}$ is a congruence family.

Now let $S_\alpha \in \mathcal{CR}$ where $\alpha \in A$ and S be a subdirect product of the S_α. Since each μ_{S_α} is idempotent separating, it follows that $\prod_{\alpha \in A} \mu_{S_\alpha}$ is also idempotent separating and therefore $\left(\prod_{\alpha \in A} \mu_{S_\alpha}\right)\big|_S$ is also idempotent separating and is contained in μ_S. The reverse containment follows from Lemma 2.3.2(ii) whence equality follows and $\mathbb{P}$(iii) is satisfied.

Finally, let $S, T \in \mathcal{CR}$ and S be a subsemigroup of T. Let $a, b \in S$. Then by [PR1, Proposition VI.2.7], we have

$$a \, \mu_T \, b \Longrightarrow a^0 = b^0 \text{ and } a^{-1}ea = b^{-1}eb \text{ for all } e \in E(T), \ e \leq a^0$$
$$\Longrightarrow a^0 = b^0 \text{ and } a^{-1}ea = b^{-1}eb \text{ for all } e \in E(S), \ e \leq a^0$$
$$\Longrightarrow a \, \mu_S \, b.$$

Thus $\mu_T\big|_S \subseteq \mu_S$ and $\mathbb{P}$(iv) is satisfied.

(i) See Theorem 1.3.4.

(iii) Let $\mathcal{U}, \mathcal{V} \in \mathcal{L}(\mathcal{CR})$. First assume that $\mathcal{U} \, \overline{\theta_{\mathcal{F}}} \, \mathcal{V}$ so that $\mathcal{U} \cap \mathcal{F} = \mathcal{V} \cap \mathcal{F}$. Then $CR_X / \rho_{\mathcal{V}}^T \in \mathcal{V} \cap \mathcal{F} = \mathcal{U} \cap \mathcal{F}$ which implies that $\rho_{\mathcal{U}} \subseteq \rho_{\mathcal{V}}^T$. Similarly $\rho_{\mathcal{V}} \subseteq \rho_{\mathcal{U}}^T$ and $\rho_{\mathcal{V}}^T \subseteq \rho_{\mathcal{U}}^T$ whence $\rho_{\mathcal{U}} \subseteq \rho_{\mathcal{V}}^T \subseteq \rho_{\mathcal{U}}^T$. This implies that $\rho_{\mathcal{U}} \, T \, \rho_{\mathcal{V}}$ and therefore $\mathcal{U} \, T \, \mathcal{V}$.

Conversely, let $\mathcal{U} \, T \, \mathcal{V}$ and $S \in \mathcal{U} \cap \mathcal{F}$. To see that $S \in \mathcal{V}$, it suffices to show that every finitely generated completely regular subsemigroup A of S belongs to $\mathcal{V}$. By Lemma 2.3.9, there exists a completely regular subsemigroup B of S with $A \subseteq B \in \mathcal{F}$ and B a finite or countably infinite. Hence there exists an epimorphism $\varphi : CR_X / \rho_{\mathcal{U}} \to B$. Now $\mathbb{M}$ is a congruence system and so, for any $a, b \in CR_X$, by Lemma 2.3.2(i),

$$a \rho_{\mathcal{U}} \, \mu \, b \rho_{\mathcal{U}} \Longrightarrow (a \rho_{\mathcal{U}})\varphi \, \mu \, (b \rho_{\mathcal{U}})\varphi.$$

But $B \in \mathcal{F}$ so that $(a\rho_\mathcal{U})\varphi = (b\rho_\mathcal{U})\varphi$ and φ induces an epimorphism $(CR_X/\rho_\mathcal{U})/\mu \to B$ where

$$(CR_X/\rho_\mathcal{U})/\mu = (CR_X/\rho_\mathcal{U})(\rho_\mathcal{U}^T/\rho_\mathcal{U}) \cong CR_X/\rho_\mathcal{U}^T = CR_X/\rho_\mathcal{V}^T \in \mathcal{V}.$$

Consequently $B \in \mathcal{V}$. Hence $S \in \mathcal{V} \cap \mathcal{F}$ and it follows that $\mathcal{U} \cap \mathcal{F} \subseteq \mathcal{V} \cap \mathcal{F}$. Reversing the roles of $\mathcal{U}$ and $\mathcal{V}$ we obtain that $\mathcal{U} \cap \mathcal{F} = \mathcal{V} \cap \mathcal{F}$ as desired.

(iv) $A \Leftrightarrow B$. This follows immediately from the definition of $\theta_\mathcal{F}$ and the fact that $T = \overline{\theta}_\mathcal{F}$.

$A \Leftrightarrow C$. This follows immediately from Theorem 2.3.8(iii) and the fact that $T = \overline{\theta}_\mathcal{F}$.

$A \Leftrightarrow D$. This follows immediately from [PR1, Corollary VII.2.9].

(v) $A = B$. This follows immediately from Theorem 2.3.8(iii) and the fact that $T = \overline{\theta}_\mathcal{F}$.

$A = C$. Let $\mathcal{V} \in L(B\mathcal{G})$. Since $B\mathcal{G} \cap \mathcal{F} = B$, it follows that

$$\mathcal{V} \cap \mathcal{F} = \mathcal{V} \cap B\mathcal{G} \cap \mathcal{F} = \mathcal{V} \cap B$$

whence $\mathcal{V}_T = \langle \mathcal{V} \cap \mathcal{F} \rangle = \mathcal{V} \cap B$. For $\mathcal{V} \notin L(B\mathcal{G})$, the claim follows from the equality $A = B$.

(vi) $A = C$. This follows from Theorem 2.3.8(iii) and the fact that $T = \overline{\theta}_\mathcal{F}$.

$B = C = D$. This follows from [PR1, Theorem IX.4.3(ii)].

$B = E$. It suffices to prove that $\mathcal{G} \circ [u = v] = [u(xu)^{-1} = v(xv)^{-1}]$ or equivalently that for $u, v \in S$, the free object on X in the variety $\mathcal{G} \circ [u = v]$,

$$S/\mu \text{ satisfies } u = v \iff S \text{ satisfies } u(xu)^{-1} = v(xv)^{-1} \text{ for } x \in X \setminus c(uv).$$

($\Rightarrow$) For any substitution of the variables in X into S, we have $u \; \mu \; v$. Then $xu \; \mu \; xv$ whence $(xu)^{-1} \; \mu \; (xv)^{-1}$ and thus $u(xu)^{-1} \; \mu \; v(xv)^{-1}$ and also $xu(xu)^{-1} \; \mu \; xv(xv)^{-1}$. Since the last two elements are idempotent, we get $xu(xu)^{-1} = xv(xv)^{-1}$. For $x = u^{-1}$, we obtain $u^0 = u^{-1}v(u^{-1}v)^{-1}$ and similarly, for $x = v^{-1}$, we get $v^{-1}u(v^{-1}u)^{-1} = v^0$. It follows that for some completely simple components S_α, S_β we have $u, v \in S_\alpha$ and $ux, xu, vx, xv \in S_\beta$. We now have $ux[u(xu)^{-1}] = ux[v(xv)^{-1}]$. In a Rees matrix semigroup, this is of the form $(j, t, \mu)(i, g, \lambda) = (j, t, \mu)(i, h, \lambda)$ since $u(xu)^{-1} \; \mu \; v(xv)^{-1}$. But then $g = h$ and thus $u(xu)^{-1} = v(xv)^{-1}$.

($\Leftarrow$) Let $x, y \in X \setminus c(uv)$ and consider any substitution of the variables into S. By the hypothesis and Lemma 2.4.2 combined with the substitution $x \to y$, we have

$$u(xu)^{-1} = v(xv)^{-1}, \quad (uy)^{-1}u = (vy)^{-1}v.$$

From the first of these equalities and the substitution $x \to u^0$ we get $u^0 = v(u^0v)^{-1}$ and, from the substitution $x \to v^0$ we get $v^0 = u(vu)^{-1}$. This implies that $u^0 = v^0$ and $u \; \mathcal{H} \; v$. Hence $xuy \; \mathcal{D} xvy$. Furthermore, $(xu)^0 = xu(xu)^{-1} = xv(xv)^{-1} = (xv)^0$ and similarly $(uy)^0 = (vy)^0$. Hence

$$xuy = xu(xu)^0y = xu(xv)^0y = xu(xv)^{-1}xvy$$

which implies that $xuy \mathrel{\mathcal{L}} xvy$. Similarly, using $(uy)^0 = (vy)^0$ we obtain $xuy \mathrel{\mathcal{R}} xvy$. Thus $xuy \mathrel{\mathcal{H}} xvy$ which, by the arbitrary nature of x and y, implies that $u \mathrel{\mu} v$.

$A = F$. This follows immediately from Lemma 1.3.7(ii).

The final claim in part (vi) follows from the fact that when $\mathcal{RB} \subseteq V$ we must have $h(u_\alpha) = h(v_\alpha)$ and $t(u_\alpha) = t(v_\alpha)$ so that the identity $u_\alpha^0 = v_\alpha^0$ is a simple consequence of the identity $(xu_\alpha y)^0 = (xv_\alpha y)^0$. The claim then follows from $A = D$.

(vii) This follows immediately from part (i) and Theorem 2.3.8(ii).

(viii) The first claim follows immediately from [PR1, Lemma I.3.4].

Now let $\mathcal{U} = \mathcal{LRO}$ and $\mathcal{V} = \mathcal{RRO}$. Let U (respectively, V) be the semigroup of all transformations on a set of two elements written on the left (respectively, right). One verifies easily that $U \in \mathcal{U} \cap \mathcal{F}$ and $V \in \mathcal{V} \cap \mathcal{F}$. Taking the nonuniversal Rees congruence on U and V, we see that both U and V admit a 2-element group with zero H as a homomorphic image. Therefore $H \in \langle \mathcal{U} \cap \mathcal{F} \rangle \cap \langle \mathcal{V} \cap \mathcal{F} \rangle$. But by [PR1, Theorem V.6.3],

$$\langle (\mathcal{U} \cap \mathcal{V}) \cap \mathcal{F} \rangle = \langle \mathcal{SG} \cap \mathcal{F} \rangle = \langle \mathcal{S} \rangle = \mathcal{S}$$

which shows that the above mapping is not an $\cap$-homomorphism when restricted to $\mathcal{L}(\mathcal{RO})$.

(ix) The first claim follows from [PR1, Lemma I.1.2]. For the remaining assertion, with the help of part (vi), we have

$$\mathcal{LZ}^T \vee \mathcal{RZ}^T = (\mathcal{G} \circ \mathcal{LZ}) \vee (\mathcal{G} \circ \mathcal{RZ}) = \mathcal{LG} \vee \mathcal{RG} = \mathcal{ReG},$$
$$(\mathcal{LZ} \vee \mathcal{RZ})^T = \mathcal{RB}^T = \mathcal{G} \circ \mathcal{RB} = \mathcal{CS}.$$

(x) By [PR1, Proposition VII.2.10(i)], we know that $K \cap T = \varepsilon$ in $\mathcal{C}(CR_X)$ and therefore also in Δ. Via the usual antiisomorphism from Δ to $\mathcal{L}(\mathcal{CR})$ this translates to the equality $K \cap T = \varepsilon$ in $\mathcal{L}(\mathcal{CR})$.

(xi) By Proposition [PR1, VII.2.10(ii)], we have

$$\rho_V = (\rho_V)_K \vee (\rho_V)_T = (\rho_V)^K \cap (\rho_V)^T$$

which, via the standard antiisomorphism, translates into $\mathcal{V} = \mathcal{V}_K \vee \mathcal{V}_T = \mathcal{V}^K \cap \mathcal{V}^T$ in terms of varieties.

(xii) For $\mathcal{V} \in \mathcal{L}(\mathcal{G})$, we have $(\mathcal{V}^T)_K = \mathcal{G}_K = \mathcal{G} = \mathcal{V}^T$. For $\mathcal{V} \in [\mathcal{RB}, \mathcal{CS}]$, we have $\mathcal{V}^T = \mathcal{CS}$ so that, by Theorem 2.1.3(v), $(\mathcal{V}^T)_K = \mathcal{CS}_K = \mathcal{CS} = \mathcal{V}^T$.

For $\mathcal{V} \in [\mathcal{ReB}, \mathcal{CR}]$, a deep result due to Kad'ourek states that $(\mathcal{V}^T)_K = \mathcal{V}^T$. The proof is beyond the scope of this book (see Kad'ourek [1]). The claim regarding τ_{CR_X} then follows from the fact that $\mathcal{CR}^T = \mathcal{CR}$ and Lemma 1.3.6(ii).

Now assume that $\mathcal{V} \notin \mathcal{L}(\mathcal{G}) \cup [\mathcal{RB}, \mathcal{CS}] \cup [\mathcal{ReB}, \mathcal{CR}]$. First suppose that $\mathcal{S} \subseteq \mathcal{V}$ but $\mathcal{ReB} \not\subseteq \mathcal{V}$. By [PR1, Proposition IX.5.2(iv)], we then have that either $\mathcal{V} \subseteq \mathcal{LLRO}$ or

$\mathcal{V} \subseteq L\mathcal{RRO}$. By duality, it suffices to consider the case $\mathcal{V} \subseteq L\mathcal{RRO}$. By the dual of [PR1, Lemma V.3.1], we have $\mathcal{RRO} = [xa = a^0xa]$ so that, by [PR1, Proposition II.7.3(iii)],

$$L\mathcal{RRO} = [y^0xy^0ay^0 = (y^0ay^0)^0y^0xy^0ay^0] = [u = v]$$

where $u = y^0xy^0ay^0$ and $v = (y^0ay^0)^0y^0xy^0ay^0$. From part (vi), first basis, we then obtain

$$(L\mathcal{RRO})^T = [u^0 = v^0, (puq)^0 = (pvq)^0] \subseteq [u^0 = v^0].$$

Any substitution of the form $y \to 1$ and $x, a \to$ distinct elements in L_2 shows that L_2^1 does not satisfy the identity $u^0 = v^0$ and therefore $L_2^1 \notin [u^0 = v^0]$. Hence, by [PR1, Theorem V.1.9(xii)], $\mathcal{ReB} \not\subseteq (L\mathcal{RRO})^T$ and, by [PR1, Proposition IX.5.2(iv)], we must have $(L\mathcal{RRO})^T \subseteq L\mathcal{O}$. Consequently, by Theorem 2.1.3(v), we have $((L\mathcal{RRO})^T)_K \subseteq (L\mathcal{O})_K = \mathcal{CS}$. Therefore, $(\mathcal{V}^T)_K \subseteq \mathcal{CS}$. But, by hypothesis, $\mathcal{S} \subseteq \mathcal{V} \subseteq \mathcal{V}^T$. Hence $(\mathcal{V}^T)_K \neq \mathcal{V}^T$.

Finally, let $\mathcal{V} \in \mathcal{L}(\mathcal{CS}) \setminus \mathcal{L}(\mathcal{G})$ and $\mathcal{RB} \not\subseteq \mathcal{V}$. Then $\mathcal{V} \in (\mathcal{L}(\mathcal{LG}) \cup \mathcal{L}(\mathcal{RG})) \setminus \mathcal{L}(\mathcal{G})$ so that $\mathcal{V}^T \in \{\mathcal{LG}, \mathcal{RG}\}$ and $(\mathcal{V}^T)_K = \mathcal{G} \neq \mathcal{V}^T$. All possible values of $\mathcal{V}$ have been considered with the outcomes as claimed.

(xiii) It will suffice to show that $\bigcup_{n \geq 0} \mathcal{J}^{(KT)^n}$ contains every completely regular semigroup with a finite number of nontrivial completely simple components (that is, $\mathcal{D}$-classes). We argue by induction on the number of nontrivial $\mathcal{D}$-classes. Let $S \in \mathcal{CR}$ have n nontrivial $\mathcal{D}$-classes. If $n = 0$, then S is a semilattice and $S \in \mathcal{J}^K$. So suppose that the claim is true for all $S \in \mathcal{CR}$ with fewer than n nontrivial $\mathcal{D}$-classes, $n \geq 1$. Then S must have a minimal nontrivial $\mathcal{D}$-class in the partial ordering of $\mathcal{D}$-classes. Let M be one such. We define two relations, λ on S and ρ on S/λ. For $a, b \in S$,

$$a \lambda b \Longleftrightarrow \text{either (i) } a = b \text{ or (ii) } a, b \in M \text{ and } a \mathcal{H} b$$
$$a\lambda \rho b\lambda \Longleftrightarrow \text{either (i) } a\lambda = b\lambda \text{ or (ii) } a, b \in M.$$

Then λ is a congruence on S contained in $\mathcal{H}$ while ρ is a congruence on S/ρ and is contained in $\tau_{S/\lambda}$. Now $(S/\lambda)/\rho$ has one fewer nontrivial $\mathcal{D}$-classes than S. Hence $(S/\lambda)/\rho \in \mathcal{S}^{(KT)^m}$, for some integer $m \geq 1$. It follows that $S \in \mathcal{S}^{(KT)^m KT} = \mathcal{S}^{(KT)^{m+1}}$. Consequently, $\bigcup_{n \geq 1} \mathcal{S}^{(KT)^n}$ contains all completely regular semigroups with a finite number of nontrivial $\mathcal{D}$-classes. Therefore $\bigvee_{n \geq 1} \mathcal{S}^{(KT)^n} = \mathcal{CR}$.

(xiv) It follows immediately from part (xii) that $\mathcal{CR}_K = \mathcal{CR}$. Since it is trivial that $\mathcal{CR}^K = \mathcal{CR}$, the kernel class of $\mathcal{CR}$ is $\{\mathcal{CR}\}$. This implies the final claim. □

In the next result we single out some simple consequences of Theorem 2.4.3(ii).

Corollary 2.4.4 (i) $\mathcal{F}$ *is closed under subdirect products.*
(ii) *Let* $S = \prod_{\alpha \in A} S_\alpha$. *Then* $S/\mu_S = \prod_{\alpha \in A} S_\alpha/\mu_{S_\alpha}$.

Proof These claims are clear.   □

In the context of $\mathcal{L}(\mathcal{BG})$ we can be much more explicit.

Theorem 2.4.5 *The following claims hold.*

(i) *For any* $\mathcal{V} \in \mathcal{L}(\mathcal{BG})$, *we have* $\mathcal{V}_T = \mathcal{V} \cap \mathcal{B}$ *and* $\mathcal{V}^T$ *is the largest cryptogroup variety* $\mathcal{U}$ *such that* $\mathcal{U} \cap \mathcal{B} = \mathcal{V} \cap \mathcal{B}$.

(ii) *The intervals* $\{[\mathcal{V}, \mathcal{G} \circ \mathcal{V}] \mid \mathcal{V} \in \mathcal{L}(\mathcal{B})\}$ *constitute the complete set of* T-*classes of varieties in* $\mathcal{L}(\mathcal{BG})$. *In particular, the following intervals are* T-*classes:*

$$[\mathcal{T}, \mathcal{G}], \quad [\mathcal{LZ}, \mathcal{LG}], \quad [\mathcal{RZ}, \mathcal{RG}], \quad [\mathcal{RB}, \mathcal{CS}], \quad [\mathcal{S}, \mathcal{SG}],$$

$$[\mathcal{RNB}, \mathcal{RNO}], \quad [\mathcal{RRB}, \mathcal{RRBG}], \quad [\mathcal{B}, \mathcal{BG}].$$

(iii) *The mapping* $\mathcal{V} \mapsto \mathcal{V}_T$ $(\mathcal{V} \in \mathcal{L}(\mathcal{BG}))$ *is a complete retraction of* $\mathcal{L}(\mathcal{BG})$ *onto* $\mathcal{L}(\mathcal{B})$.

(iv) *The mapping* $\mathcal{V} \mapsto \mathcal{V}^T$ $(\mathcal{V} \in \mathcal{L}(\mathcal{LRBG}))$ *is a complete endomorphism of* $\mathcal{L}(\mathcal{LRBG})$.

Proof (i) The first claim was established in Theorem 2.4.3(v) and is only repeated here for the sake of completeness. For the second assertion, let $S \in \mathcal{V}^T$. By the radical property of $\mathbb{M}$, we have $S/\mu \in \mathcal{F}$ and by Theorem 2.4.3(vi), we have $S/\mu \in \mathcal{V}$. Hence, $S/\mu \in \mathcal{V} \cap \mathcal{F}$. But $\mathcal{V} \cap \mathcal{F} \subseteq \mathcal{BG} \cap \mathcal{F} = \mathcal{B}$. Therefore $S/\mu \in \mathcal{B}$ so that $S \in \mathcal{BG}$. Therefore $\mathcal{V}^T$ is a variety of cryptogroups and $\mathcal{V}^T \cap \mathcal{F} = \mathcal{V} \cap \mathcal{F} = \mathcal{V} \cap \mathcal{B}$. Now let $\mathcal{U}$ be a variety of cryptogroups with $\mathcal{U} \cap \mathcal{B} = \mathcal{V} \cap \mathcal{B}$. Then

$$S \in \mathcal{U} \Longrightarrow S/\mu \in \mathcal{U} \cap \mathcal{F} \Longrightarrow S/\mu \in \mathcal{U} \cap \mathcal{B} = \mathcal{V} \cap \mathcal{B} \Longrightarrow S \in \mathcal{V}^T.$$

Thus $\mathcal{U} \subseteq \mathcal{V}^T$ and the claim holds.

(ii) For any $\mathcal{V} \in \mathcal{L}(\mathcal{B})$, by Theorem 2.4.3(v), we have $\mathcal{V}_T = \langle \mathcal{V} \cap \mathcal{F} \rangle = \langle \mathcal{V} \cap \mathcal{B} \rangle = \mathcal{V}$. Hence it follows from Theorem 2.4.3(ii)(iii) that, for any $\mathcal{V} \in \mathcal{L}(\mathcal{B})$, the T-class of $\mathcal{V}$ is $[\mathcal{V}, \mathcal{G} \circ \mathcal{V}]$, as claimed. The listed intervals are then simply instances of this fact and the verification is left as an exercise.

(iii) The first step will be to show that the mapping in the statement respects arbitrary joins.

Let $\mathcal{V}_\alpha \in \mathcal{L}(\mathcal{BG})$ where $\alpha \in A$, let $S_\alpha \in \mathcal{V}_\alpha$, S be a completely regular subsemigroup of $\prod_{\alpha \in A} S_\alpha$ and θ be a homomorphism of S onto $B \in \mathcal{B}$. Since B is a band, it follows that $\mathcal{H} \subseteq \theta \circ \theta^{-1}$ so that we have an induced epimorphism $S/\mathcal{H} \to B$. If $a = (a_\alpha)_{\alpha \in A}$ and $b = (b_\alpha)_{\alpha \in A}$ are $\mathcal{H}$-related elements in S, then necessarily $a_\alpha \mathcal{H} b_\alpha$ for all $\alpha \in A$. Hence $S/\mathcal{H}$ is isomorphic to a subsemigroup of $\prod_{\alpha \in A}(S_\alpha/\mathcal{H})$ where $S_\alpha/\mathcal{H} \in \mathcal{V}_\alpha \cap \mathcal{B}$. Accordingly, and since $\theta_{\mathcal{F}}$ is a complete homomorphism by Theorem 2.4.3(vii), we have

$$B \in \left(\bigvee_{\alpha \in A} \mathcal{V}_\alpha \right)_T = \left(\bigvee_{\alpha \in A} \mathcal{V}_\alpha \right) \cap \mathcal{B} \qquad \text{by part (i)}$$

$$= \left(\bigvee_{\alpha \in A} \mathcal{V}_\alpha \right) \cap \mathcal{B} \cap \mathcal{F} = \bigvee_{\alpha \in A} (\mathcal{V}_\alpha \cap \mathcal{F}) \cap \mathcal{B}$$

$$= \bigvee_{\alpha \in A} (\mathcal{V}_\alpha \cap \mathcal{B}\mathcal{G} \cap \mathcal{F}) \cap \mathcal{B} = \bigvee_{\alpha \in A} (\mathcal{V}_\alpha \cap \mathcal{B}) \cap \mathcal{B}$$

$$= \bigvee_{\alpha \in A} (\mathcal{V}_\alpha \cap \mathcal{B}) = \bigvee_{\alpha \in A} (\mathcal{V}_\alpha \cap \mathcal{B})$$

$$= \bigvee_{\alpha \in A} (\mathcal{V}_\alpha)_T \qquad \text{by part (i).}$$

Thus $\left(\bigvee_{\alpha \in A} \mathcal{V}_\alpha \right)_T \subseteq \bigvee_{\alpha \in A} (\mathcal{V}_\alpha)_T$ and, since the reverse containment holds trivially, equality prevails. From Theorem 2.4.3(viii), we know that the mapping $\mathcal{V} \mapsto \mathcal{V}_T$ is a complete $\vee$-endomorphism. Hence we have that the restriction of the mapping to $\mathcal{L}(\mathcal{B}\mathcal{G})$ is a complete homomorphism. Clearly $\mathcal{V}_T = \mathcal{V}$ for all $\mathcal{V} \in \mathcal{L}(\mathcal{B})$ and so the mapping is a complete retraction.

(iv) Let $\mathcal{V} \in \mathcal{L}(\mathcal{L}\mathcal{R}\mathcal{B}\mathcal{G})$. From part (i), we know that $\mathcal{V}_T = \mathcal{V} \cap \mathcal{B}$. But $\mathcal{L}\mathcal{R}\mathcal{B}\mathcal{G} \cap \mathcal{B} = \mathcal{L}\mathcal{R}\mathcal{B}$ so that, by [PR1, Theorem V.1.7], there are only five possible values for $\mathcal{V}_T$, namely $\mathcal{T}, \mathcal{L}\mathcal{Z}, \mathcal{S}, \mathcal{L}\mathcal{N}\mathcal{B}$ and $\mathcal{L}\mathcal{R}\mathcal{B}$ and therefore there are only five T-classes intersecting $\mathcal{L}(\mathcal{L}\mathcal{R}\mathcal{B}\mathcal{G})$ nontrivially: $\mathcal{T}T, \mathcal{L}\mathcal{Z}T, \mathcal{S}T, \mathcal{L}\mathcal{N}\mathcal{B}T, \mathcal{L}\mathcal{R}\mathcal{B}T$. From part (ii), we have $\mathcal{T}^T = \mathcal{G}, \mathcal{L}\mathcal{Z}^T = \mathcal{L}\mathcal{G}$, $\mathcal{S}^T = \mathcal{S}\mathcal{G}, \mathcal{L}\mathcal{N}\mathcal{B}^T = \mathcal{L}\mathcal{N}\mathcal{B}\mathcal{G}$ and $\mathcal{L}\mathcal{R}\mathcal{B}^T = \mathcal{L}\mathcal{R}\mathcal{B}\mathcal{G}$. Thus the complete set of T-classes intersecting $\mathcal{L}(\mathcal{L}\mathcal{R}\mathcal{B}\mathcal{G})$ nontrivially is

$$[\mathcal{T}, \mathcal{G}], \quad [\mathcal{L}\mathcal{Z}, \mathcal{L}\mathcal{G}], \quad [\mathcal{S}, \mathcal{S}\mathcal{G}], \quad [\mathcal{L}\mathcal{N}\mathcal{G}, \mathcal{L}\mathcal{N}\mathcal{B}\mathcal{G}], \quad [\mathcal{L}\mathcal{R}\mathcal{B}, \mathcal{L}\mathcal{R}\mathcal{B}\mathcal{G}].$$

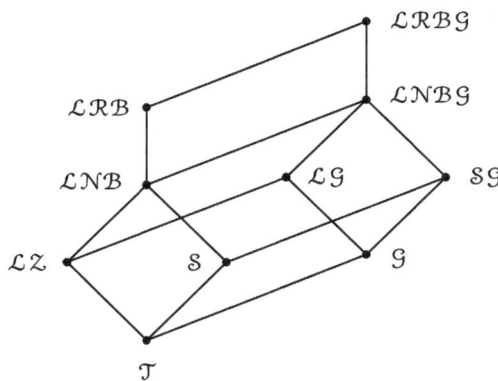

Let $\mathcal{U}, \mathcal{V} \in \mathcal{L}(\mathcal{L}\mathcal{R}\mathcal{B}\mathcal{G})$. If $\mathcal{U}T$ and $\mathcal{V}T$ are comparable T-classes, then clearly $\mathcal{U}^T \vee \mathcal{V}^T = (\mathcal{U} \vee \mathcal{V})^T$. The only incomparable T-classes are $[\mathcal{L}\mathcal{Z}, \mathcal{L}\mathcal{G}]$ and $[\mathcal{S}, \mathcal{S}\mathcal{G}]$. If $\mathcal{U} \in [\mathcal{L}\mathcal{Z}, \mathcal{L}\mathcal{G}]$ and $\mathcal{V} \in [\mathcal{S}, \mathcal{S}\mathcal{G}]$, then

$$\mathcal{U}^T \vee \mathcal{V}^T = \mathcal{LG} \vee \mathcal{SG} = \mathcal{LNBG}, \quad \mathcal{U} \vee \mathcal{V} \in [\mathcal{LNB}, \mathcal{LNBG}]$$

and thus $(\mathcal{U} \vee \mathcal{V})^T = \mathcal{LNBG} = \mathcal{U}^T \vee \mathcal{V}^T$. Consequently, the mapping $\mathcal{V} \to \mathcal{V}^T$ respects arbitrary joins in $\mathcal{L}(\mathcal{LRBG})$ while, by Theorem 2.4.3(ix), it also respects arbitrary intersections.

Therefore $\mathcal{V} \mapsto \mathcal{V}^T$ restricted to $\mathcal{L}(\mathcal{LRBG})$ is a complete homomorphism.   □

It is interesting to note that in the proof of Theorem 2.4.5(iii), we provided the argument required to establish that the mapping $\mathcal{V} \mapsto \mathcal{V} \cap \mathcal{B}$ ($\mathcal{V} \in \mathcal{L}(\mathcal{BG})$) is a complete retraction of $\mathcal{L}(\mathcal{BG})$ onto $\mathcal{L}(\mathcal{B})$. A more general result than this can be found in Chap. 3, see also [PR3, Chapter 2] for the neutrality of $\mathcal{B}$ within the lattice $\mathcal{L}(\mathcal{CR})$. For $\mathcal{L}(\mathcal{O})$ we have the following.

Proposition 2.4.6 (i) *For any* $\mathcal{V} \in \mathcal{L}(\mathcal{O})$, *we have* $\mathcal{V}^T \cap \mathcal{O} = \mathcal{V} \vee \mathcal{G}$.

(ii) *The mapping* $\mathcal{V} \mapsto \mathcal{V}^T \cap \mathcal{O}$ ($\mathcal{V} \in \mathcal{L}(\mathcal{O})$) *is a complete endomorphism of* $\mathcal{L}(\mathcal{O})$.

Proof (i) If $\mathcal{V} \in \mathcal{L}(\mathcal{ReG})$, then the claim is trivial. So assume that $\mathcal{V} \in [\mathcal{S}, \mathcal{O}]$. For any $S \in \mathcal{V}^T \cap \mathcal{O}$, we have $S/\mu \in \mathcal{V}$, while $S/\tau \in \mathcal{G} \vee \mathcal{S}$, by Theorem 2.1.3(iv)(v). But $\mu \cap \tau = \iota$ so that $S \in \mathcal{V} \vee \mathcal{S} \vee \mathcal{G}$. Thus $\mathcal{V}^T \cap \mathcal{O} \subseteq \mathcal{V} \vee \mathcal{SG} = \mathcal{V} \vee \mathcal{G}$ and the reverse containment is clear.

Let $\{\mathcal{V}_\alpha \mid \alpha \in A\} \subseteq \mathcal{L}(\mathcal{O})$. Then

$$\left(\bigvee_{\alpha \in A} \mathcal{V}_\alpha\right)^T \cap \mathcal{O} = \left(\bigvee_{\alpha \in A} \mathcal{V}_\alpha\right) \vee \mathcal{G} = \bigvee_{\alpha \in A} (\mathcal{V}_\alpha \vee \mathcal{G}) = \bigvee_{\alpha \in A} (\mathcal{V}_\alpha^T \cap \mathcal{O}).$$

Also

$$\left(\bigcap_{\alpha \in A} \mathcal{V}_\alpha\right)^T \cap \mathcal{O} = \left(\bigcap_{\alpha \in A} \mathcal{V}_\alpha^T\right) \cap \mathcal{O} \quad \text{by Theorem 2.4.3(ix)}$$

$$= \bigcap_{\alpha \in A} (\mathcal{V}_\alpha^T \cap \mathcal{O}).$$

Thus the mapping $\mathcal{V} \mapsto \mathcal{V}^T \cap \mathcal{O}$ respects arbitrary joins and intersections.   □

Given $\mathcal{V} \in \mathcal{L}(\mathcal{CR})$, the next diagram contains a few varieties which can be obtained from $\mathcal{V}$ by performing the operations of lower and upper K and T as well as some meets and joins of these varieties. Simple inspection will show that this is an inclusion diagram—not a lattice diagram.

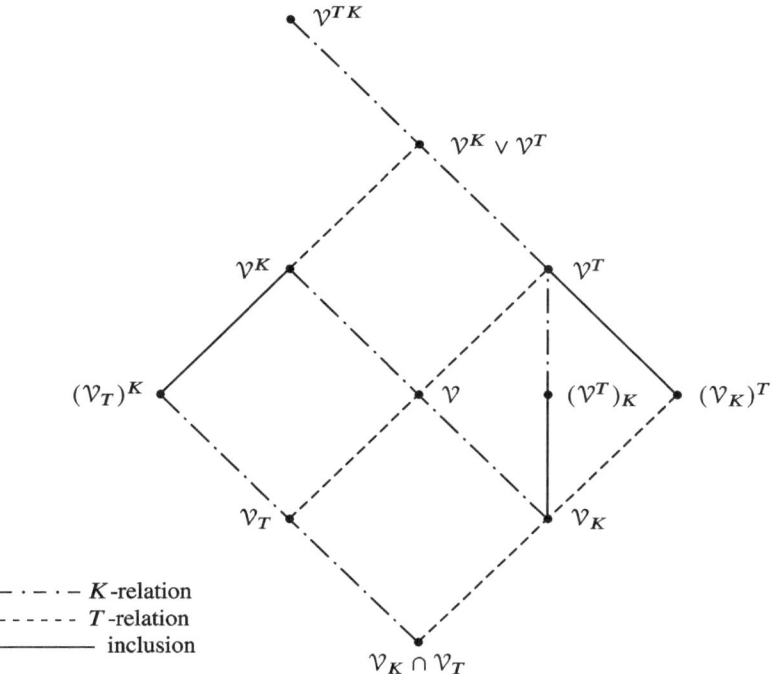

Problem 2.4.7 Given a basis for $\mathcal{V} \in \mathcal{L}(\mathcal{CR})$, construct a basis for $\mathcal{V}_T = \langle \mathcal{V} \cap \mathcal{F} \rangle$.

Reilly [1] introduced $\mathcal{V}^T$ and established that $\mathcal{V}^T$ is a variety, Theorem 2.4.3(iii) and the first basis in Theorem 2.4.3(iii). The second basis in Theorem 2.4.3(vi) is due to Kad'ourek [1]. Theorem 2.4.3(i)(iv) and the alternative characterization of the congruence T in terms of the integral members shared by two varieties are due to Petrich and Reilly [4], as well as Theorem 2.4.5. Theorem 2.4.3(vii)(viii) can be found in the same reference. Theorem 2.4.3(xiv) is due to Pastijn and Trotter [1]. A description of the free objects in $\mathcal{V}^T$ in terms of the free objects in $\mathcal{V}$, when $\mathcal{V} \in \mathcal{L}(\mathcal{B})$, can be found in Petrich and Reilly [2]. The rest of the material of this section is either folklore or is new.

2.5   Right Trace Relation

In Chap. 1 we considered the transfer of the relation T_r from the lattice of fully invariant congruences on CR_X to the lattice of varieties $\mathcal{L}(\mathcal{CR})$ via the standard antiisomorphism π. We characterize here this relation as the congruence induced by the homomorphism $\mathcal{V} \to \mathcal{V} \cap \mathcal{RF}$ where $\mathcal{RF}$ denotes the class of right fundamental completely regular semigroups. In comparison with the kernel and trace relations, the remarkable novelty here is that we are able to provide a basis of identities for $\mathcal{V}_{T_r}$, and show that the mapping $\mathcal{V} \to \mathcal{V}_{T_r}$ is a complete endomorphism of $\mathcal{L}(\mathcal{CR})$.

Recall from [PR1, Definition VI.4.8] that a completely regular semigroup S is said to be *right fundamental* if the equality relation on S is the only congruence contained in $\mathcal{R}$; that is, if $\mathcal{R}^0 = \varepsilon$. Also recall from Notation 1.3.2 that, for $\mathcal{U}, \mathcal{V} \in \mathcal{L}(\mathcal{CR})$,

$$\mathcal{U}\ T_r\ \mathcal{V} \iff \rho_{\mathcal{U}}\ T_r\ \rho_{\mathcal{V}}.$$

In addition, in [PR1, Notation V.4.1], we introduced $\mathcal{R}^*$ to denote the variety of completely regular semigroups for which $\mathcal{R}$ is a congruence.

Notation 2.5.1 We denote by $\mathcal{RF}$ the class of all right fundamental completely regular semigroups and write $\mathbb{R} = \{\mathcal{R}^0_S \mid S \in \mathcal{CR}\}$ and $\theta_{\mathcal{RF}} : \mathcal{V} \mapsto \mathcal{V} \cap \mathcal{RF}$ $(\mathcal{V} \in \mathcal{L}(\mathcal{CR}))$.

Theorem 2.5.2 *The following statements hold.*

(i) *T_r is a complete congruence on $\mathcal{L}(\mathcal{CR})$.*
(ii) *$\mathbb{R}$ is a congruence system and $\mathcal{RF}$ is its integral class.*
(iii) *$T_r = \overline{\theta}_{\mathcal{RF}}$.*
(iv) *For $\mathcal{U}, \mathcal{V} \in \mathcal{L}(\mathcal{CR})$,*

$$\mathcal{U}\ T_r\ \mathcal{V} \iff \mathcal{U} \cap \mathcal{RF} = \mathcal{V} \cap \mathcal{RF} \iff \langle \mathcal{U} \cap \mathcal{RF} \rangle = \langle \mathcal{V} \cap \mathcal{RF} \rangle$$
$$\iff \rho_{\mathcal{U}} \vee \mathcal{R} = \rho_{\mathcal{V}} \vee \mathcal{R}.$$

(v) *For $\mathcal{V} = [u_\alpha = v_\alpha]_{\alpha \in A} \in \mathcal{L}(\mathcal{CR})$, we have*

$$\mathcal{V}_{T_r} = \langle \mathcal{V} \cap \mathcal{RF} \rangle = \begin{cases} \mathcal{V} \cap \mathcal{LNB} & \text{if } \mathcal{V} \subseteq \mathcal{LRRO} \\ [s^k(u_\alpha) = s^k(v_\alpha)]_{\alpha \in A,\, k \geq 0} & \text{if } \mathcal{LRB} \subseteq \mathcal{V} \end{cases}$$

(vi) *For $\mathcal{V} = [u_\alpha = v_\alpha]_{\alpha \in A} \in \mathcal{L}(\mathcal{CR})$ and $W = \bigcup_{\alpha \in A} c(u_\alpha v_\alpha)$, we have*

$$\mathcal{V}^{T_r} = \mathcal{RG} \circ \mathcal{V} = \{S \in \mathcal{CR} \mid S/\mathcal{R}^0 \in \mathcal{V}\}$$
$$= \begin{cases} [u_\alpha x = (v_\alpha x)^0 u_\alpha x]_{\alpha \in A} & \text{if } \mathcal{S} \subseteq \mathcal{V} \\ [u_\alpha x = (v_\alpha x)^0 u_\alpha x]_{\alpha \in A} \cap \mathcal{CS} & \text{if } \mathcal{V} \subseteq \mathcal{CS} \end{cases}$$
$$= [u^0 = v^0 \mid u^0 = v^0 \text{ holds } \mathcal{V}, \text{ and } (u\zeta)^0 = (u\zeta)^0(v\zeta)^0]$$

where $x \notin W$; if $\mathcal{LRB} \subseteq \mathcal{V}$, then also $\mathcal{V}^{T_r} = [(u_\alpha x w_\alpha)^0 = (v_\alpha x w_\alpha)^0]_{\alpha \in A}$ where $x \notin W$ and, for all $\alpha \in A$, we have $w_\alpha \in \mathcal{U}$ and $c(u_\alpha v_\alpha) \subseteq c(w_\alpha)$.

(vii) *The mapping $\theta_{\mathcal{F}} : \mathcal{V} \mapsto \mathcal{V} \cap \mathcal{RF}$ $(\mathcal{V} \in \mathcal{L}(\mathcal{CR}))$ is a complete homomorphism of $\mathcal{L}(\mathcal{CR})$ onto $\mathcal{L}_{\mathbb{R}}(\mathcal{CR})$ inducing T_r.*

(viii) *The mapping $\mathcal{V} \mapsto \mathcal{V}_{T_r}$ $(\mathcal{V} \in \mathcal{L}(\mathcal{CR}))$ is a complete endomorphism of $\mathcal{L}(\mathcal{CR})$.*

(ix) *The mapping $\mathcal{V} \mapsto \mathcal{V}^{T_r}$ ($\mathcal{V} \in \mathcal{L}(\mathcal{CR})$) is a complete $\cap$-endomorphism. Its restriction to $\mathcal{L}(\mathcal{LRRO})$ is not a $\vee$-homomorphism.*

(x) $T = T_\ell \cap T_r$.

(xi) *For any $\mathcal{V} \in \mathcal{L}(\mathcal{CR})$, we have*

$$\mathcal{V}_T = \mathcal{V}_{T_\ell} \vee \mathcal{V}_{T_r}, \quad \mathcal{V}^T = \mathcal{V}^{T_\ell} \cap \mathcal{V}^{T_r},$$

$$\mathcal{V}_K \vee \mathcal{V}_{T_\ell} \vee \mathcal{V}_{T_r} = \mathcal{V} = \mathcal{V}^K \cap \mathcal{V}^{T_\ell} \cap \mathcal{V}^{T_r},$$

$$[\mathcal{V}_{T_\ell}, \mathcal{V}^{T_\ell}] \cap [\mathcal{V}_{T_r}, \mathcal{V}^{T_r}] = [\mathcal{V}_T, \mathcal{V}^T].$$

(xii) $\mathcal{V}^K = (\mathcal{V}^K)_{T_\ell} = (\mathcal{V}^K)_{T_r} = (\mathcal{V}^K)_T$. *In particular,* $\mathcal{CR}_T = \mathcal{CR}_{T_\ell} = \mathcal{CR}_{T_r} = \mathcal{CR}$.

(xiii) *For $\mathcal{V} \in \mathcal{L}(\mathcal{CR})$, we have* $(\mathcal{V}^{T_\ell})_T = (\mathcal{V}^{T_\ell})_{T_r} = \mathcal{V}^{T_\ell} \Leftrightarrow \mathcal{S} \subseteq \mathcal{V}$.

(xiv) $\mathcal{CR}$ *is finitely join irreducible.*

Notes. The T_r-classes of $\mathcal{J}$, $\mathcal{LZ}$, $\mathcal{S}$, $\mathcal{LNB}$ and $\mathcal{LRB}$ were described in Theorem 1.5.5.

Proof Recall the notational convention introduced at the beginning of this chapter. Also recall Definition 2.3.1.

(i) See Theorem 1.3.4.

(ii) Clearly $\mathbb{R}$ satisfies $\mathbb{P}$(i). That $\mathbb{R}$ satisfies $\mathbb{P}$(ii) follows immediately from [PR1, Proposition VI.4.11(i)]. Thus $\mathbb{R}$ is a congruence family. Now let $S_\alpha \in \mathcal{CR}$ where $\alpha \in A$ and S be a subdirect product of the S_α. Since each $\mathcal{R}^0_{S_\alpha}$ is left idempotent separating, so also is $\prod_{\alpha \in A} \mathcal{R}^0_{S_\alpha}$ and therefore $\prod_{\alpha \in A} \mathcal{R}^0_{S_\alpha}|_S$ must be left idempotent separating. Hence $\prod_{\alpha \in A} \mathcal{R}^0_{S_\alpha}|_S \subseteq \mathcal{R}^0_S$. The reverse containment follows from Lemma 2.3.2(ii) and we have equality. Thus $\mathbb{P}$(iii) is satisfied.

Finally, let $S, T \in \mathcal{CR}$ and S be a subsemigroup of T. By [PR1, Lemma II.3.5], we know that $\mathcal{R}_T|_S = \mathcal{R}_S$. It then follows easily that $\mathcal{R}^0_T|_S \subseteq \mathcal{R}^0_S$ and that $\mathbb{R}$ satisfies $\mathbb{P}$(iv). Thus $\mathbb{R}$ is a congruence system. It is evident from the definition of right fundamental completely regular semigroups that $\mathcal{RF}$ is the integral class of $\mathbb{R}$.

(iii) and (iv). The arguments for these parts follow the same lines as the proofs of parts (iii) and (iv) in Theorem 2.4.3 replacing T by T_r, $\mathcal{F}$ by $\mathcal{RF}$, $\mathbb{M}$ by $\mathbb{R}$ and [PR1, Corollary VII.2.9] by [PR1, Corollary VII.4.9].

(v) $A = B$. This follows immediately from Theorem 2.3.8(iii) and the fact that $T_r = \overline{\theta}_{\mathcal{RF}}$.

$B = C$. Let $\mathcal{V} \subseteq \mathcal{LRRO}$ and $S \in \mathcal{V} \cap \mathcal{RF}$. Then, by [PR1, Theorem IX.5.1(iv)], $S \in \mathcal{RG} \circ \mathcal{LNB}$ from which it follows that $\mathcal{R}$ is a congruence on S and that $S/\mathcal{R} \in \mathcal{LNB}$. But, since $S \in \mathcal{RF}$ we also have $\mathcal{R} = \mathcal{R}^0 = \varepsilon$. Therefore $S \in \mathcal{LNB}$ and $\mathcal{V} \cap \mathcal{RF} \subseteq \mathcal{LNB}$. Hence $\langle \mathcal{V} \cap \mathcal{RF} \rangle \subseteq \mathcal{V} \cap \mathcal{LNB}$. Clearly $\mathcal{V} \cap \mathcal{LNB} \subseteq \langle \mathcal{V} \cap \mathcal{RF} \rangle$. Consequently, we have $\langle \mathcal{V} \cap \mathcal{RF} \rangle = \mathcal{V} \cap \mathcal{LNB}$ and $B = C$ in this case.

We now consider the case where $\mathcal{LRB} \subseteq \mathcal{V}$ or, equivalently, $\zeta_\mathcal{V} \subseteq \zeta_{\mathcal{LRB}}$. Let $\mathcal{U} = [s^k(u_\alpha) = s^k(v_\alpha)]_{\alpha \in A, \, k \geq 0}$. For any $\alpha \in A$, we have $u_\alpha \, \zeta_\mathcal{V} \, v_\alpha$ so that $u_\alpha \, \zeta^s_\mathcal{V} \, v_\alpha$ since, by Lemma 1.5.2(i), $\zeta_\mathcal{V} \subseteq \zeta^s_\mathcal{V}$. By Lemma 1.5.2(iv), we have $\zeta^{ss}_\mathcal{V} = \zeta^s_\mathcal{V}$. Hence a simple induction argument on k will show that $s^k(u_\alpha) \zeta_v^s s^k(v_\alpha)$ for all $k \geq 0$. Therefore $\zeta_u \subseteq \zeta_v^s$ and we have $\zeta_v \subseteq \zeta_u \subseteq \zeta_v^s$.

Conversely, let $u, v \in U$ be such that $u \zeta_v^s v$. Then $u = s(w)$, $v = s(t)$ and $w \zeta_v t$ for suitable $w, t \in U$. Now ζ_v is generated by all pairs drawn from sets of the form

$$\{p, pp^{-1}p\}, \quad \{p, (p^{-1})^{-1}\}, \quad \{pp^{-1}, p^{-1}p\}, \quad (p \in U) \tag{2.5.1}$$

or

$$\{u_\alpha \varphi, v_\alpha \varphi\} \text{ for some } \alpha \in A \text{ and some endomorphism } \varphi \text{ of } U. \tag{2.5.2}$$

Note that, since $\zeta_v \subseteq \zeta_u$, all of the above pairs lie in ζ_u.

Recall from [PR1, Section I.10] that U was defined as a subsemigroup of the free semigroup on $P = X \cup \{(,)^{-1}\}$. By [PR1, Proposition I.10.7], there exist $a_i, b_i \in P^*$ and $p_i, q_i \in U$ for $i = 1, \ldots, n$ such that $\{p_i, q_i\}$ is a set of the form (2.5.1) or (2.5.2), $a_i p_i b_i$, $a_i q_i b_i \in U$ and

$$w = a_1 p_1 b_1, \quad a_1 q_1 b_1 = a_2 p_2 b_2, \quad \ldots, \quad a_n q_n b_n = t$$

where the factorizations take place in P^*. We will show that at each step

$$s(a_i p_i b_i) \zeta_u s(a_i q_i b_i) \text{ for } i = 1, \ldots, n.$$

Note that, since $\zeta_v \subseteq \zeta_{\mathcal{LRB}}$, we have $c(p_i) = c(q_i)$ whence $c(a_i p_i b_i) = c(a_i q_i b_i)$ and also $i(a_i p_i b_i) = i(a_i q_i b_i)$. In addition, $(p_i, q_i) \in \zeta_v \subseteq \zeta_u$.

Case: $c(a_i p_i b_i) = c(a_i)$. Then $s(a_i p_i b_i) = s(a_i) = s(a_i q_i b_i)$.

Case: $c(a_i p_i b_i)$ *is not contained in* $c(a_i p_i)$. Then $c(a_i p_i) = c(a_i q_i)$ and, since parentheses are matched within p_i and q_i, there exist $a_i^*, b_i^* \in P^*$ such that

$$s(a_i p_i b_i) = a_i^* p_i b_i^* \zeta_u a_i^* q_i b_i^* = s(a_i q_i b_i)$$

For example: consider $((xuy)^{-1}z)^{-1} = ((xuu^{-1}uy)^{-1}z)^{-1}$ where $x, y, z \in X \setminus c(u)$. Then

$$((xuy)^{-1}z)^{-1} = aub \qquad \text{where } a = ((x, \; b = y)^{-1}z)^{-1}$$
$$((xuu^{-1}uy)^{-1}z)^{-1} = auu^{-1}ub \qquad \text{where } a = ((x, \; b = y)^{-1}z)^{-1}$$

and

$$s(((xuy)^{-1}z)^{-1}) = a^*ub^* \qquad \text{where } a^* = (x, \; b^* = y)^{-1}$$
$$s(((xuu^{-1}uy)^{-1}z)^{-1}) = a^*uu^{-1}ub^* \quad \text{where } a^* = (x, \; b^* = y)^{-1}.$$

Case: $c(a_i p_i b_i) = c(a_i p_i)$ *is not contained in* $c(a_i)$. If $\{p_i, q_i\}$ is of the form (2.5.1), then it is evident that $a_i p_i b_i \zeta a_i q_i b_i$ and therefore, by Lemma 1.1.8, $s(a_i p_i b_i) \zeta s(a_i q_i b_i)$. So suppose that $\{p_i, q_i\}$ is of the form (2.5.2). We may assume that $p_i = u_\alpha \varphi$, $q_i = v_\alpha \varphi$ for some $\alpha \in A$ and some endomorphism φ of U. Note that we can extend φ to an endomorphism of P^* by the simple expedient of defining φ on (and)$^{-1}$ by $\varphi(() = ($ and $\varphi()^{-1}) =)^{-1}$. This extension of φ has the convenient property that $\overline{w\varphi} = \overline{w}\varphi$. Since $\zeta_V \subseteq \zeta_{\mathcal{LRB}}$, we have $i(u_\alpha) = i(v_\alpha) = x_1 \cdots x_n$, say. Thus, for suitable u_i, $v_i \in P^*$, we may write

$$u_\alpha = x_1 u_1 x_2 u_2 \cdots x_n u_n, \quad v_\alpha = x_1 v_1 x_2 v_2 \cdots x_n v_n,$$

where $x_i \notin c(x_1 u_1 x_2 u_2 \cdots x_{i-1} u_{i-1}) \cup c(x_1 v_1 x_2 v_2 \cdots x_{i-1} v_{i-1})$. Then

$$u_\alpha \varphi = (x_1\varphi)(u_1\varphi)(x_2\varphi) \cdots (x_n\varphi)(u_n\varphi).$$

Let $y \in X$ be the last variable to appear in $a_i(u_\alpha\varphi)b_i$ for the first time. This first appearance of y must occur in one of the $x_i\varphi$, say $x_{n-j+1}\varphi$, and we can write $x_{n-j+1}\varphi = wyz$ for suitable $w, z \in P^*$ with $y \notin c(w)$. Let $\#(x_{n-j+1})\varphi = m$ and y be the $(m-k+1)^{\text{st}}$ variable in $(x_{n-j+1})\varphi$ to appear for the first time. Then $s^k(x_{n-j+1}\varphi) = \overline{w}$. We also have

$$u_\alpha\varphi = (x_1\varphi)(u_1\varphi) \cdots (x_{n-j}\varphi)(u_{n-j}\varphi)(wyz)(u_{n-j+1}\varphi) \cdots (u_n\varphi)$$

so that

$$
\begin{aligned}
s(a_i u_\alpha b_i) &= \overline{a_i(x_1\varphi)(u_1\varphi) \cdots (x_{n-j}\varphi)(u_{n-j}\varphi)w} \\
&= \overline{a_i}\,\overline{(x_1\varphi)(u_1\varphi) \cdots (x_{n-j}\varphi)(u_{n-j}\varphi)\overline{w}} \\
&= \overline{a_i}\,\overline{(x_1 u_1 \cdots x_{n-j} u_{n-j})\varphi}\,\overline{w} \\
&= \overline{a_i}\,\overline{(x_1 u_1 \cdots x_{n-j} u_{n-j})}\varphi\overline{w} \\
&= \overline{a_i}(s^j(u_\alpha))\varphi\overline{w}.
\end{aligned}
$$

Similar analysis of $a_i v_\alpha b_i$ yields $s(a_i v_\alpha b_i) = \overline{a_i}(s^j(v_\alpha))\varphi\overline{w}$. Therefore, $s(a_i u_\alpha b_i)\zeta_\mathcal{U}$ $s(a_i v_\alpha b_i)$. In all cases, $s(a_i p_i b_i) \zeta_\mathcal{U} s(a_i q_i b_i)$. Hence $u = s(w) \zeta_\mathcal{U} s(t) = v$. Thus $\zeta_V^s \subseteq \zeta_\mathcal{U}$ so that $\zeta_V^s = \zeta_\mathcal{U}$ and the result follows by Theorem 1.5.6.

(vi) $A = B = C$. The proofs of these equalities follow closely the pattern of the proofs of the corresponding parts of Theorem 2.4.3(vi).

$B = D$. This simply restates [PR1, Theorem IX.4.3(iii)].

$A = E$. This is a reformulation of Lemma 1.3.7(iii).

   Now let $\mathcal{W} = \big[(u_\alpha x w_\alpha)^0 = (v_\alpha x w_\alpha)^0\big]_{\alpha \in A}$. By part (ii), we have

$$\mathcal{W}_{T_r} = \big[s^k((u_\alpha x w_\alpha)^0) = s^k((v_\alpha x w_\alpha)^0) \mid \alpha \in A, k \geq 0\big].$$

Now the identities in the above basis for $\mathcal{W}_{T_r}$ have one of the following forms:

$$(u_\alpha x w_\alpha)^0 = (v_\alpha x w_\alpha)^0, \quad u_\alpha x s^i(w_\alpha) = v_\alpha x s^i(w_\alpha),$$

$$u_\alpha x = v_\alpha x, \quad s^j(u_\alpha) = s^j(v_\alpha),$$

(although the third form may not occur). Each of these is a simple consequence of one of the identities defining $\mathcal{V}_{T_r}$ in part (v). Since the identities defining $\mathcal{V}_{T_r}$ are included in the list, it follows that $\mathcal{W}_{T_r} = \mathcal{V}_{T_r}$ and $\mathcal{W} \, T_r \, \mathcal{V}$.

Now let $\mathcal{M} \in \mathcal{V}T_r$ and $M = U/\zeta_{\mathcal{M}}$. For any $\alpha \in A$, the identity $u_\alpha = v_\alpha$ holds in $\mathcal{M}_{T_r} = \mathcal{V}_{T_r}$. This means that $u_\alpha \zeta_{\mathcal{M}} \, \mathcal{R}_M^0 \, v_\alpha \zeta_{\mathcal{M}}$. In order to simplify the notation, let $u = u_\alpha x w_\alpha$ and $v = v_\alpha x w_\alpha$. Since $\mathcal{R}_M^0$ is a congruence we have $u \zeta_{\mathcal{M}} \, \mathcal{R}^0 \, v \zeta_{\mathcal{M}}$. By the hypothesis on x and w_α, it follows from Theorem 1.1.9(ii), that $u \zeta \, \mathcal{L} \, v \zeta$ which implies that $u \zeta_{\mathcal{M}} \, \mathcal{L} \, v \zeta_{\mathcal{M}}$. Therefore $u \zeta_{\mathcal{M}} \, \mathcal{H} \, v \zeta_{\mathcal{M}}$ and $u^0 \zeta_{\mathcal{M}} = v^0 \zeta_{\mathcal{M}}$. Consequently $\mathcal{M}$ satisfies the identity $(u_\alpha x w_\alpha)^0 = (v_\alpha x w_\alpha)^0$ for all $\alpha \in A$ so that $\mathcal{M} \subseteq \mathcal{W}$. Thus $\mathcal{W} = \mathcal{V}^{T_r}$.

(vii) That $\theta_{\mathcal{F}}$ is a complete homomorphism of $\mathcal{L}(\mathcal{CR})$ onto $\mathcal{L}_{\mathbb{R}}(\mathcal{CR})$ follows from Theorem 2.3.8(ii) since, by part (ii), $\mathbb{R}$ is a congruence system with $\mathcal{F}$ as its integral class. That $\theta_{\mathcal{F}}$ induces the congruence T_r on $\mathcal{L}(\mathcal{CR})$ follows from part (iv).

(viii) This follows immediately from Theorem 1.5.8 and the standard antiisomorphism between $\Gamma(\mathrm{CR}_X)$ and $\mathcal{L}(\mathcal{CR})$.

It will simplify the proof of part (ix) if we interrupt the sequence and prove part (xiii) first.

(xiii) By duality, for the first two equalities, it suffices to consider $(\mathcal{V}^{T_r})_{T_\ell}$. Let $\mathcal{V} = [u_\alpha = v_\alpha]_{\alpha \in A}$.

First assume that $\mathcal{S} \subseteq \mathcal{V}$.

Case: $\mathcal{S} \subseteq \mathcal{V} \subseteq \mathcal{RRO}$. By Theorem 1.5.5 and [PR1, Theorem IX.5.2(iv)], $\mathcal{V}^{T_r} = \mathcal{RRO} = [xa = a^0 xa]$. Since $\mathcal{RRB} \subseteq \mathcal{RRO}$, it follows from the dual of part (v) that

$$\left(\mathcal{V}^{T_r} \right)_{T_\ell} = [e^k(xa) = e^k(a^0 xa)]_{k \geq 0} = [xa = a^0 xa] = \mathcal{V}^{T_r}$$

since all the identities with $k \geq 1$ are valid in $\mathcal{CR}$.

Case: $\mathcal{LNB} \subseteq \mathcal{V} \subseteq \mathcal{LRRO}$. By Theorem 1.5.5 and [PR1, Theorem IX.5.1(iv)],

$$\mathcal{V}^{T_r} = \mathcal{LRRO} = [yxa = (ya)^0 yxa].$$

Since $\mathcal{RRB} \subseteq \mathcal{LRRO}$, it follows from the dual of part (v) that

$$\left(\mathcal{V}^{T_r} \right)_{T_\ell} = \left[e^k(yxa) = e^k((ya)^0 yxa) \right]_{k \geq 0} = \left[yxa = (ya)^0 yxa \right] = \mathcal{V}^{T_r}$$

since all the identities with $k \geq 1$ are valid in $\mathcal{CR}$.

Case: $\mathcal{LRB} \subseteq \mathcal{V}$. By part (vi) we have $\mathcal{V}^{T_r} = \left[(u_\alpha x w_\alpha)^0 = (v_\alpha x w_\alpha)^0 \right]_{\alpha \in A}$ where, for each $\alpha \in A$, $w_\alpha \in U$, $x \notin c(w_\alpha)$, $c(u_\alpha) = c(v_\alpha) \subseteq c(w_\alpha)$. By Theorem 1.5.5(i), we have $\mathcal{RRB} \subseteq \mathcal{V}^{T_r}$ and, by the dual of part (v), we then have

$$\left(\mathcal{V}^{T_r} \right)_{T_\ell} = \left[e^k \left((u_\alpha x w_\alpha)^0 \right) = e^k \left((v_\alpha x w_\alpha)^0 \right) \right]_{\alpha \in A, \, k \geq 0}$$
$$= \left[(u_\alpha x w_\alpha)^0 = (v_\alpha x w_\alpha)^0 \right]_{\alpha \in A}$$

since $e^k\big((u_\alpha x w_\alpha)^0\big)$ is equal to $e^k\big((v_\alpha x w_\alpha)^0\big)$ for $k \geq 1$ and therefore $\big(\mathcal{V}^{T_r}\big)_{T_\ell} = \mathcal{V}^{T_r}$, as required.

Now assume that $\mathcal{V} \subseteq \mathcal{CS}$. We have $\mathcal{V}^{T_\ell} \in [\mathcal{LG}, \mathcal{CS}]$. Then $\big(\mathcal{V}^{T_\ell}\big)_{T_r} = \mathcal{LZ} \neq \mathcal{V}^{T_\ell}$.

(ix) The first assertion follows from Theorem 1.3.4 and the dual of [PR1, Lemma I.2.2].

Now consider the negative claim. We will take advantage of the description of certain T_r-intervals that can be found in Theorem 1.5.5(i) and part (xiii) just established. We have:

$$(\mathcal{LZ} \vee \mathcal{S})^{T_r} = \mathcal{LNB}^{T_r} = \mathcal{LRRO},$$
$$\mathcal{LZ}^{T_r} \vee \mathcal{S}^{T_r} = \mathcal{CS} \vee \mathcal{RRO}$$

while

$$(\mathcal{LRRO})_{T_\ell} = \mathcal{LRRO},$$
$$(\mathcal{CS} \vee \mathcal{RRO})_{T_\ell} = \mathcal{CS}_{T_\ell} \vee \mathcal{RRO}_{T_\ell} = \mathcal{RZ} \vee \mathcal{RRO} = \mathcal{RRO}.$$

Since $\mathcal{RZ} \vee \mathcal{RRO} \neq \mathcal{LRRO}$, it follows that $\mathcal{LRRO} \neq \mathcal{CS} \vee \mathcal{RRO}$ and therefore that $(\mathcal{LZ} \vee \mathcal{S})^{T_r} \neq \mathcal{LZ}^{T_r} \vee \mathcal{S}^{T_r}$ which establishes the final claim.

(x) This follows from [PR1, Corollary VII.4.4(i)] applied to $\Gamma(\mathrm{CR}_X)$ and then transferred via the usual antiisomorphism to $\mathcal{L}(\mathcal{CR})$.

(xi) From [PR1, Corollary VII.4.4(ii)] applied to $\rho_\mathcal{V} \in \Gamma(\mathrm{CR}_X)$ and the usual antiisomorphism from $\Gamma(\mathrm{CR}_X)$ to $\mathcal{L}(\mathcal{CR})$, we can deduce that $\mathcal{V}_T = \mathcal{V}_{T_\ell} \vee \mathcal{V}_{T_r}$ and $\mathcal{V}^T = \mathcal{V}^{T_\ell} \cap \mathcal{V}^{T_r}$. The remaining equalities in the first line of equalities now follow from Theorem 2.4.3(xi). The final equality then follows easily,

(xii) We first note that by Theorem 2.1.3(iii), we have $\mathcal{V}^K = \mathcal{B} \circ (\mathcal{V} \vee \mathcal{S})$ so that $\mathcal{LRB} \subseteq \mathcal{V}^K$. By Theorem 1.4.9, we have $\overline{\rho} = \zeta_{\mathcal{V}^K}$ and, by Lemma 1.5.2(i), $\overline{\rho} \subseteq \overline{\rho}^s$. Let $u, v \in \mathcal{U}$ and $u \, \overline{\rho}^s \, v$. Then there exist $w, t \in \mathcal{U}$ with $w \, \overline{\rho} \, t$, $u = s(w)$ and $v = s(t)$. But

$$w \, \overline{\rho} \, t \iff c(w) = c(t), \quad w \, \rho \, t, \quad s(w) \, \overline{\rho} \, s(t), \quad e(w) \, \overline{\rho} \, e(t).$$

Consequently, $u \, \overline{\rho} \, v$ so that $\overline{\rho}^s \subseteq \overline{\rho}$ whence $\overline{\rho}^s = \overline{\rho}$. Since $\mathcal{LRB} \subseteq \mathcal{V}^K$, it follows from Theorem 1.5.6, that $[\overline{\rho}^s] = \big(\mathcal{V}^K\big)_{T_r}$. Thus $\big(\mathcal{V}^K\big)_{T_r} = \mathcal{V}^K$ and, by duality, $\big(\mathcal{V}^K\big)_{T_\ell} = \mathcal{V}^K$. Furthermore,

$$\big(\mathcal{V}^K\big)_T = \big(\mathcal{V}^K\big)_{T_\ell} \vee \big(\mathcal{V}^K\big)_{T_r} = \mathcal{V}^K \vee \mathcal{V}^K = \mathcal{V}^K.$$

(xiv) We refer the reader to Petrich and Reilly [3] for a proof of this result.   □

In the proof of the last part of part (v), we must include the identities of the form (2.5.1). However, on applying the operator $s(*)$ to these we obtain only trivial identities of the form $s(p) = s(p)$. Thus, in the applications of this part, we need only consider the derived identitities of the form $s^k(u_\alpha) = s^k(v_\alpha)$.

Of particular significance are the sets of formulae in parts (xii) and (xiii) in the above theorem. They are all of the form $(\mathcal{V}^X)_Y = \mathcal{V}^X$ which means that the upper end in any

X-class is the lower end of its Y-class. Since we have bases for upper ends, this can be used for discovering lower ends of some Y-classes. Knowing lower ends of any Y-classes can in turn be used to determine the entire Y-class since, again, we know how to find upper ends.

In the context of $\mathcal{L}(\mathcal{R}^*)$, we can obtain more precise statements, based on results appearing in Sect. 1.5.

Theorem 2.5.3 *The following statements hold.*

(i) $\mathcal{L}(\mathcal{R}^*)$ *is a union of T_r-classes.*

(ii) *For every $\mathcal{V} \in \mathcal{L}(\mathcal{R}^*)$, we have $\mathcal{V}_{T_r} = \mathcal{V} \cap \mathcal{LRB}$ and $\mathcal{V}^{T_r}$ is the largest variety $\mathcal{U} \in \mathcal{L}(\mathcal{R}^*)$ for which $\mathcal{U} \cap \mathcal{LRB} = \mathcal{V} \cap \mathcal{LRB}$.*

(iii) *The intervals $\{[\mathcal{V}, \mathcal{RG} \circ \mathcal{V}] \mid \mathcal{V} \in \mathcal{L}(\mathcal{LRB})\}$ constitute the complete set of T_r-classes of varieties in $\mathcal{L}(\mathcal{R}^*)$. One can find this set listed explicitly in Theorem 1.5.5.*

(iv) *The mapping $\mathcal{V} \to \mathcal{V}_{T_r} = \mathcal{V} \cap \mathcal{LRB}$ ($\mathcal{V} \in \mathcal{L}(\mathcal{R}^*)$) is a complete retraction of $\mathcal{L}(\mathcal{R}^*)$ onto $\mathcal{L}(\mathcal{LRB})$.*

Proof (i) Let $\mathcal{V} \in \mathcal{L}(R^*)$. By Theorem 1.5.5, $\mathcal{V}^{T_r} \in \mathcal{L}(R^*)$ and (i) follows.

(ii) We have

$$S \in \mathcal{V}^{T_r} \cap \mathcal{LRB} \implies S \cong S/\mathcal{R}^0 \in \mathcal{V} \implies \mathcal{V}^{T_r} \cap \mathcal{LRB} \subseteq \mathcal{V} \cap \mathcal{LRB}$$

while the reverse containment is trivial so that $\mathcal{V}^{T_r} \cap \mathcal{LRB} = \mathcal{V} \cap \mathcal{LRB}$. Therefore $\mathcal{V}_{T_r} \cap \mathcal{LRB} = \mathcal{V} \cap \mathcal{LRB}$. Now let $\mathcal{U} \in \mathcal{L}(\mathcal{R}^*)$ be such that $\mathcal{U} \cap \mathcal{LRB} = \mathcal{V} \cap \mathcal{LRB}$. By Theorem 1.5.5, each T_r-class in $\mathcal{L}(\mathcal{R}^*)$ contains a unique subvariety of $\mathcal{LRB}$. Hence $\mathcal{U} T_r \mathcal{V}$ and therefore $\mathcal{U} \subseteq \mathcal{V}^{T_r}$. Consequently $\mathcal{V}^{T_r}$ is the largest variety $\mathcal{U}$ in $\mathcal{L}(\mathcal{R}^*)$ with $\mathcal{U} \cap \mathcal{LRB} = \mathcal{V} \cap \mathcal{LRB}$.

(iii) That the least member of each T_r-class containing a variety in $\mathcal{L}(\mathcal{R}^*)$ is a subvariety of $\mathcal{L}(\mathcal{LRB})$ follows from Theorem 1.5.5. By Theorem 2.5.2(vi), it then follows that the interval is of the form $[\mathcal{V}, \mathcal{RG} \circ \mathcal{V}]$ for some $\mathcal{V} \in \mathcal{L}(\mathcal{LRB})$. We can now see (once again) that the five intervals listed in Theorem 1.5.5 are T_r-classes by invoking [PR1, Theorem IX.5.1]. See diagram below in which lines of positive slope represent T_r-classes.

(iv) That $\mathcal{V}_{T_r} = \mathcal{LRB}$ for any $\mathcal{V} \in \mathcal{L}(\mathcal{R}^*)$ was proved in part (ii). Theorem 2.5.2(viii) implies that this mapping is a complete endomorphism of $\mathcal{L}(\mathcal{R}^*)$. It follows that it is a complete retraction of $\mathcal{L}(\mathcal{R}^*)$ onto $\mathcal{L}(\mathcal{LRB})$. □

Theorems 2.5.2 and 2.5.3 have a number of interesting consequences. We begin with a simple observation that we will require later in this section.

Corollary 2.5.4 *Let $\mathcal{V} \in \mathcal{L}(\mathcal{CR})$. Then $\mathcal{V} \supseteq \mathcal{S}$ if and only if $\mathcal{V}^{T_r} \supseteq \mathcal{RRB}$.*

Proof *Necessity.* By [PR1, Theorem II.5.2(ix)], $\mathcal{S} \subseteq \mathcal{V} \Rightarrow \mathcal{RRO} = \mathcal{S}^{T_r} \subseteq \mathcal{V}^{T_r}$. Since $\mathcal{RRB} \subseteq \mathcal{RRO}$, the claim follows.

Sufficiency. By contradiction, suppose that $\mathcal{V} \not\supseteq \mathcal{S}$. Then $\mathcal{V} \subseteq \mathcal{CS}$ and hence $\mathcal{V} \in [\mathcal{J}, \mathcal{RG}] \cup [\mathcal{LZ}, \mathcal{CS}]$. By Theorem 1.5.5, these two intervals are T_r-classes which implies that $\mathcal{V}^{T_r} \subseteq \mathcal{CS}$ contradicting the hypothesis that $\mathcal{V}^{T_r} \supseteq \mathcal{RRB}$. Therefore $\mathcal{V} \supseteq \mathcal{S}$ as asserted. $\square$

Note that the formula $\mathcal{V}_T = \mathcal{V}_{T_\ell} \vee \mathcal{V}_{T_r}$ may in some special cases make it possible to obtain a basis for $\mathcal{V}_T$.

The next result has a different flavor.

Proposition 2.5.5 *Let* $\mathcal{V} \in \mathcal{L}(\mathcal{CR})$. *Then* $\mathcal{V} = \mathcal{V}_{T_\ell} = \mathcal{V}_{T_r} \Leftrightarrow \mathcal{V} \in \{\mathcal{J}, \mathcal{S}, \mathcal{V}^K\}$.

Proof By Theorem 2.5.2(xii), it is clear that if $\mathcal{V} \in \{\mathcal{J}, \mathcal{S}, \mathcal{V}^K\}$, then $\mathcal{V} = \mathcal{V}_{T_\ell} = \mathcal{V}_{T_r}$. So consider the converse and let $\mathcal{V} = \mathcal{V}_{T_\ell} = \mathcal{V}_{T_r}$.

Case: $\mathcal{V} \subseteq \mathcal{CS}$. By Theorem 1.5.5, we have $\mathcal{V}_{T_r} \in \{\mathcal{J}, \mathcal{LZ}\}$ so that $\mathcal{V} = (\mathcal{V}_{T_r})_{T_\ell} = \mathcal{J}$.

Case: $\mathcal{V} \in [\mathcal{S}, L\mathcal{LRO}] \cup [\mathcal{S}, L\mathcal{RRO}]$. If $\mathcal{V} \in \mathcal{L}(L\mathcal{RRO})$, then

$$\mathcal{V} = (\mathcal{V}_{T_r})_{T_\ell} \in \{L\mathcal{NB}_{T_\ell}, \mathcal{S}_{T_\ell}\} \qquad \text{by Theorem 1.5.5}$$
$$= \mathcal{S} \qquad\qquad\qquad\qquad \text{by dual of Theorem 1.5.5}$$

and the claim follows by duality.

Case: $\mathcal{V} \supseteq \mathcal{ReB}$. Let $\rho = \zeta_\mathcal{V}$. By Theorem 1.4.9, it suffices to show that $\overline{\rho} = \rho$. Let $u, v \in U$ and $u \rho v$. Since $\mathcal{ReB} \subseteq \mathcal{V}$, we have $\rho \subseteq \zeta_{L\mathcal{RB}} \cap \zeta_{\mathcal{RRB}}$ whence $i(u) = i(v)$ and $f(u) = f(v)$. By Theorem 1.5.6 and its dual, $\rho = \rho^s = \rho^e$. Now $s(u) \, \rho^s \, s(v)$ and $e(u) \, \rho^e \, e(v)$ so that $s(u) \, \rho \, s(v)$ and $e(u) \, \rho \, e(v)$. A simple induction argument based on $\#(u) = \#(v)$ will now show that $w(u) \, \rho \, w(v)$ for any $w \in \{s, e\}^+$. By the definition of $\overline{\rho}$, this implies that $u \, \overline{\rho} \, v$. Therefore $\rho = \overline{\rho}$ and thus $\mathcal{V} = \mathcal{V}^K$.

By [PR1, Corollary IX.5.3(ii) and its dual], the above cases exhaust all of $\mathcal{L}(\mathcal{CR})$. $\square$

We conclude this section with two useful corollaries to Theorem 2.5.2 and Proposition 2.5.5.

Corollary 2.5.6 *Let* $S \in \mathcal{CR}$. *Then* $\langle S \rangle \, T_r \, \langle S/\mathcal{R}^0 \rangle$.

Proof By Theorem 2.5.2(vi), we have $S \in \langle S/\mathcal{R}^0 \rangle^{T_r}$ so that $\langle S/\mathcal{R}^0 \rangle \subseteq \langle S \rangle \subseteq \langle S/\mathcal{R}^0 \rangle^{T_r}$. The claim follows. $\square$

A remarkable feature of the next result is the fact that although the statement is exclusively concerned with the relation K the proof depends heavily on the relations T_ℓ and T_r.

Corollary 2.5.7 *Let* $\mathcal{V} = [u_\alpha = v_\alpha]_{\alpha \in A} \in \mathcal{L}(\mathcal{CR})$. *Then*

$$\mathcal{V}^K = [w_\alpha x u_\alpha x w_\alpha = w_\alpha x v_\alpha x w_\alpha]_{\alpha \in A}$$

where $x \in X$ *and for all* $\alpha \in A$, *the element* $w_\alpha \in U$ *is such that* $c(u_\alpha v_\alpha) \subseteq c(w_\alpha)$ *and* $x \notin c(w_\alpha)$.

Proof Let $\mathcal{W} = [w_\alpha x u_\alpha x w_\alpha = w_\alpha x v_\alpha x w_\alpha]_{\alpha \in A}$. Clearly any regular band satisfies the defining identities of $\mathcal{W}$ so that $\mathcal{ReB} \subseteq \mathcal{W}$. For all $\alpha \in A$ and $k \geq 1$, we have

$$s^k(w_\alpha x u_\alpha x w_\alpha) = s^{k-1}(w_\alpha) = s^k(w_\alpha x v_\alpha x w_\alpha)$$

so that the corresponding identities are all trivial. Theorem 2.5.2(v) now yields

$$\begin{aligned}
\mathcal{W}_{T_r} &= \left[s^k(w_\alpha x u_\alpha x w_\alpha) = s^k(w_\alpha x v_\alpha x w_\alpha) \right]_{\alpha \in A, \, k \geq 0} \\
&= \left[w_\alpha x u_\alpha x w_\alpha = w_\alpha x v_\alpha x w_\alpha \right]_{\alpha \in A} = \mathcal{W},
\end{aligned}$$

and dually $\mathcal{W}_{T_\ell} = \mathcal{W}$. It follows from Proposition 2.5.5 that $\mathcal{W}^K = \mathcal{W}$.

For any $\alpha \in A$, we have $w_\alpha x u_\alpha x w_\alpha \; \zeta_\mathcal{W} \; w_\alpha x v_\alpha x w_\alpha$ by the definition of $\mathcal{W}$. By Theorem 1.2.12 (with $\lambda = \zeta_\mathcal{W}$), this implies that $u_\alpha \zeta \; (\zeta_\mathcal{W}/\zeta)^K v_\alpha \zeta$. Hence $\zeta_\mathcal{V}/\zeta \subseteq (\zeta_\mathcal{W}/\zeta)^K = \zeta_{\mathcal{W}_K}/\zeta$. Therefore $\zeta_\mathcal{V} \subseteq \zeta_{\mathcal{W}_K}$ and thus $\mathcal{W}_K \subseteq \mathcal{V}$. Since clearly $\mathcal{V} \subseteq \mathcal{W}$, we conclude that $\mathcal{V} \, K \, \mathcal{W}$. Consequently $\mathcal{V}^K = \mathcal{W}^K = \mathcal{W}$. $\qquad \square$

Concerning Theorem 2.5.2, part (i) can be found in Pastijn [5] and Polák [2], parts (ii), (iii), (iv), (vii) are due to Petrich and Reilly [4], Part (v) is due to Polák [2], parts of (vi) are due to Jones [8], Reilly [1] and Kad'ourek [1] while (viii) is due to Pastijn [5] and Polák [2]. The positive part of (ix) is due to Petrich and Reilly [4] and Polák [2]. Parts (x), (xi), (xii) and (xiii) are due to Polák [2]. See also Pastijn and Trotter [1] for part of (ii). For Corollary 2.5.7, see Jones [8] and Polák [2].

2.6   Right Kernel Relation

The analysis here is parallel to those in the preceding two sections. Only here we consider the relation K_r which is now studied directly on the lattice $\mathcal{L}(\mathcal{CR})$. Recall the definition and basic properties of K_r from [PR1, Section VII.4]. The relation K_r on $\mathcal{L}(\mathcal{CR})$ is an important tool, indeed our only tool, for investigating the detailed structure of K-classes in $\mathcal{L}(\mathcal{CR})$.

Recall from [PR1, Definition VI.1.15] that $\mathcal{RD}$ denotes the class of all right E-disjunctive completely regular semigroups and that these are the semigroups in which $\tau \cap \mathcal{R}^0 = \varepsilon$, that is, the equality is the only idempotent pure congruence contained in Green's relation $\mathcal{R}$. Also recall from Notation 1.3.2 that, for $\mathcal{U}, \mathcal{V} \in \mathcal{L}(\mathcal{CR})$,

$$\mathcal{U} \; K_r \; \mathcal{V} \Longleftrightarrow \rho_{\mathcal{U}} \; K_r \; \rho_{\mathcal{V}}.$$

The class $\mathcal{RD}$ now plays the role of the classes $\mathcal{F}$ and $\mathcal{RF}$ in the preceding two sections.

Notation 2.6.1 We write $\mathbb{RD} = \{\tau_S \cap \mathcal{R}_S^0 \mid S \in \mathcal{CR}\}$.

The following theorem encompasses all the basic facts concerning the relation K_r on $\mathcal{L}(\mathcal{CR})$ and $\mathbb{RD}$.

Theorem 2.6.2 *The following statements hold.*

(i) $K_r = K \cap T_r$ *is a complete congruence on* $\mathcal{L}(\mathcal{CR})$.
(ii) $\mathbb{RD}$ *is a congruence system and* $\mathcal{RD}$ *is its integral class.*
(iii) $K_r = \bar{\theta}_{\mathcal{RD}}$.
(iv) *For* $\mathcal{U}, \mathcal{V} \in \mathcal{L}(\mathcal{CR})$,

$$\mathcal{U} \; K_r \; \mathcal{V} \Longleftrightarrow \mathcal{U} \cap \mathcal{RD} = \mathcal{V} \cap \mathcal{RD} \Longleftrightarrow \langle \mathcal{U} \cap \mathcal{RD} \rangle = \langle \mathcal{V} \cap \mathcal{RD} \rangle.$$

(v)   *For any* $\mathcal{V} \in \mathcal{L}(\mathcal{CR})$, *we have* $\mathcal{V}_{K_r} = \mathcal{V}_K \vee \mathcal{V}_{T_r} = \langle \mathcal{V} \cap \mathcal{RD} \rangle$.
(vi) *For* $\mathcal{V} = [u_\alpha = v_\alpha]_{\alpha \in A} \in \mathcal{L}(\mathcal{CR})$, *we have*

$$\mathcal{V}^{K_r} = \mathcal{V}^K \cap \mathcal{V}^{T_r} = \mathcal{RZ} \circ \mathcal{V} = \{S \in \mathcal{CR} \mid S/(\tau \cap \mathcal{R}^0) \in \mathcal{V}\}$$

$$= \begin{cases} [xu_\alpha y(xv_\alpha y)^{-1} \in E, \; u_\alpha x = (v_\alpha x)^0 u_\alpha x, \; v_\alpha x = (u_\alpha x)^0 v_\alpha x]_{\alpha \in A} & \text{if } S \subseteq \mathcal{V} \\ [u_\alpha x = v_\alpha x]_{\alpha \in A} & \text{if } \mathcal{V} \subseteq \mathcal{CS} \end{cases} \qquad \text{where } x, y \notin \bigcup_{\alpha \in A} c(u_\alpha v_\alpha)$$

$$= [u = u^0, \; v^0 = w^0 \mid w^0 v^0 \; \zeta \; v^0 \; \text{and } u = u^0, \; v^0 = w^0 \; \text{hold in } \mathcal{V}].$$

If $\mathcal{LRB} \subseteq \mathcal{V}$, *then we also have* $\mathcal{V}^{K_r} = [u_\alpha x w_\alpha = v_\alpha x w_\alpha]_{\alpha \in A}$ *where the* $w_\alpha \in U$
are such that $c(u_\alpha v_\alpha) \subseteq c(w_\alpha)$ *and* $x \notin \bigcup_{\alpha \in A} c(w_\alpha)$.
(vii) *The mapping* $\theta_{\mathcal{RD}} : \mathcal{V} \mapsto \mathcal{V} \cap \mathcal{RD}$ $(\mathcal{V} \in \mathcal{L}(\mathcal{CR}))$ *is a complete homomorphism of*
$\mathcal{L}(\mathcal{CR})$ *onto* $\mathcal{L}_{\tau \cap \mathcal{R}^0}(\mathcal{CR})$ *which induces* K_r.
(viii) *The mapping* $\mathcal{V} \to \mathcal{V}_{K_r}$ $(\mathcal{V} \in \mathcal{L}(\mathcal{CR}))$ *is a complete* $\vee$-*endomorphism of* $\mathcal{L}(\mathcal{CR})$. *Its*
restriction to $\mathcal{L}(\mathcal{CSHA})$ *is not an* $\cap$-*homomorphism.*
(ix) *The mapping* $\mathcal{V} \to \mathcal{V}^{K_r}$ $(\mathcal{V} \in \mathcal{L}(\mathcal{CR}))$ *is a complete* $\cap$-*endomorphism of* $\mathcal{L}(\mathcal{CR})$. *Its*
restriction to $\mathcal{L}(\mathcal{B} \cap \mathcal{LRRO})$ *is not a* $\vee$-*homomorphism.*
(x)   *For* $\mathcal{V} \in \mathcal{L}(\mathcal{CR})$, *we have* $\mathcal{V} = \mathcal{V}_{K_\ell} \vee \mathcal{V}_{K_r} = \mathcal{V}^{K_\ell} \cap \mathcal{V}^{K_r}$
(xi) $K_\ell \cap K_r = \varepsilon$.
(xii) *For* $\mathcal{V} \in [\mathcal{S}, \mathcal{CR}]$, *we have* $\mathcal{V}^{K_\ell} = (\mathcal{V}^{K_\ell})_{K_r} = (\mathcal{V}^{K_\ell})_{T_r}$.
(xiii) *For* $\mathcal{V} \in [\mathcal{J}, \mathcal{CS}]$, *we have* $\mathcal{V}^{K_\ell K_r} = \mathcal{RB} \vee \mathcal{V}$.
(xiv) *Let* $\mathcal{V} \in \mathcal{L}(\mathcal{CR})$, $\sigma = T_x T_y \cdots T_z$, $\tau = K_x K_y \cdots K_z$, $|\sigma| = |\tau|$ *and* $x, y, z \in \{\ell, r\}$
and no two consecutive terms in either σ *or* τ *are equal. Then* $\mathcal{V}^\tau = \mathcal{V}^\sigma \cap \mathcal{V}^K$.
(xv) *For* $\mathcal{V} \in [\mathcal{S}, \mathcal{CR}]$, *we have* $\mathcal{V}^K = \bigvee_{w \in \{K_\ell, K_r\}^*} \mathcal{V}^w$.
(xvi) *For* $\mathcal{V} \in [\mathcal{S}, \mathcal{CR}]$, *we have that* $\mathcal{V} \neq \mathcal{V}^K \Leftrightarrow$ *either* $\mathcal{V} \neq \mathcal{V}^{K_\ell}$ *or* $\mathcal{V} \neq \mathcal{V}^{K_r}$.

Proof Recall the convention introduced at the beginning of this chapter and Definition 2.3.1. It is convenient to establish part (ii) first.

(ii) Clearly $\mathbb{P}$(i) is satisfied. Let $\lambda \in \mathcal{C}(S/(\tau_S \cap \mathcal{R}^0))$ be idempotent pure and $\lambda \subseteq \mathcal{R}$. Let $\gamma \in \mathcal{C}(S)$ be such that $\gamma/(\tau_S \cap \mathcal{R}^0) = \lambda$. It is straightforward to see that γ is idempotent pure (so that $\gamma \subseteq \tau_S$) and that $\gamma \subseteq \mathcal{R}$ (so that $\gamma \subseteq \mathcal{R}^0$). Therefore $\gamma \subseteq \tau_S \cap \mathcal{R}^0$ but, by the definition of γ, we have $\tau_S \cap \mathcal{R}^0 \subseteq \gamma$ so that $\gamma = \tau_S \cap \mathcal{R}^0$. Thus $\lambda = \varepsilon$ and there are no nontrivial idempotent pure congruences on $S/(\tau_S \cap \mathcal{R}^0))$.

Now let $\rho \in \mathcal{C}(S)$ be such that S/ρ has no nontrivial idempotent pure congruences contained in $\mathcal{R}$. By [PR1, Lemma VII.1.9], since $\tau_S \cap \mathcal{R}^0 \subseteq \tau_S \cap \mathcal{D}$, we have $\ker(\tau_S \cap \mathcal{R}^0) \vee \rho = \ker \rho$. Hence $\ker((\tau_S \cap \mathcal{R}^0) \vee \rho)/\rho = E(S/\rho)$ so that $((\tau_S \cap \mathcal{R}^0) \vee \rho)/\rho \subseteq \tau_{S/\rho}$. Let $a, b \in S$. Then

$$ a \ (\tau_S \cap \mathcal{R}^0) \ b \Longrightarrow a \ \mathcal{R} \vee \rho \ b \Longrightarrow a\rho \ \mathcal{R} \ b\rho \quad \text{by [PR1, Theorem VI.5.1(ii)].} $$

Therefore $((\tau_S \cap \mathcal{R}^0) \vee \rho)/\rho \subseteq \mathcal{R}$ which implies that $((\tau_S \cap \mathcal{R}^0) \vee \rho)/\rho \subseteq \mathcal{R}^0$. Thus $((\tau_S \cap \mathcal{R}^0) \vee \rho)/\rho \subseteq \tau_{S/\rho} \cap \mathcal{R}^0$. Consequently, by the hypothesis, $((\tau_S \cap \mathcal{R}^0) \vee \rho)/\rho = \epsilon$ which implies that $\tau_S \cap \mathcal{R}^0) \subseteq \rho$. Hence $\tau_S \cap \mathcal{R}^0$ is the least congruence on S such that the quotient has no nontrivial idempotent pure congruences contained in $\mathcal{R}$. Thus $\mathbb{P}$(ii) is satisfied.

Now consider $\mathbb{P}$(iii). Let S be a subdirect product of the semigroups S_α, where $\alpha \in A$. By Lemma 2.3.2(ii), we have $\tau_S \cap \mathcal{R}^0_S \subseteq \left(\prod_{\alpha \in A} \left(\tau_{S_\alpha} \cap \mathcal{R}^0_{S_\alpha} \right) \right)\big|_S$. On the other hand, the fact that $\tau_{S_\alpha} \cap \mathcal{R}^0_{S_\alpha}$ is idempotent pure for all $\alpha \in A$ implies that $\left(\prod_{\alpha \in A} \left(\tau_{S_\alpha} \cap \mathcal{R}^0_{S_\alpha} \right) \right)\big|_S$ is idempotent pure so that $\left(\prod_{\alpha \in A} \left(\tau_{S_\alpha} \cap \mathcal{R}^0_{S_\alpha} \right) \right)\big|_S \subseteq \tau_S$. Moreover, $\left(\tau_{S_\alpha} \cap \mathcal{R}^0_{S_\alpha} \right) \subseteq \mathcal{R}^0 \subseteq \mathcal{R}$ for all $\alpha \in A$, implies that $\prod_{\alpha \in A} \left(\tau_{S_\alpha} \cap \mathcal{R}^0_{S_\alpha} \right) \subseteq \mathcal{R}$ and therefore we have $\left(\prod_{\alpha \in A} \left(\tau_{S_\alpha} \cap \mathcal{R}^0_{S_\alpha} \right) \right)\big|_S \subseteq \mathcal{R}$. Hence $\left(\prod_{\alpha \in A} \left(\tau_{S_\alpha} \cap \mathcal{R}^0_{S_\alpha} \right) \right)\big|_S \subseteq \tau_S \cap \mathcal{R}^0$ and equality prevails. Thus $\mathbb{P}$(iii) is satisfied.

$\mathbb{P}$(iv) Let $S, T \in \mathcal{CR}$ and S be a subsemigroup of T. Then

$$ \tau_T\big|_S \text{ idempotent pure} \Longrightarrow \tau_T\big|_S \subseteq \tau_S, $$
$$ \mathcal{R}^0_T\big|_S \subseteq \mathcal{R}_S \Longrightarrow \mathcal{R}^0_T\big|_S \subseteq \mathcal{R}^0_S. $$

Therefore $\left(\tau_T \cap \mathcal{R}^0_T \right)\big|_S \subseteq \tau_S \cap \mathcal{R}^0_S$. This completes the proof that $\{ \tau_S \cap \mathcal{R}^0_S \mid S \in \mathcal{CR} \}$ is a congruence system. Clearly $\mathcal{RD}$ is its integral class.

(i) There is a small point of clarification required here due to the fact that we are using the same notation $(K, T, \ldots)$ to denote both relations on $\mathcal{C}(CR_X)$ and relations on $\mathcal{L}(\mathcal{CR})$. We defined K_r on $\mathcal{C}(S)$, for any completely regular semigroup S by $K_r = K \cap T_r$ [PR1, Notation VII.4.11] and defined K_r on $\mathcal{L}(\mathcal{CS})$ as the translation of K_r on $\mathcal{C}(CR_X)$ to $\mathcal{L}(\mathcal{CR})$ under the standard anti-isomorphism. So K_r on $\mathcal{L}(\mathcal{CR})$ is not equal to $K \cap T_r$ simply by definition. However, we do have

$$\mathcal{U} \, K_r \, \mathcal{V} \Longleftrightarrow \rho_{\mathcal{U}} \, K_r \, \rho_{\mathcal{V}}$$
$$\Longleftrightarrow \rho_{\mathcal{U}} \, K \cap T_r \, \rho_{\mathcal{V}}$$
$$\Longleftrightarrow \rho_{\mathcal{U}} \, K \, \rho_{\mathcal{V}} \text{ and } \rho_{\mathcal{U}} \, T_r \, \rho_{\mathcal{V}}$$
$$\Longleftrightarrow \mathcal{U} \, K \cap T_r \, \mathcal{V}.$$

Therefore we have $K_r = K \cap T_r$ on $\mathcal{L}(\mathcal{CR})$. The equality is simply the definition of K_r. By Theorem 1.3.4, the relation K_r is a complete congruence on Δ. Therefore, since the relation K_r on $\mathcal{L}(\mathcal{CR})$ is defined via the standard antiisomorphism between Δ and $\mathcal{L}(\mathcal{CR})$, it follows that K_r is a complete congruence on $\mathcal{L}(\mathcal{CR})$.

(iii) and (iv) The proofs of these parts follow word for word the proofs of parts (iii) and (iv) in Theorem 2.4.3, replacing T, $\mathcal{F}$ and $\mathbb{M}$ by K_r, $\mathcal{RD}$ and $\mathbb{RD}$, respectively.

(v) The first equality follows from Lemma 1.3.7(iv) and the second from Lemma 1.3.8(iii) with $\mathcal{P} = \mathcal{RD}$.

(vi) $A = B$. This follows from the dual of Lemma 1.3.7(iv).

$A = D$. This follows from Theorem 2.3.8(iii) with $\mathcal{P} = \mathcal{RD}$.

$C = D = E$. These follow from [PR1, Theorem IX.4.3(iii)].

$E = F$. This follows from the equality $\mathcal{V}^{K_r} = \mathcal{V}^K \cap \mathcal{V}^{T_r}$ combined with Theorems 2.1.3(iv) and 2.5.2(vi).

Now assume that $\mathcal{LRB} \subseteq \mathcal{V}$ and that $x \in X$, $w_\alpha \in U$ are as in the statement of the theorem. As a direct consequence of Corollary 2.5.7 and Theorem 2.5.2(vi), we have

$$\mathcal{V}^{K_r} = \left[w_\alpha x u_\alpha x w_\alpha = w_\alpha x v_\alpha x w_\alpha, \ (u_\alpha x w_\alpha)^0 = (v_\alpha x w_\alpha)^0 \right]_{\alpha \in A}.$$

Let $\mathcal{W}$ denote the variety in the statement of the theorem, that is,

$$\mathcal{W} = [u_\alpha x w_\alpha = v_\alpha x w_\alpha]_{\alpha \in A}.$$

Trivially, we have $\mathcal{W} \subseteq \mathcal{V}^{K_r}$.

For the opposite inclusion, we drop all the subscripts and assume that

$$wxuxw = wxvxw, \quad (uxw)^0 = (vxw)^0 \tag{2.6.1}$$

holds, where $x \in X$, $w \in U$, $c(uv) \subseteq c(w)$ and $x \notin c(w)$. Then $wx \, \mathcal{D} \, uxw \, \mathcal{D} \, vxw$ so that, in the Rees representation of their $\mathcal{D}$-class, we may set

$$wx = (i, g, \lambda), \quad uxw = (j, h, \mu), \quad vxw = (k, t, \nu).$$

The second part of (2.6.1) yields $j = k$ and $\mu = \nu$ so that the first part of (2.6.1) gives

$$(i, g, \lambda)(j, h, \mu) = (i, g, \lambda)(j, t, \mu)$$

whence immediately $h = t$. Therefore $uxw = vxw$, as required.

(vii) The proof of this part follows the proof of Theorem 2.4.3(vii) almost word for word, with the substitution of $\tau_S \cap \mathcal{R}^0_S$ and $\mathcal{R}\mathcal{D}$ for μ_S and $\mathcal{F}$, respectively.

(viii) The first assertion follows from Theorem 1.3.4 and [PR1, Lemma I.2.2]. For the negative assertion we have, by [PR1, Theorem VIII.5.5(i)],

$$\mathcal{CSH}\mathcal{A}_2 \cap \mathcal{CSH}\mathcal{A}_3 = \mathcal{CSH}(\mathcal{A}_2 \cap \mathcal{A}_3) = \mathcal{CSH}\mathcal{J} = \mathcal{RB}$$

and it follows easily from part (i) that $\mathcal{RB}_{K_r} = \mathcal{LZ}$. On the other hand, by [PR1, Proposition VIII.3.9], $(\mathcal{CSH}\mathcal{A}_2)_{K_r} \cap (\mathcal{CSH}\mathcal{A}_3)_{K_r} = \mathcal{CSH}\mathcal{A}_2 \cap \mathcal{CSH}\mathcal{A}_3 = \mathcal{RB}$, thereby establishing the negative claim.

(ix) The first assertion follows from Theorem 1.3.4 and [PR1, Lemma I.2.2].
    For the negative assertion, we have

$$
\begin{aligned}
\mathcal{LZ}^{K_r} \vee \mathcal{S}^{K_r} &= (\mathcal{RZ} \circ \mathcal{LZ}) \vee (\mathcal{RZ} \circ \mathcal{S}) && \text{by part (vi)} \\
&= (\mathcal{RZ} \vee \mathcal{LZ}) \vee (\mathcal{B} \cap \mathcal{RR}\mathcal{O}) && \text{by [PR1, Theorems IX.4.3(iv)} \\
&&& \text{and IX.5.1(iii)]} && (2.6.2) \\
&= \mathcal{RB} \vee \mathcal{RRB} && \\
&= [x^2 = x, \ yxa = yaxa] && \text{by [PR1, Theorem V.1.7].}
\end{aligned}
$$

However,

$$
\begin{aligned}
(\mathcal{LZ} \vee \mathcal{S})^{K_r} &= \mathcal{LNB}^{K_r} \\
&= \mathcal{LNB}^K \cap \mathcal{LNB}^{T_r} && \text{by part (ii)} \\
&= \mathcal{B} \cap \mathcal{LRR}\mathcal{O} && \text{by Theorems 2.1.13(v) and 1.5.5(I)} \\
&= [x^2 = x, \ yxa = (ya)^0 yxa] && \text{by [PR1, Theorem IX.5.1(iv)]} \\
&= [x^2 = x, \ yxa = yayxa].
\end{aligned}
$$

$$(2.6.3)$$

The five element semigroup $\{1, 2, 3, 4, 5\}$ with the multiplication table below and the substitution $a \to 2$, $x \to 3$, $y \to 4$ establishes that the varieties in (2.6.2) and (2.6.3) are distinct:

	1	2	3	4	5
1	1	1	1	4	5
2	1	2	2	4	5
3	1	3	3	4	5
4	1	1	5	4	5
5	1	5	5	4	5

Thus the negative assertion follows.

(x) Let $\rho = \rho_V$. Then

$$\rho_{K_\ell} \vee \rho_{K_r} = \rho_K \vee \rho_{T_\ell} \vee \rho_{T_r} \qquad \text{by [PR1, Lemma VII.4.13]}$$
$$= \rho_K \vee \rho_T \qquad \text{by [PR1, Corollary VII.4.4(ii)]}$$
$$= \rho \qquad \text{by [PR1, Proposition VII.2.10(ii)].}$$

Via the usual antiisomorphism from $\Gamma(CR_X)$ to $\mathcal{L}(\mathcal{CR})$ we have $\mathcal{V} = \mathcal{V}^{K_\ell} \cap \mathcal{V}^{K_r}$. That $\mathcal{V} = \mathcal{V}_{K_\ell} \vee \mathcal{V}_{K_r}$ follows similarly.

(xi) This follows immediately from (x).

(xii) We have

$$(\mathcal{V}^{K_\ell})_{T_r} = (\mathcal{V}^K \cap \mathcal{V}^{T_\ell})_{T_r} \qquad \text{by part (vi)}$$
$$= (\mathcal{V}^K)_{T_r} \cap (\mathcal{V}^{T_\ell})_{T_r} \quad \text{by Theorem 2.5.2(viii)}$$
$$= \mathcal{V}^K \cap \mathcal{V}^{T_\ell} \qquad \text{by Theorem 2.5.2(xii)(xiii)}$$
$$= \mathcal{V}^{K_\ell}.$$

Since $K_r \subseteq T_r$, the claim holds.

(xiii) We have, by [PR1, Theorem IX.4.3(iv) and its dual],

$$\mathcal{V}^{K_\ell K_r} = \mathcal{RZ} \circ (\mathcal{LZ} \circ \mathcal{V}) \subseteq \mathcal{RZ} \circ (\mathcal{RB} \vee \mathcal{V})$$
$$\subseteq \mathcal{RB} \vee (\mathcal{RB} \vee \mathcal{V}) = \mathcal{RB} \vee \mathcal{V} \subseteq \mathcal{J}^{K_\ell K_r} \vee \mathcal{V} \subseteq \mathcal{V}^{K_\ell K_r}.$$

Thus equality prevails.

(xiv) We argue by induction on $|\sigma| = |\tau|$. Without loss of generality, we may assume that $x = \ell$, $y = r$, $z = r$. By the dual of part (vi), we have $\mathcal{V}^{K_r} = \mathcal{V}^{T_r} \cap \mathcal{V}^K$ so that the claim holds when $|\sigma| = |\tau| = 1$. Now assume that $|\sigma| = |\tau| > 1$ and that the claim holds for shorter words σ, τ. Then

$$\mathcal{V}^{K_\ell K_r \cdots K_r} = (\mathcal{V}^{K_\ell K_r \cdots K_\ell})^{K_r}$$
$$= (\mathcal{V}^{K_\ell K_r \cdots K_\ell})^{T_r} \cap (\mathcal{V}^{K_\ell K_r \cdots K_\ell})^K$$
$$= (\mathcal{V}^{K_\ell K_r \cdots K_\ell})^{T_r} \cap \mathcal{V}^K$$
$$= (\mathcal{V}^{T_\ell T_r \cdots T_\ell} \cap \mathcal{V}^K)^{T_r} \cap \mathcal{V}^K \qquad \text{by Induction Hypothesis}$$
$$= \mathcal{V}^{T_\ell T_r \cdots T_\ell T_r} \cap \mathcal{V}^{K T_r} \cap \mathcal{V}^K \qquad \text{by Theorem 2.6.2(ix)}$$
$$= \mathcal{V}^{T_\ell T_r \cdots T_\ell T_r} \cap \mathcal{V}^K,$$

as required.

(xv) This follows immediately from Theorem 1.4.9, Lemma 1.6.2(iv) and Theorem 1.6.3 and its dual.

(xvi) We have

$$V \neq V^K \implies \text{there exists } w \in \{K_\ell, K_r\}^+ \text{ such that } V \neq V^w$$
$$\implies \text{either } V \neq V^{K_\ell} \text{ or } V \neq V^{K_r}$$
$$\implies V \neq V^K.$$

$\square$

Next, we consider the result of applying combinations of the operators $V \mapsto V^{K_\ell}$ and $V \mapsto V^{T_r}$

Theorem 2.6.3 *The following statements hold.*

(i) *For $V \in [\mathcal{J}, \mathcal{CS}]$, we have $V^{K_\ell T_r} = \mathcal{CS}$.*
(ii) *For $S \in \mathcal{CR}$ with only a finite number of components, there exists a positive integer k with $S \in \mathcal{S}^{(K_\ell T_r)^k}$.*
(iii) *For $V \in [\mathcal{S}, \mathcal{CR})$ we have*

$$V \subset V^{K_\ell T_r} \subset V^{(K_\ell T_r)^2} \subset \cdots \subset \mathcal{CR}, \quad \bigvee_{n \geq 0} V^{(K_\ell T_r)^n} = \mathcal{CR}.$$

Proof (i) We have $\mathcal{LZ} \subseteq V^{K_\ell}$ and, by Theorem 1.5.5, $\mathcal{CS} \subseteq V^{K_\ell T_r} \subseteq \mathcal{CS}$ so that the claim holds.

(ii) If S has only one component then $S \in \mathcal{CS} = \mathcal{J}^{K_\ell T_r} \subseteq \mathcal{S}^{K_\ell T_r}$ and the claim holds. So now assume that S has $m + 1$ components and that the claim holds for all completely regular semigroups with m or fewer components. Then for each completely regular subsemigroup T of S which is a union of at most m of the components of S there exists an integer k (dependent on T) such that $T \in \mathcal{S}^{(K_\ell T_r)^k}$. However, there are only finitely many such subsemigroups and so there exists an integer k for which all such subsemigroups T lie in $\mathcal{S}^{(K_\ell T_r)^k}$.

Let M denote the least component (least ideal) of S and define ρ on S by

$$a \, \rho \, b \iff \begin{cases} a, b \in M & \text{and } a \, \mathcal{R} \, b \\ a = b & \text{otherwise.} \end{cases}$$

Then ρ is a congruence on S and $\rho \subseteq \mathcal{R}$. Let γ denote the Rees congruence on S/ρ determined by the ideal M/ρ of S/ρ. Since $M/\rho \in \mathcal{LZ}$ we have that γ is idempotent pure and $\gamma \subseteq \mathcal{L}$. Also $(S/\rho)/\gamma \cong S/M$, the Rees quotient of S determined by the ideal M.

For each component D of S/M covering the zero component, let $\overline{D}$ be the union of all components equal to D or above it in the $\mathcal{D}$-class order. Note that $\overline{D}$ is a completely regular subsemigroup of S consisting of a union of some of the components of S excluding M so that $\overline{D}$ has fewer components than S and $\overline{D} \in \mathcal{S}^{(K_\ell T_r)^k}$.

On S/M define a relation ρ_D by $a \;\rho_D\; b \Leftrightarrow a = b$ or $a, b \notin \overline{D}$. Then ρ_D is a congruence on S/M and $(S/M)/\rho_D \cong \overline{D}^0$. In addition, the intersection of all the congruences of the form ρ_D is the equality relation on S/M.

We conclude that S/M is a subdirect product of the semigroups of the form $\overline{D}^0$. Since each $\overline{D}$ and, therefore, each $\overline{D}^0$ lies in $\mathcal{S}^{(K_\ell T_r)^k}$, it follows that $S/M \in \mathcal{S}^{(K_\ell T_r)^k}$. From the fact that γ is idempotent pure and $\gamma \subseteq \mathcal{L}$, it follows that $S/\rho \in \mathcal{S}^{(K_\ell T_r)^k K_\ell}$. Therefore $S \in \mathcal{S}^{(K_\ell T_r)^k K_\ell T_r} = \mathcal{S}^{(K_\ell T_r)^{k+1}}$, as required.

(iii) In order to establish that $\bigvee_{n\geq 0} \mathcal{V}^{(K_\ell T_r)^n} = \mathcal{CR}$, it suffices to show that $\bigvee_{n\geq 0} \mathcal{V}^{(K_\ell T_r)^n}$ contains every finitely generated completely regular semigroup. Let S be any such semigroup. Then S has only a finite number of components and therefore, by part (ii) there exists an integer k with

$$ S \in \mathcal{S}^{(K_\ell T_r)^k} \subseteq \mathcal{V}^{(K_\ell T_r)^k} \subseteq \bigvee_{n\geq 0} \mathcal{V}^{(K_\ell T_r)^n}. $$

Hence $\mathcal{CR} = \bigvee_{n\geq 0} \mathcal{V}^{(T_\ell T_r)^n}$.

Now suppose that $\mathcal{V}^{T_r} = \mathcal{CR}$. By Theorem 2.5.2(vi), we then have that $CR_X/\mathcal{R}^0 \in \mathcal{V}$. But, by the dual of Corollary 1.1.11, $\mathcal{R}^0 = \varepsilon$ on CR_X which would imply that $CR_X \in \mathcal{V}$, a contradiction. Hence, for any proper subvariety $\mathcal{V}$ of $\mathcal{CR}$, $\mathcal{V}^{T_r}$ and, by duality, $\mathcal{V}^{T_\ell}$ are both proper subvarieties. Therefore if $\mathcal{V}$ is properly contained in $\mathcal{CR}$, so also are all the varieties $\mathcal{V}^{(K_\ell T_r)^k}$. If, for some integer m, $\mathcal{V}^{(K_\ell T_r)^m} = \mathcal{V}^{(K_\ell T_r)^{m+1}}$, then clearly $\mathcal{V}^{(K_\ell T_r)^m} = \mathcal{V}^{(K_\ell T_r)^k}$ for all $k \geq m$. But this would imply that $\mathcal{CR} = \bigvee_{n\geq 0} \mathcal{V}^{(K_\ell T_r)^n} = \mathcal{V}^{(K_\ell T_r)^m}$ contradicting the fact, established above, that $\mathcal{V}^{(K_\ell T_r)^m}$ is a proper subvariety of $\mathcal{CR}$. Thus the containments are all proper and the claims hold.   □

Corollary 2.6.4 *Let $\mathcal{V} \in \mathcal{L}(\mathcal{CR})$.*

(i) $\mathcal{V}^{T_\ell} = \mathcal{V}^{T_r} \Leftrightarrow \mathcal{V} \in [\mathcal{RB}, \mathcal{CS}] \cup \{\mathcal{CR}\}$.   (ii) $\mathcal{V}^{K_\ell} = \mathcal{V}^{T_r} \Leftrightarrow \mathcal{V} \in \{\mathcal{CS}, \mathcal{CR}\}$.

Proof (i) Let $\mathcal{V}^{T_\ell} = \mathcal{V}^{T_r}$. Assume $\mathcal{S} \subseteq \mathcal{V}$. By Theorem 2.6.3(iii),

$$ \mathcal{CR} = \bigvee_{n\geq 0} \mathcal{V}^{(K_\ell T_r)^n} = \bigvee_{n\geq 0} \mathcal{V}^{(T_\ell T_r)^n} = \mathcal{V}^{T_\ell} = \mathcal{V}^{T_r}. $$

But if $\mathcal{V}$ is a proper subvariety of $\mathcal{CR}$ then, again by Theorem 2.6.3(iii), so also are $\mathcal{V}^{T_\ell}$ and $\mathcal{V}^{T_r}$. Hence we must have $\mathcal{V} = \mathcal{CR}$. Now let $\mathcal{V} \subseteq \mathcal{CS}$. By Theorem 1.5.5 and its dual, the only possibility is for $\mathcal{V} \in [\mathcal{RB}, \mathcal{CS}]$ where $\mathcal{V}^{T_\ell} = \mathcal{V}^{T_r} = \mathcal{CS}$. The converse is clear.

(ii) For $\mathcal{S} \subseteq \mathcal{V}$, the argument is as in part (i) with K_ℓ replaced by T_r. Now let $\mathcal{V} \subseteq \mathcal{CS}$. By Theorem 1.5.5, $\mathcal{V}^{T_r} \in \{\mathcal{RG}, \mathcal{CS}\}$. On the other hand $\mathcal{V}^{K_\ell} = \mathcal{LZ} \circ \mathcal{V} = \mathcal{LZ} \vee \mathcal{V}$ by [PR1, Theorem IX.4.3(iv)]. So the only possible $\mathcal{V}$ with $\mathcal{V}^{K_\ell} = \mathcal{V}^{K_r}$ is $\mathcal{CS}$. Thus the claim holds.   □

In certain special contexts it is possible to give more complete or more precise results than in Theorem 2.6.2.

Theorem 2.6.5 *The following statements hold.*

(i) $(L\mathcal{O})\mathcal{R}^* = [xya = (xyxa)^0(xya)^0(yxa)^0xya]$.
(ii) *The following intervals constitute the complete set of K_r-classes containing varieties in $(L\mathcal{O})\mathcal{R}^*$:*

> *for* $\mathcal{V} \in \mathcal{L}(\mathcal{G})$:
>
> $\{\mathcal{V}, \mathcal{RZ} \vee \mathcal{V}\}$, $\{\mathcal{LZ} \vee \mathcal{V}, \mathcal{RB} \vee \mathcal{V}\}$, $[\mathcal{S} \vee \mathcal{V}, \mathcal{RRO}(H\mathcal{V})]$,
>
> $[\mathcal{LNB} \vee \mathcal{V}, L\mathcal{RRO}(H\mathcal{V})]$, $[\mathcal{LRB} \vee \mathcal{V}, \mathcal{OR}^*(H\mathcal{V})]$,
>
> *for* $\mathcal{V} \in \mathcal{L}(\mathcal{CS}) \setminus \mathcal{L}(\mathcal{ReG})$:
>
> $\{\mathcal{V}\}$, $[\mathcal{LNB} \vee \mathcal{V}, (L\mathcal{RRO})D\mathcal{V}]$, $[\mathcal{LRB} \vee \mathcal{V}, (L\mathcal{O})\mathcal{R}^*D\mathcal{V}]$.

These intervals form a partition of $(L\mathcal{O})\mathcal{R}^*$.
(iii) *The mapping $\mathcal{V} \mapsto \mathcal{V}_{K_r}$ $(\mathcal{V} \in \mathcal{L}(\mathcal{RRO}))$ is a complete retraction of $\mathcal{L}(\mathcal{RRO})$ onto $\mathcal{L}(\mathcal{SG})$.*
(iv) *The mapping $\mathcal{V} \to \mathcal{V}^{K_r}$ $(\mathcal{V} \in \mathcal{L}(\mathcal{CS}))$ is a complete endomorphism of $\mathcal{L}(\mathcal{CS})$.*

Proof (i) Let $\mathcal{W} = [xya = (xyxa)^0(xya)^0(yxa)^0xya]$. For any $a, x, y \in S \in (L\mathcal{O})\mathcal{R}^*$, we have

$$
\begin{aligned}
xya &= (xyxa)^0xya \quad \text{by the dual of [PR1, Lemma V.4.2]} \\
&= (xyxa)^0\left[(yxa)^0(xya)^0\right]xya \\
&= (xyxa)^0\left[(yxa)^0(xya)^0\right]\left[(yxa)^0(xya)^0\right]xya \quad \text{by [PR1, Lemma II.7.4]} \\
&= (xyxa)^0(xya)^0(yxa)^0(xya).
\end{aligned}
$$

Thus $(L\mathcal{O})\mathcal{R}^* \subseteq \mathcal{W}$. Conversely, let $S \in \mathcal{W}$ and consider $e, f, g \in E(S)$, $e \geq f, e \geq g$. Make the substitution $a \to e, x \to f, y \to g$ in the identity. Then

$$
\begin{aligned}
fg &= (fgf)^0(fg)^0(gf)^0fg \\
&= (fg)^0(gf)^0g \quad \text{by [PR1, Lemma II.2.2(i)]} \\
&= (fg)^0g(fg)^0 \quad \text{by [PR1, Lemma II.4.4(ii)]} \\
&= (fg)^0 \in E(S).
\end{aligned}
$$

Thus $S \in L\mathcal{O}$. In addition, it follows immediately from the identity that S satisfies $xya = (xyxa)^0xya$ so that $S \in \mathcal{R}^*$. Thus $\mathcal{W} \subseteq (L\mathcal{O})\mathcal{R}^*$ and equality prevails.

(ii) By Theorem 1.5.5 we have all the T_r-classes contained in $\mathcal{R}^*$:

$$[\mathcal{T}, \mathcal{RG}], \quad [\mathcal{LZ}, \mathcal{CS}], \quad [\mathcal{S}, \mathcal{RRO}], \quad [\mathcal{LNB}, \mathcal{LRRO}], \quad [\mathcal{LRB}, \mathcal{R}^*]. \tag{2.6.4}$$

These intersected with $\mathcal{L}(\mathcal{O})$ give the classes:

$$[\mathcal{T}, \mathcal{RG}], \quad [\mathcal{LZ}, \mathcal{ReG}], \quad [\mathcal{S}, \mathcal{RRO}], \quad [\mathcal{LNB}, \mathcal{O}(\mathcal{LRRO})], \quad [\mathcal{LRB}, \mathcal{OR}^*]. \tag{2.6.5}$$

In (2.6.4), the first and the third intervals are contained in $\mathcal{L}(\mathcal{O})$; the remaining ones intersected with $\mathcal{L}(L\mathcal{O})$ are:

$$[\mathcal{LZ}, \mathcal{CS}], \quad [\mathcal{LNB}, \mathcal{LRRO}], \quad [\mathcal{LRB}, (L\mathcal{O})\mathcal{R}^*)]. \tag{2.6.6}$$

For each variety $\mathcal{V}$ in (2.6.5), by Theorem 2.1.3(iii), we have $\mathcal{V}_K = \mathcal{V} \cap \mathcal{G}$ and $\mathcal{V}^K = \mathcal{O}H\mathcal{V}$ and for each $\mathcal{V}$ in (2.6.6) not in $\mathcal{L}(\mathcal{O})$, we get $\mathcal{V}_K = \mathcal{V} \cap \mathcal{CS}$ and $\mathcal{V}^K = (L\mathcal{O})D\mathcal{V}$. We thus have all the needed parameters and it remains to apply Theorem 2.6.2(v)(vi), namely

$$\mathcal{V}_{K_r} = \mathcal{V}_K \vee \mathcal{V}_{T_r}, \quad \mathcal{V}^{K_r} = \mathcal{V}^K \cap \mathcal{V}^{T_r}.$$

We do this in the form of the following table for the intervals listed in (2.6.5) and (2.6.6).

	$\mathcal{V}_K$	$\mathcal{V}_{T_r}$	$\mathcal{V}_{K_r}$	$\mathcal{V}^K$	$\mathcal{V}^{T_r}$	$\mathcal{V}^{K_r}$
1	$\mathcal{V} \cap \mathcal{G}$	$\mathcal{T}$	$\mathcal{V} \cap \mathcal{G}$	$\mathcal{O}H\mathcal{V}$	$\mathcal{RG}$	$\mathcal{RZ} \vee (\mathcal{V} \cap \mathcal{G})$
2	$\mathcal{V} \cap \mathcal{G}$	$\mathcal{LZ}$	$\mathcal{LZ} \vee (\mathcal{V} \cap \mathcal{G})$	$\mathcal{O}H\mathcal{V}$	$\mathcal{CS}$	$\mathcal{RB} \vee (\mathcal{V} \cap \mathcal{G})$
3	$\mathcal{V} \cap \mathcal{G}$	$\mathcal{S}$	$\mathcal{S} \vee (\mathcal{V} \cap \mathcal{G})$	$\mathcal{O}H\mathcal{V}$	$\mathcal{RRO}$	$\mathcal{RRO}(H\mathcal{V})$
4	$\mathcal{V} \cap \mathcal{G}$	$\mathcal{LNB}$	$\mathcal{LNB} \vee (\mathcal{V} \cap \mathcal{G})$	$\mathcal{O}H\mathcal{V}$	$\mathcal{O}(\mathcal{LRRO})$	$\mathcal{LRRO}(H\mathcal{V})$
5	$\mathcal{V} \cap \mathcal{G}$	$\mathcal{LRB}$	$\mathcal{LRB} \vee (\mathcal{V} \cap \mathcal{G})$	$\mathcal{O}H\mathcal{V}$	$\mathcal{OR}^*$	$\mathcal{OR}^*(H\mathcal{V})$
6	$\mathcal{V} \cap \mathcal{CS}$	$\mathcal{LZ}$	$\mathcal{V} \cap \mathcal{CS}$	$(L\mathcal{O})D\mathcal{V}$	$\mathcal{CS}$	$\mathcal{V}$
7	$\mathcal{V} \cap \mathcal{CS}$	$\mathcal{LNB}$	$\mathcal{LNB} \vee (\mathcal{V} \cap \mathcal{CS})$	$(L\mathcal{O})D\mathcal{V}$	$\mathcal{LRRO}$	$(\mathcal{LRRO})D\mathcal{V}$
8	$\mathcal{V} \cap \mathcal{CS}$	$\mathcal{LRB}$	$\mathcal{LRB} \vee (\mathcal{V} \cap \mathcal{CS})$	$(L\mathcal{O})D\mathcal{V}$	$(L\mathcal{O})\mathcal{R}^*$	$(L\mathcal{O})\mathcal{R}^*D\mathcal{V}$

The first five cases follow the intervals in (2.6.5); to get the intervals in the statement of the proposition it suffices to put $\mathcal{U} = \mathcal{V} \cap \mathcal{G}$. For the next three cases, we follow the intervals in (2.6.5) from which we subtract the lattice $\mathcal{L}(\mathcal{O})$; to get the intervals in the statement of the proposition it suffices to put $\mathcal{U} = \mathcal{V} \cap \mathcal{CS}$. The verification of the correctness of the table is left as an exercise.

(iii) From part (ii) we see that the K_r-classes within $\mathcal{L}(\mathcal{RRO})$ are of the form

$$\{\mathcal{U}, \mathcal{RZ} \vee \mathcal{U}\}, \quad \{\mathcal{S} \vee \mathcal{U}, \mathcal{RRO}H\mathcal{U}\} \quad \text{for } \mathcal{U} \in \mathcal{L}(\mathcal{G}).$$

Now let $\mathcal{U}, \mathcal{U}', \mathcal{V}, \mathcal{V}' \in \mathcal{L}(\mathcal{G})$. Then the following tables provide the pattern of intersections within $\mathcal{L}(\mathcal{RRO})$, which proves that $\mathcal{V} \to \mathcal{V}_{K_r}|_{\mathcal{L}(\mathcal{RRO})}$ is a (complete) $\cap$-homomorphism. Combined with Theorem 2.6.2(vii), this proves that the mapping is a complete homomorphism.

$\cap$	$\mathcal{U}'$	$\mathcal{S} \vee \mathcal{V}'$
$\mathcal{U}$	$\mathcal{U} \cap \mathcal{U}'$	$\mathcal{U} \cap \mathcal{V}'$
$\mathcal{S} \vee \mathcal{V}$	$\mathcal{V} \cap \mathcal{U}'$	$\mathcal{S} \vee (\mathcal{U} \cap \mathcal{V}')$

$\cap$	$\mathcal{RZ} \vee \mathcal{U}'$	$\mathcal{RROHV}'$
$\mathcal{RZ} \vee \mathcal{U}$	$\mathcal{RZ} \vee (\mathcal{U} \cap \mathcal{U}')$	$\mathcal{RZ} \vee (\mathcal{U} \cap \mathcal{V}')$
$\mathcal{RROHV}$	$\mathcal{RZ} \vee (\mathcal{V} \cap \mathcal{U}')$	$\mathcal{RROH}(\mathcal{V} \cap \mathcal{V}')$

(iv) For $\mathcal{V} \in \mathcal{L}(\mathcal{CS})$, by Theorem 2.6.2(vi) and [PR1, Theorem IX.4.3(iv)], we have $\mathcal{V}^{K_r} = \mathcal{RZ} \circ \mathcal{V} = \mathcal{RZ} \vee \mathcal{V}$ and clearly the mapping $\mathcal{V} \to \mathcal{RZ} \vee \mathcal{V}$ respects arbitrary joins. Combined with Theorem 2.6.2(ix), this establishes the claim. $\square$

The five element semigroup used in Theorem 2.6.2(ix) is taken from Lee, Rhodes and Steinberg [1]. Otherwise the results in this section are either new or based on Reilly [2].

2.7   Review

We now give a common frame for the principal results in this chapter in an abstract setting. The pattern that emerges reflects some underlying connections and has merits in its own right.

Fix a subclass $\mathcal{A}$ of $\mathcal{CR}$ and define a relation, say A, on $\mathcal{L}(\mathcal{CR})$ by the rule

$$\mathcal{U} \ A \ \mathcal{V} \ \text{ if } \ \mathcal{U} \cap \mathcal{A} = \mathcal{V} \cap \mathcal{A}. \tag{2.7.1}$$

The nature of $\mathcal{A}$ will depend on what kind of relation we want A to be. Hence we actually start with A and search for $\mathcal{A}$ (which, a priori, may not exist). For example, does $\mathcal{A}$ exist if A is defined by $\mathcal{U} \ A \ \mathcal{V} \Leftrightarrow \mathcal{U} \vee \mathcal{B} = \mathcal{V} \vee \mathcal{B}$? Also, it does not mean that A uniquely determines $\mathcal{A}$. In fact, for $A = K$ we have a whole interval of varieties $\mathcal{A}$ fitting the requirement that A and $\mathcal{A}$ satisfy condition (2.7.1). In other cases under study, we found only one fitting $\mathcal{A}$ which does not mean that there are no others.

We write $A \sim \mathcal{A}$ to indicate that the relation in (2.7.1) holds. In all cases under review for A and $\mathcal{A}$, due to previously established notation, we do not have matching letters. Actually,

$$K \sim \mathcal{V} \ \text{ for } \ \mathcal{PD} \subseteq \mathcal{V} \subseteq \mathcal{PE}, \ T \sim \mathcal{F}, \ T_r \sim \mathcal{RF}, \ K_r \sim \mathcal{RD}. \tag{2.7.2}$$

The first use we obtain for this representation of A is that for any $\mathcal{V} \in \mathcal{L}(\mathcal{CR})$, we have the handy formula

$$\mathcal{V}_A = \langle \mathcal{V} \cap \mathcal{A} \rangle, \tag{2.7.3}$$

where $A \sim \mathcal{A}$ in all cases. However, $\mathcal{V}^A$ takes on values not related directly to $\mathcal{A}$. But for the cases listed in (2.7.2), we have the Mal'cev products

$$\mathcal{V}^K = \mathcal{B} \circ (\mathcal{V} \vee \mathcal{S}), \quad \mathcal{V}^T = \mathcal{G} \circ \mathcal{V}, \quad \mathcal{V}^{T_r} = \mathcal{RG} \circ \mathcal{V}, \quad \mathcal{V}^{K_r} = \mathcal{RZ} \circ \mathcal{V} \tag{2.7.4}$$

and thus arrive at a new association $K \approx \mathcal{B}, T \approx \mathcal{G}, T_r \approx \mathcal{RG}, K_r \approx \mathcal{RZ}$. All these new entities are varieties and, except in the first case, for $A \approx \mathcal{A}'$,

$$\mathcal{V}^A = \mathcal{A}' \circ \mathcal{V}, \tag{2.7.5}$$

compare with (2.7.3) above. As a consequence, for all these cases, we have for any $\mathcal{U}, \mathcal{V} \in \mathcal{L}(\mathcal{CR})$, $\mathcal{U} A \mathcal{V} \Leftrightarrow \mathcal{A}' \circ \mathcal{U} = \mathcal{A}' \circ \mathcal{V}$ with a small adjustment for $A = K$.

Historically, all of this development started with the trace relation T on the congruence lattice of a regular semigroup (Reilly and Scheiblich [1]). Then came the kernel relation on the same objects and finally T_r (Pastijn and Petrich [1]). They were transferred to $\mathcal{L}(\mathcal{CR})$ both directly and later via the Polák Theorem for varieties above semilattices. The relation K_r was defined directly on $\mathcal{L}(\mathcal{CR})$.

Of all these relations, K is the most difficult to deal with. It is related to the classes $\mathcal{PD}$ and $\mathcal{PE}$ (recall the expression for $\mathcal{V}_K$ as in (2.7.3)). In their turn, these two classes are based on the class $\mathcal{E}$ of E-disjunctive completely regular semigroups, where the latter are defined by means of the congruence τ. Since τ is the principal congruence on a (completely regular) semigroup relative to its set of idempotents, this makes K difficult to study. In particular, it is the only relation of those mentioned above which does not saturate $\mathcal{CS}$, so it is not "local" in any sense.

Next we review an interesting analogy between the lattice $\mathcal{C}(S)$ of congruences on a completely regular semigroup and $\mathcal{L}(\mathcal{CR})$. Below $\lambda, \rho \in \mathcal{C}(S)$ and $\mathcal{U}, \mathcal{V} \in \mathcal{L}(\mathcal{CR})$. We have

$$\lambda \, K \, \rho \Longleftrightarrow \lambda \cap \mathcal{H} = \rho \cap \mathcal{H}, \qquad \mathcal{U} \, K \, \mathcal{V} \Longleftrightarrow \mathcal{U} \cap \mathcal{PE} = \mathcal{V} \cap \mathcal{PE},$$

$$\lambda \, T \, \rho \Longleftrightarrow \lambda \cap \Theta = \rho \cap \Theta, \qquad \mathcal{U} \, T \, \mathcal{V} \Longleftrightarrow \mathcal{U} \cap \mathcal{F} = \mathcal{V} \cap \mathcal{F},$$

$$\lambda \, T_r \, \rho \Longleftrightarrow \lambda \cap \Theta_r = \rho \cap \Theta_r, \qquad \mathcal{U} \, T_r \, \mathcal{V} \Longleftrightarrow \mathcal{U} \cap \mathcal{RF} = \mathcal{V} \cap \mathcal{RF},$$

$$\lambda \, K_r \, \rho \Longleftrightarrow \lambda \cap \mathcal{H} \cap \Theta_r = \rho \cap \mathcal{H} \cap \Theta_r, \qquad \mathcal{U} \, K_r \, \mathcal{V} \Longleftrightarrow \mathcal{U} \cap \mathcal{RD} = \mathcal{V} \cap \mathcal{RD}.$$

Furthermore,

$$\rho_K = (\rho \cap \mathcal{H})^*, \qquad\qquad \mathcal{V}_K = \langle \mathcal{V} \cap \mathcal{PE} \rangle,$$
$$\rho_T = (\rho \cap \Theta)^*, \qquad\qquad \mathcal{V}_T = \langle \mathcal{V} \cap \mathcal{F} \rangle,$$
$$\rho_{T_r} = (\rho \cap \Theta_r)^*, \qquad\qquad \mathcal{V}_{T_r} = \langle \mathcal{V} \cap \mathcal{RF} \rangle,$$
$$\rho_{K_r} = (\rho \cap \mathcal{H} \cap \Theta_r)^*, \qquad\qquad \mathcal{V}_{K_r} = \langle \mathcal{V} \cap \mathcal{RD} \rangle.$$

As a bonus, we have

$$\lambda \, T \, \rho \iff \lambda \vee \mathcal{H} = \rho \vee \mathcal{H},$$
$$\lambda \, T_r \, \rho \iff \lambda \vee \mathcal{R} = \rho \vee \mathcal{R},$$

$$\rho^T = (\rho \vee \mathcal{H})^0; \qquad\qquad \rho = \rho^T \iff \rho \supseteq \mathcal{H},$$
$$\rho^{T_r} = (\rho \vee \mathcal{R})^0; \qquad\qquad \rho = \rho^{T_r} \iff \rho \supseteq \mathcal{R}.$$

See [PR1, Chapter VII: Theorems 1.2, 2.2, 4.2 together with Corollaries 1.8, 2.9, 4.9].

In view of these results, we arrive at some curious correspondences:

$$\text{for congruences} \quad \mathcal{H} \quad \Theta \quad \Theta_r \quad \mathcal{H} \cap \Theta_r,$$
$$\text{for varieties} \quad \mathcal{PE} \quad \mathcal{F} \quad \mathcal{RF} \quad \mathcal{RD}.$$

Radical congruence systems and their integral classes, introduced in Sect. 2.3, find useful applications in subsequent sections of this chapter. Indeed the concepts that they embody have been useful in shedding light on structures beyond completely regular semigroups, for instance, see Pastijn and Trotter [4], Reilly and Zhang [5]. The stage is then set for an intensive study of the kernel and trace relations directly on the lattice $\mathcal{L}(\mathcal{CR})$. This is the subject of four sections of this chapter: the kernel, trace, right trace and right kernel relations. The contents of these sections are similar: a mapping which induces the respective relations, except for the kernel where there are some irregularities, the expressions for the lower and upper ends of the relevant intervals, with a basis for the upper end, and a sundry collection of results pertaining to the properties of these ends in terms of the mappings induced by them. In particular, the mappings $\mathcal{V} \to \mathcal{V}^K$ and $\mathcal{V} \to \mathcal{V}_{T_x}$ ($\mathcal{V} \in \mathcal{L}(\mathcal{CR})$, $x \in \{\ell, r\}$) are complete endomorphisms of $\mathcal{L}(\mathcal{CR})$ while a similar statement fails for all other choices of lower or upper ends. This exhibits the powerful tools provided by the constructions in Sects. 2.4, 2.5, 2.6. To this circle of results belongs also the fact that for $\mathcal{V} \in [\mathcal{LRB}, \mathcal{CR}]$, and given a basis of identities of $\mathcal{V}$, simple constructions of bases for $\mathcal{V}^K$ and $\mathcal{V}_{T_r}$ are possible.

Bands

The first significant lattice of varieties of semigroups to be completely classified was the lattice of varieties of bands. Since bands are also completely regular semigroups (in which $x = x^0$) it is appropriate for us to consider this lattice here.

However, in view of the historical importance of bands for semigroup theory, we will begin with a brief history of the early progress on this topic.

In 1954, MacLean [1] published a note on idempotent semigroups. This was followed in 1957, 1958 by four papers by Kimura, [1, 2, 3, 4]. He published a list of all varieties of bands with bases of identities in up to three variables without proof (he told the first author of this book that the proofs got lost in the mail). At a mathematical meeting in Russia, in 1969, Birjukov announced a diagram of "idempotent semigroups". In 1970, Birjukov [1] published a proof and gave a diagram of the general lattice of varieties of bands. In the same year, Gerhard [1] published from his doctoral dissertation the lattice of "idempotent semigroups" in terms of fully invariant congruences. By this time Clifford had promoted the name "band" for semigroups of idempotents which became generally adopted. In 1970/1971 Fennemore [1, 2, 3], from his doctoral dissertation, described the lattice $\mathcal{L}(\mathcal{B})$ in terms of the identities of each band variety.

The description of $\mathcal{L}(\mathcal{B})$ is now credited to all three of the authors Birjukov, Gerhard and Fennemore. A relatively short explanation of the structure of $\mathcal{L}(\mathcal{B})$ with a new proof was published by Gerhard and Petrich [8] in 1989. The development of the theory of varieties of bands provided in this chapter takes advantage of the tools introduced in Chap. 2 for the study of the lattice of varieties of completely regular semigroups and the techniques introduced by Polák [1, 2]. Indeed this chapter serves as a ramp up to the theory developed by Polák as the lattice of varieties of bands plays a very special role in that theory.

We begin more generally by studying the fully invariant congruences ζ_V on $C R_X$ corresponding to varieties V containing $\mathcal{S}$. This enables us to obtain a solution to the word

© The Author(s), under exclusive license to Springer Nature Switzerland AG 2024   101
M. Petrich and N. R. Reilly, *Completely Regular Semigroup Varieties*, Synthesis Lectures
on Mathematics & Statistics, https://doi.org/10.1007/978-3-031-42891-3_3

problem for $\zeta_\mathcal{V}$ that is somewhat unwieldy in general but reduces to a useful form in many instances.

We then proceed to an intermediate description of the lattice $[\mathcal{S}, \mathcal{B}]$ in terms of mappings from a relatively free semigroup into a particular four element lattice.

From that we extract a detailed description of $\mathcal{L}(\mathcal{B})$. This introduces us to a useful tool in the study of $\mathcal{L}(\mathcal{CR})$ known as a ladder. We also see how to describe any variety in $[\mathcal{S}, \mathcal{B})$ starting from just three varieties: $\mathcal{S}$, $\mathcal{LNB}$ and $\mathcal{RNB}$.

We obtain a basis of identities for each variety in $\mathcal{L}(\mathcal{B})$. In fact, we provide two bases of identities for each variety. These bases are built on two systems of words each consisting of three sequences of words. We also determine the K_ℓ- and K_r-classes within $\mathcal{L}(\mathcal{B})$.

Finally, we show that the mapping $\mathcal{V} \to \mathcal{V} \cap \mathcal{B}$ is a complete retraction of $\mathcal{L}(\mathcal{CR})$ onto $\mathcal{L}(\mathcal{B})$. This mapping induces a complete congruence $\mathbf{B}$, say, on $\mathcal{L}(\mathcal{CR})$. Consequently, for any $\mathcal{V} \in \mathcal{L}(\mathcal{CR})$, the $\mathbf{B}$-class of $\mathcal{V}$ is an interval $[\mathcal{V}_\mathbf{B}, \mathcal{V}^\mathbf{B}]$ where, clearly, $\mathcal{V}_\mathbf{B} = \mathcal{V} \cap \mathcal{B}$. It is then fascinating to discover that the varieties of the form $\mathcal{V}^\mathbf{B}$, for $\mathcal{V} \in [\mathcal{S}, \mathcal{B})$, can be captured by a process very similar to that used for the elements of $[\mathcal{S}, \mathcal{B})$ themselves and that bases of identities for these varieties can also be obtained using analogous sequences of words written in the type $(\cdot, {}^{-1})$ as opposed to just semigroup type.

As suggested above, the study of the lattice of varieties of bands can be approached both from the semigroup theoretical point of view ignoring the fact that they are also varieties of completely regular semigroups subject to the additional law $x^2 = x$. The former approach is also the traditional one as the lattice of varieties of bands (as semigroups) was the subject of intense investigation in the nineteen sixties, seventies and eighties where the techniques depended heavily on computation with identities. The approach used in this chapter is quite different from that. It employs ideas developed for the investigation of completely regular semigroup and their varieties. By using these ideas it is possible to highlight the central role that $\mathcal{L}(\mathcal{B})$ plays in any deep investigation of $\mathcal{L}(\mathcal{CR})$ and makes it possible to bring this fact out into the light of day when working with the Polák theorem in the next chapter. In doing so, this chapter will provide a useful preparation for the Polák theorem.

3.1   Analysis of Fully Invariant Congruences

Except for the last corollary, this section deals with the interval $[\mathcal{S}, \mathcal{CR}]$ in full generality. We consider here the effect of composition of the operators $\mathcal{V} \to \mathcal{V}_P$, $\mathcal{V} \to \mathcal{V}^P$ ($P \in \{K, T_\ell, T_r\}$) and the companion operators on the associated fully invariant congruences. We first introduce a partial order on the free semigroup $\{T_\ell, T_r\}^+$ modulo the congruence ρ generated by $\{(T_\ell, T_\ell^2), (T_r, T_r^2)\}$. This partially ordered set is the domain of the functions in the lattice of functions used to describe $\mathcal{L}(\mathcal{B})$ in later sections. The final theorem of this section gives an important characterization of any congruence of the form $\zeta_\mathcal{V}$, for $\mathcal{V} \in [\mathcal{S}, \mathcal{CR}]$, in terms of the congruences associated with varieties obtained from $\mathcal{V}$ by applying operators of the form $\mathcal{V} \to \mathcal{V}_P$, $P \in \{K, T_\ell, T_r\}$. Since these varieties are smaller than $\mathcal{V}$ we thereby obtain a characterization of $\zeta_{\mathcal{V}}$ in terms of the fully invariant congruences associated with smaller and, possibly, better understood varieties. This theorem is the foundation stone for this and the next chapter.

Definition 3.1.1 Let Θ denote the set of all (nonempty) words over the alphabet $\{T_\ell, T_r\}$ of the form $P_1 P_2 \cdots P_n$ where $P_i \in \{T_\ell, T_r\}$ and $P_i \neq P_{i+1}$ for $i = 1, 2, \ldots, n-1$ with the multiplication

$$(P_1 P_2 \cdots P_m)(Q_1 Q_2 \cdots Q_n) = \begin{cases} P_1 P_2 \cdots P_m Q_2 Q_3 \cdots Q_n & \text{if } P_m = Q_1 \\ P_1 P_2 \cdots P_m Q_1 Q_2 \cdots Q_n & \text{otherwise.} \end{cases}$$

In order to emphasize the presence of the condition $P_i \neq P_{i+1}$, we shall refer to $P_1 P_2 \ldots P_n$ as being in *canonical form*. Clearly Θ is a semigroup but is not a subsemigroup of $\{T_\ell, T_r\}^+$. Indeed Θ is just an alternative description of $\{T_\ell, T_r\}^+/\rho$ where ρ is as described above. From the latter however, Θ inherits the following parameters: for $\tau = P_1 P_2 \cdots P_n$, we have:

(i) the length of τ, $|\tau| = n$,   (iii) the tail of τ, $t(\tau) = P_n$,

(ii) the head of τ, $h(\tau) = P_1$,   (iv) the mirror image τ, $\overline{\tau} = P_n \cdots P_2 P_1$.

Note that the meaning given to the bar notation here differs from the usage in Chap. 1. However, there should be no confusion since it will be used consistently in this chapter in the sense of "reverse" or "mirror image".

We adjoin $\emptyset$ to Θ as an identity to form Θ^1 and set the length $|\emptyset| = 0$. The relation $\leq$ defined by $\sigma \leq \tau$ if $|\sigma| > |\tau|$ or $\sigma = \tau$ ($\sigma, \tau \in \Theta^1$) is easily seen to be a partial order which can be depicted as

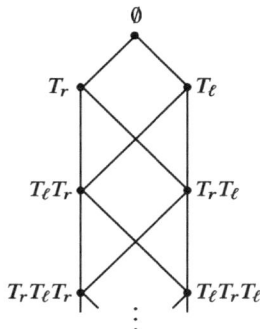

For any nonempty set A and any mapping $\psi \in A^{\Theta^1}$, there is a useful diagrammatic representation of ψ. We begin with a graph displaying the underlying partially ordered set Θ^1. We now replace the label θ, say, at each vertex by $\theta\psi$ to obtain a diagram with labels $c = \emptyset\psi, a_1 = T_r\psi, b_1 = T_\ell\psi$ and so on. We will refer to this as the *ladder (representation)* L_ψ of ψ. In the reverse direction, we might start with a nonempty set A and a labelled graph G of the above form where $a, a_i, b_i \in A$ for $0 \le i$. We will refer to such a labelled diagram/graph as a *ladder over A* . We can associate with each such labelled graph, G say, an element $\psi_G \in A^{\Theta^1}$ in the obvious way:

$$\emptyset\psi_G = a_0, \quad T_r\psi_G = a_1, \quad T_\ell\psi_G = b_1, \quad T_\ell T_r\psi_G = a_2, \ldots$$

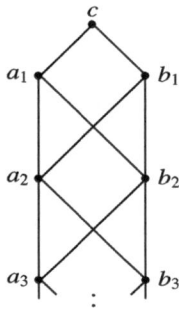

Clearly, the mappings $\psi \mapsto L_\psi$ and $G \mapsto \psi_G$ are inverse bijections between A^{Θ^1} and the set of ladders over A. For each claim about A^{Θ^1} or $\psi \in A^{\Theta^1}$, there is a corresponding claim that can be made about the set of ladders over A or about L_ψ and vice versa. We will essentially consider ψ and L_ψ (or G and L_G) as manifestations of the same object. The concept of ladders, introduced by Polák [1], turns out to be extremely helpful in visualizing, formulating and proving results about A^{Θ^1} and we will employ them extensively, using the language of functions or ladders as seems most suitable to the issue at hand.

Recall the definitions of U and Δ from Chap. 1.

Definition 3.1.2 Let $\Pi = \{s, e\}^+$ and $\varphi_{\Pi\Theta} : \Pi \to \Theta$ be the homomorphism defined by $\varphi_{\Pi\Theta} : s \mapsto T_r, \ e \mapsto T_\ell$. We extend $\varphi_{\Pi\Theta}$ to a homomorphism of Π^1 to Θ^1 by setting $1\varphi_{\Pi\Theta} = \emptyset$. For any $\pi \in \Pi^1$, we write $\pi^* = \pi\varphi_{\Pi\Theta}$.

We define two actions $V \mapsto V_\tau, V \mapsto V^\tau$ of the semigroup Θ^1 on $\mathcal{L}(\mathcal{CR})$ inductively as follows: for $V \in \mathcal{L}(\mathcal{CR})$, we set $V^\emptyset = V_\emptyset = V$ while for $\tau = \tau_1 \cdots \tau_n$ where $\tau_i \in \{T_\ell, T_r\}$ for $1 \le i \le n$ we define

$$V_\tau = (V_{\tau_1 \cdots \tau_{n-1}})_{\tau_n}, \quad V^\tau = (V^{\tau_1 \cdots \tau_{n-1}})^{\tau_n}.$$

We extend the definitions of s and e to include: $s(\emptyset) = \emptyset = e(\emptyset)$. For $u \in U$, any binary relation ρ on U, $\pi_i \in \{s, e\}$, $1 \le i \le n$, $\pi = \pi_1 \cdots \pi_n$, we define $u\emptyset = u$, $\rho^\emptyset = \rho$ and then inductively we define

$$u\pi = (u\pi_1 \cdots \pi_{n-1})\pi_n, \quad \rho^\pi = (\rho^{\pi_1 \cdots \pi_{n-1}})^{\pi_n}.$$

Clearly $\varphi_{\Pi\Theta}$ is surjective and the basic properties of the mappings $V \mapsto V_\tau$, $V \mapsto V^\tau$ are explored below.

Note that Definition 3.1.2 is slightly in conflict with our practice heretofore since we are going to have elements of Π^1 operate on the right, whereas previously we wrote $s(u)$ and $e(u)$. This is motivated by the desire to have Π^1 and Θ^1 operate on the same side. However, we will continue to write $s(u)$ and $e(u)$ when convenient.

Lemma 3.1.3 *Let* $V \in [\mathcal{S}, \mathcal{CR}]$, P_i, $Q_j \in \{T_\ell, T_r\}$ *for* $1 \le i \le m$ *and* $1 \le j \le n$, *and let* $P_1 \cdots P_m$ *and* $Q_1 \cdots Q_n$ *be in canonical form. Then*

$$(V^{P_1 \cdots P_m})_{Q_1 \cdots Q_n} = \begin{cases} V^{P_1 \cdots P_{m-n}} & \text{if } P_m = Q_1, \ m > n & (3.1.1) \\ V_{Q_m \cdots Q_n} & \text{if } P_m = Q_1, \ m \le n & (3.1.2) \\ V^{P_1 \cdots P_{m-n+1}} & \text{if } P_m \ne Q_1, \ m > n - 1 & (3.1.3) \\ V_{Q_{m+1} \cdots Q_n} & \text{if } P_m \ne Q_1, \ m \le n - 1. & (3.1.4) \end{cases}$$

Proof The proof is by induction on $m + n$. The case $m + n = 2$, that is $m = n = 1$, is covered by Theorem 2.5.2(xiii). So suppose that $m + n \ge 3$.

If $P_m = Q_1$, then with $W = V^{P_1 \cdots P_{m-1}}$, we have

$$\begin{aligned}
(V^{P_1 \cdots P_m})_{Q_1 \cdots Q_n} &= \left((W^{Q_1})_{Q_1} \right)_{Q_2 \cdots Q_n} \\
&= (W_{Q_1})_{Q_2 \cdots Q_n} \\
&= (V^{P_1 \cdots P_{m-1}})_{Q_1 Q_2 \cdots Q_n}
\end{aligned}$$

and the result follows from (3.1.3) and (3.1.4) by the induction hypothesis.

If $P_m \ne Q_1$, then

$$\begin{aligned}
(V^{P_1 \cdots P_m})_{Q_1 \cdots Q_n} &= \left((W^{P_m})_{Q_1} \right)_{Q_2 \cdots Q_n} \\
&= (W^{P_m})_{Q_2 \cdots Q_n} \quad \text{by Theorem 2.5.2(xiii)} \\
&= (V^{P_1 \cdots P_n})_{Q_2 \cdots Q_n}
\end{aligned}$$

and the result follows from (3.1.1) and (3.1.2) by the induction hypothesis.   □

Lemma 3.1.4 *Let $\tau \in \Theta$.*

(i) *The mapping $\mathcal{V} \mapsto \mathcal{V}_\tau$ ($\mathcal{V} \in \mathcal{L}(\mathcal{CR})$) is a complete endomorphism of $\mathcal{L}(\mathcal{CR})$. It maps $[\mathcal{S}, \mathcal{CR}]$ into itself.*

(ii) *The mapping $\mathcal{V} \mapsto \mathcal{V}^\tau$ ($\mathcal{V} \in \mathcal{L}(\mathcal{CR})$) is a complete $\cap$-endomorphism on $\mathcal{L}(\mathcal{CR})$. It maps $[\mathcal{S}, \mathcal{CR}]$ into itself.*

(iii) *For all $\mathcal{V} \in \mathcal{L}(\mathcal{CR})$ we have $\mathcal{V} \subseteq (\mathcal{V}_\tau)^{\overline{\tau}}$.*

Proof (i) The first claim is an immediate consequence of the definition of $\mathcal{V}_\tau$ and Theorem 2.5.2(viii) and its dual. The second claim follows from the first and the observation that $\mathcal{S}_{T_r} = \mathcal{S}_{T_\ell} = \mathcal{S}$.

(ii) This is an immediate consequence of the definition of $\mathcal{V}^\tau$ and Theorem 2.5.2(ix) and its dual.

(iii) We argue by induction on $|\tau|$, where τ is written in canonical form. If $|\tau| = 1$, say $\tau = P \in \{T_\ell, T_r\}$, then $\mathcal{V} \subseteq \mathcal{V}^P = (\mathcal{V}_P)^P$.

So suppose that $|\tau| = n$ and that the claim is true for elements of Θ with shorter canonical forms. Let $\tau = \sigma P$ where $P \in \{T_\ell, T_r\}$ and σP is written in canonical form. Then $|\sigma| < |\tau|$ so that

$$(\mathcal{V}_\tau)^{\overline{\tau}} = (\mathcal{V}_{\sigma P})^{P\overline{\sigma}} = \left(((\mathcal{V}_\sigma)_P)^P \right)^{\overline{\sigma}}$$

$$= \left((\mathcal{V}_\sigma)^P \right)^{\overline{\sigma}} \supseteq (\mathcal{V}_\sigma)^{\overline{\sigma}} \quad \text{by part (ii)}$$

$$\supseteq \mathcal{V}$$

by the induction hypothesis. $\square$

The next result clarifies the correspondence between the action of $\pi \in \Pi$ on a fully invariant congruence $\zeta_\mathcal{V}$ and the corresponding action of $\pi\varphi_{\Pi\Theta}$ on $\mathcal{V}$. But first we label a particular set of three varieties that will play a special role in the ensuing discussion.

Notation 3.1.5 Let $\mathbb{N}_3 = \{\mathcal{LNB}, \mathcal{S}, \mathcal{RNB}\}$.

In the discussion to follow, we will frequently encounter expressions of the form $\zeta_\mathcal{V}^\tau$ where $\mathcal{V} \in \mathcal{L}(\mathcal{CR})$, $\tau \in \Theta^1$. This should always be interpreted to mean $(\zeta_\mathcal{V})^\tau$.

Lemma 3.1.6 *Let $\mathcal{V} \in [\mathcal{S}, \mathcal{CS}]$, $\pi \in \Pi$ and $\tau = \pi\varphi_{\Pi\Theta}$.*

(i) *The following statements are equivalent:*

(a) *either $\pi = \emptyset$ or $\mathcal{V}_\tau \notin \mathbb{N}_3$;*

(b) *$\zeta_\mathcal{V}^\pi = \zeta_{\mathcal{V}_\tau}$;*

(c) $\zeta_{\mathcal{V}}^{\pi} \in \Delta$;

(d) *either* $\pi = \emptyset$

    or $\pi = \pi_1 s$ where $\pi_1 \in \Pi^1$ and $\mathcal{V}_{\tau_1} \supseteq \mathcal{LRB}$;

    or $\pi = \pi_1 e$ where $\pi_1 \in \Pi^1$ and $\mathcal{V}_{\tau_1} \supseteq \mathcal{RRB}$.

(ii) *We have*

$$\widehat{\zeta_{\mathcal{V}}^{\pi}} = \begin{cases} \zeta_{\mathcal{LZ}} & \text{if } \mathcal{V}_\tau = \mathcal{LNB} & (3.1.5) \\ \omega & \text{if } \mathcal{V}_\tau = \mathcal{S} & (3.1.6) \\ \zeta_{\mathcal{RZ}} & \text{if } \mathcal{V}_\tau = \mathcal{RNB}. & (3.1.7) \end{cases}$$

Proof Recall the notation $\pi^* = \pi\varphi_{\Pi\Theta}$ from Definition 3.1.2.

(i) (a) $\Rightarrow$ (b). The claim is trivial for $\pi = \emptyset$. So assume that $\mathcal{V}_\tau \not\subseteq \mathbb{N}_3$. By duality, it suffices to consider the case $t(\pi) = s$, $t(\tau) = T_r$. We argue by induction on $|\pi| = n$. If $n = 1$, then by hypothesis $\mathcal{V}_{T_r} \not\subseteq \mathbb{N}_3$ and, by Theorem 1.5.5(ii), $\mathcal{LRB} \subseteq \mathcal{V}_{T_r}$ so that by Theorem 1.5.6, we obtain $\zeta_{\mathcal{V}}^{\pi} = \zeta_{\mathcal{V}}^{s} = \zeta_{\mathcal{V}_{T_r}} = \zeta_{\mathcal{V}_\tau}$. Now let $n > 1$, $\pi = \pi_1 s$ and $\tau_1 = \pi_1^*$ so that $\tau = \tau_1 T_r$. But $\mathcal{V}_\tau \subseteq \mathcal{V}_{\tau_1}$. Consequently $\mathcal{V}_{\tau_1} \not\subseteq \mathbb{N}_3$ and, by the induction hypothesis, $\zeta_{\mathcal{V}}^{\pi_1} = \zeta_{\mathcal{V}_{\tau_1}}$. Since $(\mathcal{V}_{\tau_1})_{T_r} = \mathcal{V}_\tau \not\subseteq \mathbb{N}_3$, we have from the case $n = 1$, that $(\zeta_{\mathcal{V}_{\tau_1}})^s = \zeta_{\mathcal{V}_{\tau_1 T_r}} = \zeta_{\mathcal{V}_\tau}$ and therefore that $\zeta_{\mathcal{V}}^{\pi} = (\zeta_{\mathcal{V}}^{\pi_1})^s = (\zeta_{\mathcal{V}_{\tau_1}})^s = \zeta_{\mathcal{V}_\tau}$. as required.

(b) $\Rightarrow$ (c). From (b), we have $\zeta_{\mathcal{V}}^{\pi} = \zeta_{\mathcal{V}_\tau} \in \Delta$.

(c) $\Rightarrow$ (d). Suppose that $\pi \neq \emptyset$. First assume that $\pi = \pi_1 s$ where $\pi_1 \in \Pi^1$. Then $\mathcal{V} \not\subseteq \mathcal{LRRO}$ since otherwise, by Theorem 1.5.6, $\zeta_{\mathcal{V}_{\pi_1}}^{s} \notin \Delta$ so that $\zeta_{\mathcal{V}}^{\pi} \notin \Delta$, a contradiction. Therefore $\mathcal{V}_{\pi_1} \supseteq \mathcal{LRB}$. By duality, the claim holds.

(d) $\Longrightarrow$ (a). Suppose that $\pi \neq \emptyset$ and let $\pi = \pi_1 s$ where $\pi_1 \in \Pi^1$. Then $\mathcal{V}_{\pi_1} \supseteq \mathcal{LRB}$. Hence $\zeta_{\mathcal{V}_\pi} = (\zeta_{\mathcal{V}_{\pi_1}})^s \supseteq (\zeta_{\mathcal{LRB}})^s = \zeta_{\mathcal{LRB}}$. Therefore $\mathcal{V}_\pi \not\subseteq \mathbb{N}_3$. By duality the claim holds.

(ii) We consider the three cases.

Case 1: $\mathcal{V}_\tau = \mathcal{LNB}$. By the dual of Theorem 1.5.5, it is not possible to have $t(\tau) = T_\ell$. Therefore $t(\pi) = s$ and $t(\tau) = T_r$. Let $\pi = \pi_1 ss \cdots s$ where either $\pi_1 = \emptyset$ or $t(\pi_1) = e$.

Subcase: $\pi_1 = \emptyset$. Then $\pi = s \cdots s$ and $\tau = \pi^* = T_r$. Since $\mathcal{V}_{T_r} = \mathcal{LNB}$, we have $\mathcal{V} \in [\mathcal{LNB}, \mathcal{LRRO}]$ so that $\zeta_{\mathcal{V}} \in [\zeta_{\mathcal{LRRO}}, \zeta_{\mathcal{LZ}}]$. By Lemma 1.5.2(iv), it follows that $\widehat{\zeta_{\mathcal{V}}^{s}} = \zeta_{\mathcal{LZ}}$. Hence $\widehat{\zeta_{\mathcal{V}}^{\pi}} = \widehat{\zeta_{\mathcal{V}}^{ss \cdots s}} = \zeta_{\mathcal{LZ}}$.

Subcase: $\pi_1 \neq \emptyset$ and $t(\pi_1) = e$. Let $\tau_1 = \pi_1^*$. Then $t(\tau_1) = T_\ell$. Since $\mathcal{V}_{\tau_1 T_r} = \mathcal{V}_\tau = \mathcal{LNB}$, we have $\mathcal{V}_{\tau_1} \in [\mathcal{LNB}, \mathcal{LRRO}]$. Thus $\mathcal{V}_{\tau_1} \neq \mathcal{S}$, $\mathcal{LRB}$ and $\mathcal{V}_{\tau_1} \neq \mathcal{LNB}$ since $\mathcal{V}_{\tau_1}$ is the least element in its T_ℓ-class but $\mathcal{LNB}_{T_\ell} = \mathcal{S}$. Hence $\mathcal{V}_{\tau_1} \not\subseteq \mathbb{N}_3$ and, by the first case, we have $\zeta_{\mathcal{V}}^{\pi_1} = \zeta_{\mathcal{V}_{\tau_1}}$. Since $\mathcal{V}_{\tau_1} \in [\mathcal{LNB}, \mathcal{LRRO}]$, it follows that $\zeta_{\mathcal{V}}^{\pi_1} \in [\zeta_{\mathcal{LRRO}}, \zeta_{\mathcal{LNB}}]$. Therefore $(\zeta_{\mathcal{V}}^{\pi_1})^s = \zeta_{\mathcal{LZ}}$, by Lemma 1.5.2(iv). Consequently, again by Lemma 1.5.2(iv)

$$\widehat{(\zeta_{\mathcal{V}})^{\pi}} = \widehat{(\zeta_{\mathcal{V}}^{\pi_1})^{s \cdots s}} = \zeta_{\mathcal{LZ}}.$$

Case 2: $\mathcal{V}_\tau = \mathcal{S}$. We argue by induction on $|\pi|$. By duality, we may assume that $t(\pi) = s$ and $t(\tau) = T_r$. If $|\pi| = 1$ (or even if $\pi = ss \cdots s$), then $\tau = T_r$ (or $\tau = T_r T_r \cdots T_r = T_r$) and, by Theorem 1.5.5, we get $\mathcal{V} \in [\mathcal{S}, \mathcal{RRO}]$. Hence, by Lemma 1.5.2(iv), we obtain $\widehat{\zeta^s_\mathcal{V}} = \omega$.

So now assume that $|\pi| > 1$. Indeed, by the preceding argument, we may assume that $\pi = \pi_1 ss \cdots s$ where $\pi_1 \neq \emptyset$ and $t(\pi_1) = e$. Let $\tau_1 = \pi_1^*$. By Theorem 1.5.5, we have $\mathcal{V}_{\tau_1} \in [\mathcal{S}, \mathcal{RRO}]$.

If $\mathcal{V}_{\tau_1} = \mathcal{S}$, then $\widehat{\zeta^{\pi_1}_\mathcal{V}} = \omega$ by the induction hypothesis and so $\widehat{\zeta^\pi_\mathcal{V}} = (\widehat{\zeta^{\pi_1}_\mathcal{V}})^{s \cdots s} = \omega$. If $\mathcal{V}_{\tau_1} = \mathcal{RNB}$ then, by the dual of Case 1 and Lemma 1.5.2(iv), it follows that

$$\widehat{\zeta^\pi_\mathcal{V}} = (\widehat{\zeta^{\pi_1}_\mathcal{V}})^{s \cdots s} = (\widehat{\zeta_{\mathcal{RZ}}})^{s \cdots s} = \omega.$$

So suppose that $\mathcal{V}_{\tau_1} \notin \{\mathcal{S}, \mathcal{RNB}\}$. Since $t(\pi_1) = e$ we also have $t(\tau_1) = T_\ell$. Hence $(\mathcal{V}_{\tau_1})_{T_\ell} = \mathcal{V}_{\tau_1}$. If $\mathcal{V} \subseteq LL\mathcal{RO}$ then, by the dual of Theorem 1.5.5, we would have $\mathcal{V}_{\tau_1} \in \{\mathcal{S}, \mathcal{RNB}\}$. Since this is not the case, $\mathcal{V}_{\tau_1} \notin \mathcal{L}(LL\mathcal{RO})$ and so, by the dual of [PR1, Corollary IX.5.3(i)], we get $\mathcal{RRB} \subseteq \mathcal{V}_{\tau_1}$. Consequently, by Case 1, $\zeta^{\pi_1}_\mathcal{V} = \zeta_{\mathcal{V}_{\tau_1}}$. Since $\mathcal{V}_{\tau_1} \in [\mathcal{S}, \mathcal{RRO}]$, it follows from Lemma 1.5.2(iv) that

$$\widehat{\zeta^\pi_\mathcal{V}} = (\widehat{\zeta^{\pi_1}_\mathcal{V}})^{s \cdots s} = (\widehat{\zeta_{\mathcal{V}_{\tau_1}}})^{s \cdots s} = \omega.$$

Case 3: $\mathcal{V}_r = \mathcal{RNB}$. This is the dual of Case 1.   □

The next theorem provides what appears to be a complicated characterization of $\zeta_\mathcal{V}$ for $\mathcal{V} \in [\mathcal{S}, \mathcal{CR}]$. Yet it is the key to the proof of Theorem 4.4.1(to come) and also provides a method for relating the word problems for $\mathcal{V}$ to that for certain smaller varieties.

Theorem 3.1.7 *Let* $\mathcal{V} \in [\mathcal{S}, \mathcal{CR}]$, $u, v \in U$ *and* $\pi^* = \pi\varphi_{\Pi\Theta}$ ($\pi \in \Pi^1$).
(i) *We have* $u \zeta_\mathcal{V} v \Rightarrow u\pi (\zeta_\mathcal{V})^\pi v\pi$ *for all* $\pi \in \Pi^1$.
(ii) *We have*

$$u \zeta_\mathcal{V} v \Longleftrightarrow \begin{cases} u\pi ((\zeta_\mathcal{V})^{\pi K} \cap \eta) \, v\pi & \text{for all } \pi \in \Pi^1 \text{ such that} \\ & \text{either } \pi = \emptyset \text{ or } \mathcal{V}_{\pi*} \notin \mathbb{N}_3 \quad\quad (3.1.8) \\ u\pi (\zeta_\mathcal{V})^\pi \, v\pi & \text{for all } \pi \in \Pi \text{ such that } \mathcal{V}_{\pi*} \in \mathbb{N}_3. \quad (3.1.9) \end{cases}$$

Remark. It is very important to recognize that, whereas $\mathcal{V}_{T_\ell T_\ell} = \mathcal{V}_{T_\ell}$, $\mathcal{V}_{T_r T_r} = \mathcal{V}_{T_r}$, $\rho^{ss} = \rho^s \in \Delta$ for $\rho \subseteq \zeta_{L\mathcal{RB}}$ and $\rho^{ee} = \rho^e \in \Delta$ for $\rho \subseteq \zeta_{R\mathcal{RB}}$, we must not identify ss with s or ee with e since these operators are distinct when applied to words. We will see the importance of this when we apply Lemma 3.1.6 to the word problem for $\mathcal{ReB}$, for instance.

It is also important to recognize that it may happen that, when $\mathcal{V}_{\pi*} \in \mathbb{N}_3$, we may have that $(\zeta_\mathcal{V})^\pi$ is not a fully invariant congruence. However, when that occurs, recall that the definitions of ρ^s, ρ^e in Notation 1.5.1, include not just fully invariant congruences ρ but any binary relation ρ on U. With this convention, we can avoid having to involve the fully invariant congruence generated by $(\zeta_V)^s$, or $(\zeta_V)^e$ in these cases.

Proof (i) Let $u \; \zeta_V \; v$. From the definitions of ρ^s and ρ^e, we see that $s(u) \; (\zeta_V)^s \; s(v)$ and $e(u) \; (\zeta_V)^e \; e(v)$; that is, for $\pi \in \{s, e\}$, we have $u\pi \; \zeta_V^\pi \; v\pi$. A simple induction argument based on $|\pi|$ will now establish that

$$u\pi \; \zeta_V^\pi \; v\pi \qquad (\pi \in \Pi^1) \tag{3.1.10}$$

and (i) holds.

(ii) *Direct part.* We consider three cases.

Case: $\mathcal{V}_{\pi^*} \in \mathbb{N}_3$. The claim in (3.1.9) follows immediately from (3.1.10).

Case: $\pi = \emptyset$. Since $\mathcal{S} \subseteq \mathcal{V}$, we have $\zeta_V \subseteq \eta$ so that $u \; \eta \; v$. Also $\zeta_V \subseteq (\zeta_V)^K$, whence $u \; (\zeta_V)^K \; v$ and the claim follows.

Case: $\pi \neq \emptyset$ and $\mathcal{V}_{\pi^*} \notin \mathbb{N}_3$. By Lemma 1.1.6, $\zeta_V^\pi = \zeta_{(\mathcal{V}_{\pi^*})}$ and, by Theorem 1.5.5 and its dual, since $\mathcal{S} \subseteq \mathcal{V}$ we also have $\mathcal{S} \subseteq \mathcal{V}_{\pi^*}$. Hence $\zeta_V^\pi = \zeta_{(\mathcal{V}_{\pi^*})} \subseteq \eta$. Additionally, $\zeta_V^\pi \subseteq (\zeta_V)^{\pi K}$ so that from (3.1.10), we can deduce that $u\pi \; (\zeta_V)^{\pi K} \cap \eta \; v\pi$ and the claim is verified in this case also.

Converse. Now let $u, v \in U$ and ζ_V satisfy the conditions in (3.1.8) and (3.1.9). With $\pi = \emptyset$, we obtain

$$u \; \left(\zeta_V^K \cap \eta\right) \; v \tag{3.1.11}$$

so that $c(u) = c(v)$ and we can argue by induction on $\#(u)$.

Let $\#(u) = 1$. Then $c(u) = c(v) = \{x\}$ for some $x \in X$, and $u\zeta \; \mathcal{H} \; v\zeta$. By (3.1.11), $u \; \zeta_V^K \; v$ so that $u\zeta \; (\zeta_V/\zeta)^K \; v\zeta$ and thus

$$u\zeta \; \left((\zeta_V/\zeta)^K \cap \mathcal{H}\right) \; v\zeta. \tag{3.1.12}$$

But, by [PR1, Corollary VII.1.8], $(\zeta_V/\zeta)^K \cap \mathcal{H} = (\zeta_V/\zeta) \cap \mathcal{H}$ and therefore (3.1.12) implies that $u\zeta \; (\zeta_V/\zeta) \; v\zeta$ whence $u \; \zeta_V \; v$, as required.

Now suppose that $\#(u) > 1$ and that the claim holds for smaller values of $\#(u)$. Using this induction hypothesis we will first show that $s(u) \; (\zeta_V)^s \; s(v)$.

Case: $\mathcal{V}_{T_r} \in \mathbb{N}_3$. We have immediately from the hypothesis in (3.1.9) that $s(u) \; \zeta_V^s \; s(v)$.

Case: $\mathcal{V}_{T_r} \notin \mathbb{N}_3$. By Lemma 3.1.6. The argument will be more transparent if we write $\mathcal{U} = \mathcal{V}_{T_r}$. Then $\zeta_\mathcal{U} = \zeta_{\mathcal{V}_{T_r}} = \zeta_V^s$.

We wish to show first that $s(u)$, $s(v)$ and $\zeta_\mathcal{U}$ also satisfy the conditions in (3.1.8) and (3.1.9). To this end, let $\pi \in \Pi^1$ and $\tau = \pi^*$. We break the argument into three subcases.

Subcase: $\mathcal{V}_{T_r} \notin \mathbb{N}_3, \pi = \emptyset$. Since $\mathcal{U}_{T_r} = (\mathcal{V}_{T_r})_{T_r} = \mathcal{V}_{T_r} \notin \mathbb{N}_3$, we may invoke (3.1.8) with $s \in \Pi$ to obtain $s(u) \; ((\zeta_V)^{sK} \cap \eta) \; s(v)$ or

$$s(u) \; \left(\zeta_\mathcal{U}^K \cap \eta\right) \; s(v). \tag{3.1.13}$$

This tells us that (3.1.8) is satisfied by $s(u)$, $s(v)$ and ζ_u in the case $\pi = \emptyset$.

Subcase: $V_{T_r} \not\subseteq N_3$, $\pi \neq \emptyset$, $\tau = \pi^*$, $V_{T,\tau} \in N_3$. From the hypothesis in (3.1.9) we have

$$s(u)\pi = u(s\pi) \ (\zeta_v)^{s\pi} \ v(s\pi) = s(v)\pi$$

that is,

$$s(u)\pi \ (\zeta_u)^\pi \ s(v)\pi \tag{3.1.14}$$

so that (3.1.9) is satisfied for $s(u)$, $s(v)$ and ζ_u.

Subcase: $V_{T_r} \not\subseteq N_3$, $\pi \neq \emptyset$, $\tau = \pi^*$, $V_{T,\tau} \not\subseteq N_3$. Then by the hypothesis in (3.1.8) we get

$$s(u)\pi = u(s\pi) \ \left((\zeta_v)^{s\pi K} \cap \eta\right) \ v(s\pi) = s(v)\pi,$$

whence

$$s(u)\pi \ \left((\zeta_u)^{\pi K} \cap \eta\right) \ s(v)\pi. \tag{3.1.15}$$

Combining (3.1.13) and (3.1.15), we get

$$s(u)\pi \ \left((\zeta_u)^{\pi K} \cap \eta\right) \ s(v)\pi \quad \text{if} \ \ \pi = \emptyset \ \ \text{or} \ \ U_r \not\subseteq N_3. \tag{3.1.16}$$

By (3.1.14) and (3.1.16), we see that $s(u)$, $s(v)$ and ζ_u satisfy the conditions (3.1.8) and (3.1.9). By (3.1.13), $\#(s(u)) = \#(s(v)) < \#(u)$, so that the induction hypothesis yields that $s(u) \ \zeta_u \ s(v)$, as required. In other words, $s(u) \ \zeta_v^s \ s(v)$.

By duality, $e(u) \ \zeta_v^e \ e(v)$. By (3.1.8), with $\pi = \emptyset$, we have $u \ \left(\zeta_v^K \cap \eta\right) v$ and thus

$$u \ \zeta_v^K \ v, \quad c(u) = c(v), \quad s(u) \ \zeta_v^s \ s(v), \quad e(u) \ \zeta_v^e \ e(v).$$

By Lemma 1.5.9 with $\lambda = \zeta_v^K$, we can now conclude that $u \ \zeta_v \ v$. $\square$

Because $\mathcal{L}(\mathcal{B})$ constitutes a single K-class, Theorem 3.1.7 simplifies slightly in the context of subvarieties of $\mathcal{B}$.

Corollary 3.1.8 *Let* $V \in [S, \mathcal{B}]$ *and* $u, v \in U$. *Then*

$$u \ \zeta_v \ v \iff \begin{cases} \text{(i)} \ \ u\pi \ \eta \ v\pi & \text{for } \pi = \emptyset \text{ and } \pi \in \Pi \text{ such that } V_{\pi^*} \not\subseteq N_3, \\ \text{(ii)} \ \ u\pi \ \zeta_v^\pi \ v\pi & \text{for all } \pi \in \Pi \text{ with } V_{\pi^*} \in N_3. \end{cases}$$

Proof Immediate. $\square$

Theorem 3.1.7 and Corollary 3.1.8 can be viewed as providing a solution to the word problem for the varieties involved. However, the usefulness of the solution depends on the nature of V.

There is a limit on the applicability of Theorem 3.1.7 to the solution of word problems on account of (3.1.8). For any $V \in [S, C\mathcal{R}]$ with $V_K = V$ and any $u, v \in U$, we see that (3.1.8) includes the condition $u \; \zeta_{\mathcal{V}_K} \; v$, that is, $u \; \zeta_{\mathcal{V}} \; v$ so that no simplification is obtained from Theorem 3.1.7.

However, whenever $\mathcal{V}_K$ is a proper subvariety of $\mathcal{V}$, Theorem 3.1.7 does describe $\zeta_{\mathcal{V}}$ in terms of congruences $\zeta_{\mathcal{W}}$ for certain varieties $\mathcal{W}$ that are properly contained in $\mathcal{V}$ making it possible in some instances to obtain a usable solution to the word problem for $\mathcal{V}$.

When applying Theorem 3.1.7 to the word problem for particular varieties, it is helpful to have the following information at hand. From Lemma 1.5.2(iv) and its dual,

$$\mathcal{V} \in [\mathcal{T}, \mathcal{LRO}] \implies \widehat{\zeta_{\mathcal{V}}^{e}} = \omega$$

$$\mathcal{V} \in [\mathcal{T}, \mathcal{RRO}] \implies \widehat{\zeta_{\mathcal{V}}^{s}} = \omega$$

$$\mathcal{V} \in [\mathcal{LZ}, \mathcal{LRRO}] \implies \widehat{\zeta_{\mathcal{V}}^{s}} = \zeta_{\mathcal{LZ}}$$

$$\mathcal{V} \in [\mathcal{RZ}, \mathcal{LLRO}] \implies \widehat{\zeta_{\mathcal{V}}^{e}} = \zeta_{\mathcal{RZ}}.$$

In particular,

$$\widehat{\zeta_{\mathcal{S}}^{e}} = \widehat{\zeta_{\mathcal{S}}^{s}} = \omega, \quad \widehat{\zeta_{\mathcal{LNB}}^{e}} = \widehat{\zeta_{\mathcal{LRB}}^{e}} = \widehat{\zeta_{\mathcal{RNB}}^{s}} = \widehat{\zeta_{\mathcal{RRB}}^{s}} = \omega,$$

$$\widehat{\zeta_{\mathcal{LNB}}^{s}} = \zeta_{\mathcal{LZ}}, \quad \widehat{\zeta_{\mathcal{RNB}}^{e}} = \zeta_{\mathcal{RZ}},$$

$$\widehat{\zeta_{\mathcal{LRB}}^{s}} = \zeta_{\mathcal{LRB}}, \quad \widehat{\zeta_{\mathcal{RRB}}^{e}} = \zeta_{\mathcal{RRB}}$$

where the equalities in the last line follow from Lemma 1.5.2(vi) and its dual.

It is important to get a proper understanding of how Theorem 3.1.7 works since it is the linchpin for what is to follow in this chapter and the next. Towards this end we consider some examples. At first glance, it would appear that in order to apply Theorem 3.1.7 we may be required to calculate infinitely many congruences of the forms $(\zeta_{\mathcal{V}})^{\pi K}$ or $(\zeta_{\mathcal{V}})^{\pi}$. However, in many instances it will turn out that all but a small, or at least finite, number of these congruences will be equal to ω and therefore make no contribution to the value of $\zeta_{\mathcal{V}}$.

It is not immediately evident, but there is a big divide in the application of Theorem 3.1.7 depending on whether the variety $\mathcal{V}$ contains $\mathcal{B}$ or not. If $\mathcal{B} \subseteq \mathcal{V}$, then since $\mathcal{B}_{\tau} = \mathcal{B}$ for all $\tau \in \Theta$ there is no $\pi \in \Pi$ such that $\mathcal{V}_{\pi^*} \in \mathbb{N}_3$ and that simplifies the description of $\zeta_{\mathcal{V}}$ in Theorem 3.1.7 considerably. So we begin with some illustrations in that simpler context.

Examples (i) *The word problem for* $\mathcal{B}$. This is probably the simplest illustration of Theorem 3.1.7. Since $\mathcal{B}_{\tau} = \mathcal{B}$ for all $\tau \in \Theta$, we deduce from Theorem 3.1.7 that, for $u, v \in U$,

$$u \; \zeta_{\mathcal{B}} \; v \iff u\pi \; ((\zeta_{\mathcal{B}})^{\pi K} \cap \eta) \; v\pi \text{ for all } \pi \in \Pi^{1}$$

$$\iff u\pi \; (\omega \cap \eta) \; v\pi \text{ for all } \pi \in \Pi^{1}$$

$$\iff u\pi \; \eta \; v\pi \text{ for all } \pi \in \Pi^{1}.$$

(ii) *The word problem for* $\mathcal{O}$. The required basic facts regarding $\mathcal{O}$ are:

$$\mathcal{O}_K = \mathcal{G}, \; \mathcal{O}^K = \mathcal{O}, \; \mathcal{O}_{T_\ell} = \mathcal{O}_{T_r} = \mathcal{O}.$$

From Theorem 3.1.7, we then obtain that, for any $u, v \in U$,

$$u \; \zeta_{\mathcal{O}} \; v \iff u\left((\zeta_{\mathcal{O}})^K \cap \eta\right) vu\pi \left((\zeta_{\mathcal{O}})^{\pi K} \cap \eta\right) v\pi \text{ for all } \pi \in \Pi^1$$
$$\iff u \;(\zeta_{\mathcal{G}} \cap \eta) \; vu\pi \;(\zeta_{\mathcal{O}_{\tau K}} \cap \eta) \; v\pi \qquad \text{for all } \pi \in \Pi^1, \tau \in \Theta$$
$$\iff u\pi \;(\zeta_{\mathcal{G}} \cap \eta) \; v\pi \qquad\qquad \text{for all } \pi \in \Pi^1.$$

(iii) $L\mathcal{O}$: We have $L\mathcal{O}_K = \mathcal{CS}$, $L\mathcal{O} = \mathcal{CS}^K$, $L\mathcal{O}_\pi = L\mathcal{O}$ for all $\pi \in \Pi^1$. Hence $L\mathcal{O}_{\pi^*} \notin \mathbb{N}_3$ for all $\pi \in \Pi^1$. Therefore, for $u, v \in U$,

$$u \; \zeta_{L\mathcal{O}} \; v \iff u\pi \; \zeta_{L\mathcal{O}}^{\pi K} \cap \eta \; v\pi \quad \text{for all } \; \pi \in \Pi^1$$
$$\iff u\pi \; \zeta_{L\mathcal{O}}^{K} \cap \eta \; v\pi \quad \text{for all } \; \pi \in \Pi^1$$
$$\iff u\pi \; \zeta_{\mathcal{CS}} \cap \eta \; v\pi \quad \text{for all } \; \pi \in \Pi^1 \text{ with } |\pi| < \#(u).$$

(iv) *The word problem for* $\mathcal{BG}$. The relevant facts for $\mathcal{BG}$ are:

$$\mathcal{BG}_K = (\mathcal{B}^T)_K = \mathcal{B}^T = \mathcal{BG}, \quad \mathcal{BG}_{T_\ell} = \mathcal{BG}_{T_r} = \mathcal{B}.$$

The characterization of $\zeta_{\mathcal{BG}}$ in Theorem 3.1.7, includes $u\;(\zeta_{\mathcal{BG}}^K \cap \eta)\;v$ which reduces to $u\;(\zeta_{\mathcal{BG}} \cap \eta)\;v$. Since this involves $\zeta_{\mathcal{BG}}$ it is of no help in finding a solution to the word problem for $\mathcal{BG}$.

Solutions to the word problems for examples that are low in the lattice $\mathcal{L}(\mathcal{CR})$ can mostly be obtained by ad hoc methods and Theorem 3.1.7 really only comes into play above $\mathcal{ReB}$. The example $\mathcal{ReB}$ itself illustrates how Theorem 3.1.7 works for varieties that contain $\mathcal{ReB}$ but not $\mathcal{B}$.

(v) *The word problem for* $\mathcal{ReB}$. Combining the fact that $\mathcal{ReB} = \mathcal{LRB} \vee \mathcal{RRB}$ with [PR1, Lemma V.1.4] and its dual, we already have a simple solution to the word problem for $\mathcal{ReB}$: for $u, v \in U, u \; \zeta_{\mathcal{ReB}} \; v \Leftrightarrow i(u) = i(v)$ and $f(u) = f(v)$. The following alternative proof of this fact will serve to illustrate how Theorem 3.1.7 can be applied to varieties in the interval $[\mathcal{ReB}, \mathcal{B})$. First a few preliminaries. We have

$$\mathcal{ReB}_K = \mathcal{J}, \quad \mathcal{ReB}_{T_\ell} = \mathcal{RRB}, \quad \mathcal{ReB}_{T_r} = \mathcal{LRB}.$$

Consequently, $\mathcal{ReB}_\tau = \mathcal{S}$ for all $\tau \in \Theta$, $|\tau| \geq 2$. Also, by Theorem 1.5.6 and its dual, we have

$$(\zeta_{\mathcal{ReB}})^s = \zeta_{\mathcal{ReB}_{T_r}} = \zeta_{\mathcal{LRB}}, \; (\zeta_{\mathcal{ReB}})^e = \zeta_{\mathcal{ReB}_{T_\ell}} = \zeta_{\mathcal{RRB}}.$$

From Corollary 1.5.10, we then have

$$u \; \zeta_{\mathcal{ReB}} \; v \iff u\left((\zeta_{\mathcal{ReB}})^K \cap \eta\right) v,$$
$$s(u) \;(\zeta_{\mathcal{LRB}} \cap \eta)\; s(v), \; e(u) \;(\zeta_{\mathcal{RRB}} \cap \eta)\; e(v),$$
$$\iff c(u) = c(v),$$
$$s(u) \;(\zeta_{\mathcal{LRB}} \cap \eta)\; s(v), \; e(u) \;(\zeta_{\mathcal{RRB}} \cap \eta)\; e(v).$$

Therefore the above characterization of $\zeta_{\mathcal{R}e\mathcal{B}}$ simplifies as follows

$$u \; \zeta_{\mathcal{R}e\mathcal{B}} \; v \iff c(u) = c(v), \; c(s(u)) = c(s(v)), \; c(e(u)) = c(e(v)),$$
$$s(u) \; \zeta_{\mathcal{L}\mathcal{R}\mathcal{B}} \; s(v), \; e(u) \; \zeta_{\mathcal{R}\mathcal{R}\mathcal{B}} \; s(v)$$
$$\iff i(u) = i(v) \text{ and } f(u) = f(v).$$

We conclude this section with a deeper analysis of the relation $(\zeta_{\mathcal{V}})^s$ for $\mathcal{V} \in [\mathcal{S}, \mathcal{L}\mathcal{R}\mathcal{R}\mathcal{O}]$ than we were able to provide in Lemma 1.5.2(vii). This will be required in the next section. This is also the delayed proof of Lemma 1.5.2(vii).

Lemma 3.1.9 *Let* $\mathcal{U} \in [\mathcal{S}, \mathcal{R}\mathcal{R}\mathcal{O}]$ *and* $\mathcal{V} \in [\mathcal{L}\mathcal{N}\mathcal{B}, \mathcal{L}\mathcal{R}\mathcal{R}\mathcal{O}]$; *then* $(\zeta_{\mathcal{U}})^s = (\zeta_{\mathcal{S}})^s \notin \Delta$ *and* $(\zeta_{\mathcal{V}})^s = (\zeta_{\mathcal{L}\mathcal{N}\mathcal{B}})^s \notin \Delta$. *Also* $\zeta_{\mathcal{S}}^{s^k}, \zeta_{\mathcal{L}\mathcal{N}\mathcal{B}}^{s^k} \notin \Delta$ *for all* $k \in \mathbb{N}$.

Proof For the first part, it suffices to show that $(\zeta_{\mathcal{R}\mathcal{R}\mathcal{O}})^s = (\zeta_{\mathcal{S}})^s$. Clearly $(\zeta_{\mathcal{R}\mathcal{R}\mathcal{O}})^s \subseteq (\zeta_{\mathcal{S}})^s$. For the reverse containment, let $u, v \in U, u \; \zeta_{\mathcal{S}} \; v$ so that $c(u) = c(v)$ and consider $s(u), s(v)$. By Theorem 2.5.2(vi) and since $\mathcal{R}\mathcal{R}\mathcal{O} = \mathcal{S}^{Tr}$, we have

$$ux \; \zeta_{\mathcal{R}\mathcal{R}\mathcal{O}} \; (vx)^0 ux \text{ where } x \in X \setminus c(u).$$

Let φ denote the endomorphism of U that maps x to $\sigma(u)$ and leaves all other elements of X unchanged. Then $u\sigma(u) \; \zeta_{\mathcal{R}\mathcal{R}\mathcal{O}} \; (v\sigma(u))^0 u\sigma(u)$. Consequently,

$$s(u) = s(u\sigma(u)) \; (\zeta_{\mathcal{R}\mathcal{R}\mathcal{O}})^s \; s((v\sigma(u))^0 u\sigma(u)) = s(v).$$

Therefore, $(\zeta_{\mathcal{S}})^s \subseteq (\zeta_{\mathcal{R}\mathcal{R}\mathcal{O}})^s$ and equality prevails.

Now suppose that $(\zeta_{\mathcal{S}})^s$ is a fully invariant congruence on U. By Lemma 1.5.2(iv), we must then have $(\zeta_{\mathcal{S}})^s = \omega$. Therefore, for distinct $x, y, p, q \in X$, there must exist $u, v \in U$ such that $u \; \zeta_{\mathcal{S}} \; v, xy = s(u), pq = s(v)$. Then $\#(u) = \#(xy) + 1 = \#(pq) + 1 = 3$. However, $c(u) = c(v) \supseteq \{x, y, p, q\}$ so that $\#(u) \geq 4$ and we have a contradiction. Therefore $(\zeta_{\mathcal{S}})^s$ is not a fully invariant congruence.

In the case of $[\mathcal{L}\mathcal{N}\mathcal{B}, \mathcal{L}\mathcal{R}\mathcal{R}\mathcal{O}]$, it suffices to show that $(\zeta_{\mathcal{L}\mathcal{R}\mathcal{R}\mathcal{O}})^s = (\zeta_{\mathcal{L}\mathcal{N}\mathcal{B}})^s$. It is clear that $(\zeta_{\mathcal{L}\mathcal{R}\mathcal{R}\mathcal{O}})^s \subseteq (\zeta_{\mathcal{L}\mathcal{N}\mathcal{B}})^s$. To establish the reverse containment, consider $s(u), s(v)$ where $u \; \zeta_{\mathcal{L}\mathcal{N}\mathcal{B}} \; v$. In particular, $c(u) = c(v)$. By Theorem 1.5.5, $\mathcal{L}\mathcal{N}\mathcal{B}^{Tr} = \mathcal{L}\mathcal{R}\mathcal{R}\mathcal{O}$ and, by Theorem 2.5.2(vi), for any $x \in X \setminus c(u)$, we have $ux \; \zeta_{\mathcal{L}\mathcal{R}\mathcal{R}\mathcal{O}} \; (vx)^0 ux$. Now define an endomorphism φ on U by its action on the elements of X: $x\varphi = \sigma(u)$ and φ leaves all other elements of X unchanged. Then

$$u\sigma(u) = (ux)\varphi \; (\zeta_{\mathcal{L}\mathcal{R}\mathcal{R}\mathcal{O}})^s \; ((vx)^0 ux)\varphi = (v\sigma(u))^0 u\sigma(u)$$

where

$$s(u\sigma(u)) = s(u), s(v\sigma(u))^0 u\sigma(u)) = s(v).$$

Thus $s(u)$ $(\zeta_{L\mathcal{RRO}})^s$ $s(v)$. Consequently $(\zeta_{L\mathcal{NB}})^s \subseteq (\zeta_{L\mathcal{RRO}})^s$ and equality prevails: $(\zeta_{L\mathcal{NB}})^s = (\zeta_{L\mathcal{RRO}})^s$.

Now suppose that $(\zeta_{L\mathcal{NB}})^s$ is a fully invariant congruence. By Lemma 1.5.2(iv), we then know that $(\zeta_{L\mathcal{NB}})^s = \zeta_{L\mathcal{Z}}$. Let x, y, z, p, q be distinct elements of X. Clearly $xyz\ \zeta_{L\mathcal{Z}}\ xpq$. Suppose by way of contradiction, that xyz $(\zeta_{L\mathcal{NB}})^s xpq$. Then there must exist $u, v \in U$ such that $s(u) = xyz$, $s(v) = xpq$ and $u\ \zeta_{L\mathcal{NB}}\ v$. It follows that $\#(u) = \#(v) = 4$. But $c(u) = c(v) \supseteq c(s(u)) \cup c(s(v)) = \{x, y, z, p, q\}$ which has five elements. Therefore $(\zeta_{L\mathcal{NB}})^s$ is not a fully invariant congruence on U. We have established the final claim for the case $k = 1$. A very slight modification of that argument will cover the general case. $\square$

With the exception of Lemma 3.1.9, this section is mainly a modified version of parts of Polák [1].

3.2   A Representation of the Interval $[\mathcal{S}, \mathcal{CR}]$

This is the second stage of our development. In this section, we construct an epimorphism from $[\mathcal{S}, \mathcal{CR}]$ onto a lattice of mappings of Θ^1 into the following four element lattice.

Definition 3.2.1 Let $D = \{1, L, R, T\}$ be the four element lattice

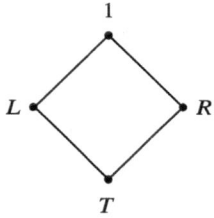

and let Φ_D denote the set of all mappings $\psi : \Theta^1 \to D$ satisfying the following conditions:

(B1) $\emptyset\psi = 1$,
(B2) ψ is order preserving,
(B3) if $\tau\psi = L$, then $t(\tau) = T_r$ and if $\tau\psi = R$, then $t(\tau) = T_\ell$.

Since D is a complete distributive lattice, so also is the direct product D^{Θ^1} under the componentwise order. As a subset of D^{Θ^1}, Φ_D inherits its partial order relation and more, as we will see below.

Lemma 3.2.2 *Let $\sigma, \tau \in \Theta$, $\psi \in \Phi_D$ and $\tau\psi = L$.*

$$\text{(i) } \sigma = \tau T_\ell \implies \sigma \psi = T. \qquad \text{(ii) } |\sigma| \geq |\tau| + 2 \implies \sigma \psi = T.$$

Proof This follows easily from conditions (B2) and (B3). □

Proposition 3.2.3 Φ_D *is a complete sublattice of* D^{Θ^1}.

Proof Let $\{\psi_\alpha \mid \alpha \in A\}$ be a nonempty subset of Φ_D and $\xi = \bigwedge_{\alpha \in A} \psi_\alpha$, $\varphi = \bigvee_{\alpha \in A} \psi_\alpha$. Since each ψ_α satisfies the conditions (B1) and (B2), so also do ξ and φ. Now suppose that $\tau \xi = L$. Since L is a meet irreducible element in D, it follows that there must be at least one $\alpha \in A$ with $\tau \psi_\alpha = L$. By (B3) applied to ψ_α, we then know that $t(\tau) = T_r$. By duality, ξ satisfies (B3) and $\xi \in \Phi_D$. Similarly,

$$\begin{aligned}
\tau \varphi = L \implies L = \tau \bigvee_{\alpha \in A} \psi_\alpha = \bigvee_{\alpha \in A} \tau \psi_\alpha \\
\implies \theta \psi_\alpha \leq L && \text{for all } \alpha \in A \text{ and} \\
\tau \psi_\beta = L && \text{for at least one } \beta \in A \\
\implies t(\tau) = T_r && \text{since } \psi_\beta \text{ satisfies (B3).}
\end{aligned}$$

By duality, φ satisfies (B3) and $\varphi \in \Phi_D$. □

Notation 3.2.4 For $\tau \in \Theta$, we denote by $\tau(K_\ell, K_r)$ the word in $\{K_\ell, K_r\}^+$ obtained from τ by replacing each occurrence of T_ℓ (respectively, T_r) in τ by K_ℓ (respectively, K_r).

We require a preliminary lemma.

Lemma 3.2.5 *Let* $\mathcal{V} \in \mathcal{L}(\mathcal{CR})$, $\tau \in \Theta^1$. *If* $\mathcal{B} \subseteq \mathcal{V}$, *then* $\mathcal{V}^\tau \cap \mathcal{B} = \mathcal{B}$. *Otherwise,*

$$\mathcal{V}^\tau \cap \mathcal{B} = (\mathcal{V} \cap \mathcal{B})^{\tau(K_\ell, K_r)}.$$

Proof The claim in the case where $\mathcal{B} \subseteq \mathcal{V}$ is obvious. Now assume that $\mathcal{B} \not\subseteq \mathcal{V}$. Clearly $(\mathcal{V} \cap \mathcal{B})^{\tau(K_\ell, K_r)} \subseteq \mathcal{V}^\tau \cap \mathcal{B}$. In order to prove the reverse containment, we argue by induction on the length of τ. The claim is trivially true for $\tau = \emptyset$. Now assume that the claim is true for shorter values of τ and assume, without loss of generality, that $\tau = \tau_1 T_r$ for some $\tau_1 \in \Theta^1$. Let $S \in \mathcal{V}^\tau \cap \mathcal{B} = (\mathcal{V}^{\tau_1})^{T_r} \cap \mathcal{B}$. By Theorem 2.5.2(vi), $S/R^0 \in \mathcal{V}^{\tau_1}$. Since S is a band, R^0 must be a congruence over $\mathcal{RZ}$ which implies that $R^0 = \tau_S \cap R^0$. Hence $S/(\tau_S \cap R^0) \in \mathcal{V}^{\tau_1} \cap \mathcal{B} = (\mathcal{V} \cap \mathcal{B})^{\tau_1(K_\ell, K_r)}$ by the induction hypothesis. By Theorem 2.6.2(vi) this implies that

$$S \in \mathcal{RZ} \circ \mathcal{V}^{\tau_1(K_\ell, K_r)} = \mathcal{V}^{\tau_1(K_\ell, K_r)K_r} = \mathcal{V}^{\tau(K_\ell, K_r)}.$$

Thus $\mathcal{V}^\tau \cap \mathcal{B} \subseteq (\mathcal{V} \cap \mathcal{B})^{\tau(K_\ell, K_r)}$ and equality prevails. □

Notation 3.2.6 For any word $\tau \in \Theta$, we denote by τ^d the word obtained from τ by replacing every occurrence of T_ℓ in τ by T_r and every occurrence of T_r in τ by T_ℓ.

For example, if $\tau = T_r T_\ell T_r$, then $\tau^d = T_\ell T_r T_\ell$. Note that if $|\tau| = n$, then τ and τ^d are the only words in Θ of length n. We will require the following basic result from lattice theory.

Lemma 3.2.7 Let L and M be lattices and $\theta : L \to M$, $\varphi : M \to L$ be mappings such that

(i) θ and φ are order preserving.   (ii) $\theta\varphi = \iota_L$.   (iii) φ is one-to-one.

Then θ and φ are mutually inverse isomorphisms.

Proof Condition (ii) implies that θ is injective and that φ is surjective. Let $b \in M$. Then by part (ii), we get $(b\varphi\theta)\varphi = (b\varphi)\theta\varphi = b\varphi$. But φ is one-to-one and so $(b\varphi)\theta = b$. Hence $\varphi\theta = \iota_M$, θ is also surjective and θ, φ are both bijections and are clearly inverse bijections.

Now let $a, b \in L$. Since θ is order preserving, we have $(a \vee b)\theta \geq a\theta \vee b\theta$. Suppose that $(a \vee b)\theta > a\theta \vee b\theta$. Then

$$
\begin{aligned}
a \vee b = (a \vee b)\theta\varphi &> (a\theta \vee b\theta)\varphi \\
&\geq a\theta\varphi \vee b\theta\varphi \qquad \text{since } \varphi \text{ is order preserving} \\
&= a \vee b
\end{aligned}
$$

which is a contradiction. Hence $(a \vee b)\theta = a\theta \vee b\theta$. Dually, $(a \wedge b)\theta = a\theta \wedge b\theta$ and so θ is a lattice isomorphism. Similarly φ is a lattice isomorphism so that θ and φ are mutually inverse isomorphisms. □

We are now ready for the main result of this section.

Theorem 3.2.8 For $\mathcal{V} \in [\mathcal{S}, \mathcal{CR}]$ define a mapping $\delta_{\mathcal{V}} : \Theta^1 \to D$ by

$$
\tau\delta_{\mathcal{V}} = \begin{cases}
1 & \text{if } \tau = \emptyset \text{ or } \mathcal{V}_\tau \notin \mathbb{N}_3 \\
L & \text{if } \mathcal{V}_\tau = \mathcal{LNB}, \tau \neq \emptyset \\
R & \text{if } \mathcal{V}_\tau = \mathcal{RNB}, \tau \neq \emptyset \\
T & \text{if } \mathcal{V}_\tau = \mathcal{S}, \qquad \tau \neq \emptyset.
\end{cases}
$$

For each $\psi \in \Phi_D$, define $\psi\xi_D \in \mathcal{L}(\mathcal{B})$ by

$$
\psi\xi_D = \left(\bigcap_{\tau\psi=1} \mathcal{B} \right) \cap \left(\bigcap_{\tau\psi=L} \mathcal{LNB}^{\overline{\tau}} \right) \cap \left(\bigcap_{\tau\psi=R} \mathcal{RNB}^{\overline{\tau}} \right) \cap \left(\bigcap_{\tau\psi=T} \mathcal{S}^{\overline{\tau}} \right).
$$

(i) The mapping $\delta : \mathcal{V} \to \delta_{\mathcal{V}}$ is a complete homomorphism of $[\mathcal{S}, \mathcal{CR}]$ onto Φ_D.
(ii) The restriction δ^* of δ to $[\mathcal{S}, \mathcal{B}]$ and the mapping $\xi_D : \psi \to \psi\xi_D$ are mutually inverse isomorphisms between the lattices $[\mathcal{S}, \mathcal{B}]$ and Φ_D.

(iii) *For* $V \in [8, C\mathcal{R}]$, $V\delta\xi_D = V \cap B$ *and the mapping* $\delta\xi_D$ *is a complete retraction of* $[8, C\mathcal{R}]$ *onto* $[8, B]$.

(iv) *Let* $V \in [8, C\mathcal{R}]$. *Then*

$$V \cap B = B \cap \left(\bigcap_{V_\tau = \mathcal{LNB}} \mathcal{LNB}^{\bar{\tau}} \right) \cap \left(\bigcap_{V_\tau = \mathcal{RNB}} \mathcal{RNB}^{\bar{\tau}} \right) \cap \left(\bigcap_{V_\tau = 8} 8^{\bar{\tau}} \right)$$

$$= \begin{cases} B & \text{if } B \subseteq V \\ \left(\bigcap_{V_\tau = \mathcal{LNB}} \mathcal{LNB}^{\bar{\tau}(K_\ell, K_r)} \right) \cap \left(\bigcap_{V_\tau = \mathcal{RNB}} \mathcal{RNB}^{\bar{\tau}(K_\ell, K_r)} \right) \cap \left(\bigcap_{V_\tau = 8} 8^{\bar{\tau}(K_\ell, K_r)} \right) \\ \qquad\qquad\qquad\qquad otherwise. \end{cases}$$

(v) *The mapping* $\mu_B : V \mapsto V \cap B$ *is a complete retraction of* $\mathcal{L}(C\mathcal{R})$ *onto* $\mathcal{L}(B)$.

Proof The steps of the proof are as follows:

(A) $\delta_V \in \Phi_D$.
(B) δ is order preserving.
(C) δ is a complete homomorphism.
(D) ξ_D is order preserving.
(E) The restriction δ^* of δ to $[8, B]$ is injective.
(F) $\xi_D \delta^* = \iota_{\Phi_D}$.

The proof will then follow by parts.

(A) $\delta_V \in \Phi_D$.

(B1) is satisfied by definition. Let $\sigma, \tau \in \Theta^1$ and $\sigma < \tau$, that is, $|\sigma| > |\tau|$.

Case: $\tau = \emptyset$. Then $\tau \chi_V = 1 \geq \sigma \chi_V$.

Case: $\tau \in \Theta$, $V_\tau \notin \mathbb{N}_3$. Again $\tau \chi_V = 1 \geq \sigma \chi_V$.

Case: $\tau \in \Theta$, $V_\tau \in \{8, \mathcal{LNB}, \mathcal{RNB}\}$. If $V_\tau = 8$, then also $V_\sigma = 8$ and $\tau\delta_V = T = \sigma\delta_V$. If $V_\tau = \mathcal{LNB}$, then $V_\sigma \subseteq V_\tau$ so that $V_\sigma \in \{8, \mathcal{LNB}\}$. Thus $\tau\delta_V = L$ while $\sigma\delta_V \in \{T, L\}$ which implies that $\tau\delta_V \geq \sigma\delta_V$. A dual argument applies when $\tau\delta_V = \mathcal{RNB}$. Thus (B2) is satisfied.

Now let $\tau \in \Theta$ be such that $\tau\delta_V = L$. Then, by the definition of δ_V, we have $V_\tau = \mathcal{LNB}$. Suppose that $t(\tau) = T_\ell$ and let $\tau = \tau_1 T_\ell$ where $\tau_1 \in \Theta^1$. Then $V_\tau = (V_{\tau_1})_{T_\ell}$ is the least element of its T_ℓ-class. But $\mathcal{LNB}$ is not the least element of its T_ℓ-class. Thus we have a contradiction and we must have $t(\tau) = T_r$. By duality (B3) is satisfied and therefore $\delta_V \in \Phi_D$.

(B) δ *is order preserving.* Let $\mathcal{U}, V \in [8, C\mathcal{R}]$ and $\mathcal{U} \subseteq V$. First $\emptyset\delta_\mathcal{U} = 1 = \emptyset\delta_V$. Now let $\tau \in \Theta$. Then $\mathcal{U}_\tau \subseteq V_\tau$. If $\mathcal{U}_\tau \notin \mathbb{N}_3$, then $V_\tau \notin \mathbb{N}_3$ and $\tau\delta_\mathcal{U} = 1 = \tau\delta_V$. So suppose that $\mathcal{U}_\tau = \mathcal{LNB}$. Then $\mathcal{LNB} = \mathcal{U}_\tau \subseteq V_\tau$ so that either $\tau\delta_V = 1$ or $\tau\delta_V = L$. In either case $\tau\delta_V \geq L = \tau\delta_\mathcal{U}$. A dual argument applies to the case $\mathcal{U}_\tau = \mathcal{RNB}$. For the last case we have

$\mathcal{U}_\tau = \mathcal{S} \Rightarrow \tau\delta u = T \leq \tau\delta v$. Thus $\tau\delta u \leq \tau\delta v$ for all $\tau \in \Theta^1$; in other words, $\delta u \leq \delta v$ and δ is order preserving.

(C) δ *is a complete homomorphism.* Let $\{\mathcal{U}_\alpha \mid \alpha \in A\}$ be a family of varieties in $[\mathcal{S}, \mathcal{CR}]$. To establish that $\bigvee_{\alpha \in A} \delta u_\alpha = \delta \bigvee_{\alpha \in A} u_\alpha$, let $\tau \in \Theta^1$.

Case: $\tau \bigvee_{\alpha \in A} \delta u_\alpha = 1$. Then $\bigvee_{\alpha \in A} \tau\delta u_\alpha = 1$ and there are two possibilities. Either there exists $\beta \in A$ with $\tau\delta u_\beta = 1$ or there exist $\beta, \gamma \in A$ with $\tau\delta u_\beta = L$, $\tau\delta u_\gamma = R$. However, because the δu_α satisfy (B3), the latter is impossible. Since δ is order preserving, we have $\tau\delta \bigvee_{\alpha \in A} u_\alpha \geq \tau\delta u_\beta$ so that $\tau\delta \bigvee_{\alpha \in A} u_\alpha = 1$.

Case: $\tau \bigvee_{\alpha \in A} \delta u_\alpha = L$. Then $\bigvee_{\alpha \in A} \tau\delta u_\alpha = L$ so that $\tau\delta u_\alpha \in \{T, L\}$ for all $\alpha \in A$ and for at least one $\beta \in A$ we have $\tau\delta u_\beta = L$. In other words, $(\mathcal{U}_\alpha)_\tau \in \{\mathcal{S}, \mathcal{LNB}\}$ and there exists $\beta \in A$, such that $(\mathcal{U}_\beta)_\tau = \mathcal{LNB}$. Hence $\left(\bigvee_{\alpha \in A} \mathcal{U}_\alpha\right)_\tau = \bigvee_{\alpha \in A} (\mathcal{U}_\alpha)_\tau = \mathcal{LNB}$ so that $\tau\delta \bigvee_{\alpha \in A} u_\alpha = L$.

Case: $\tau \bigvee_{\alpha \in A} \delta u_\alpha = T$. Then necessarily, $\tau\delta u_\alpha = T$ for all $\alpha \in A$ so that $(\mathcal{U}_\alpha)_\tau = \mathcal{S}$ for all $\alpha \in A$. Hence $(\bigvee_{\alpha \in A} \mathcal{U}_\alpha)_\tau = \bigvee_{\alpha \in A} (\mathcal{U}_\alpha)_\tau = \mathcal{S}$ and we have $\tau\delta \bigvee_{\alpha \in A} u_\alpha = T$. Thus $\tau \bigvee_{\alpha \in A} \delta u_\alpha = \tau\delta \bigvee_{\alpha \in A} u_\alpha$ for all τ so that $\bigvee_{\alpha \in A} \delta u_\alpha = \delta \bigvee_{\alpha \in A} u_\alpha$.

Now consider $\bigwedge_{\alpha \in A} \delta u_\alpha$.

Case: $\tau(\bigwedge_{\alpha \in A} \delta u_\alpha) = 1$. This implies that $\tau\delta u_\alpha = 1$ or, equivalently, $(\mathcal{U}_\alpha)_\tau \notin \mathbb{N}_3$, for all $\alpha \in A$. Now assume that $t(\tau) = T_r$. Since $(\mathcal{U}_\alpha)_\tau$ is then the least element in its T_r-class, we must have $\mathcal{LRB} \subseteq (\mathcal{U}_\alpha)_\tau$ and therefore $\mathcal{LRB} \subseteq \bigwedge_{\alpha \in A} (\mathcal{U}_\alpha)_\tau = (\bigwedge_{\alpha \in A} \mathcal{U}_\alpha)_\tau$. Thus $\tau\delta \bigwedge_{\alpha \in A} u_\alpha \notin \mathbb{N}_3$ so that $\tau\delta \bigwedge_{\alpha \in A} u_\alpha = 1$. A dual argument applies when $t(\tau) = T_\ell$.

Case: $\tau(\bigwedge_{\alpha \in A} \delta u_\alpha) = L$. This implies that $\tau\delta u_\alpha \in \{1, L\}$ for all $\alpha \in A$ and that $\tau\delta u_\beta = L$ for at least one $\beta \in A$. The latter implies that $t(\tau) = T_r$. Hence

$$(\mathcal{U}_\alpha)_\tau \in [\mathcal{LRB}, \mathcal{CR}] \cup \{\mathcal{LNB}\}$$

and, in particular, $(\mathcal{U}_\beta)_\tau = \mathcal{LNB}$. Consequently $(\bigwedge_{\alpha \in A} \mathcal{U}_\alpha)_\tau = \bigwedge_{\alpha \in A} (\mathcal{U}_\alpha)_\tau = \mathcal{LNB}$. Therefore $\tau\delta \bigwedge_{\alpha \in A} u_\alpha = L = \tau(\bigwedge_{\alpha \in A} \delta u_\alpha)$.

Case: $\tau(\bigwedge_{\alpha \in A} \delta u_\alpha) = T$. Since $t(\tau)$ must equal either T_ℓ or T_r, it is not possible for there to be $\beta, \gamma \in A$ with $\tau\delta u_\beta = L$, $\tau\delta u_\gamma = R$. Hence there must exist $\beta \in A$ with $\tau\delta u_\beta = T$ and $(\mathcal{U}_\beta)_\tau = \mathcal{S}$. But then $(\bigwedge_{\alpha \in A} \mathcal{U}_\alpha)_\tau = \bigwedge_{\alpha \in A} (\mathcal{U}_\alpha)_\tau = \mathcal{S}$ so that $\tau\delta \bigwedge_{\alpha \in A} u_\alpha = T$. Hence we have $\bigwedge_{\alpha \in A} \delta u_\alpha = \delta \bigwedge_{\alpha \in A} u_\alpha$ and δ is a complete homomorphism.

(D) ξ_D *is order preserving.* Let $\psi_1, \psi_2 \in \Phi_D$ and $\psi_1 \leq \psi_2$. For any $\psi \in \Phi_D$, it will be convenient to refer to

$$A = \bigcap_{\tau\psi=1} \mathcal{B}, \quad B = \bigcap_{\tau\psi=L} \mathcal{LNB}^\tau, \quad C = \bigcap_{\tau\psi=R} \mathcal{RNB}^\tau, \quad D = \bigcap_{\tau\psi=T} \mathcal{S}^\tau$$

as the *component* of $\psi\xi_D$. Note that, for any $\tau \in \Theta^1$, τ contributes to exactly one of the components A, B, C, D. Let $\tau \in \Theta^1$ and consider its contribution to $\psi_1\xi_D$ and $\psi_2\xi_D$.

Case: $\tau\psi_1 = 1$. Then $\tau\psi_2 = 1$ and τ contributes to the A component in both $\tau\psi_1$ and $\tau\psi_2$.

Case: $\tau \psi_1 = L$. Then $\tau \psi_2 \in \{1, L\}$ and τ contributes to the B component in $\psi_1 \xi_D$ and either the A or B component in $\tau \psi_2$.

Case: $\tau \psi_1 = R$. This is the dual of the preceding case.

Case: $\tau \psi_1 = T$. Then τ contributes to the D component in $\psi_1 \xi_D$.

Noting that $S^{\tau} \subseteq B \cap LNB^{\tau} \cap RNB^{\tau}$ we see that the contribution of τ to the components of $\psi_2 \xi_D$ is, in all cases, at least as great as the contribution of τ to the components of $\psi_1 \xi_D$. Thus $\psi_1 \xi_D \subseteq \psi_2 \xi_D$ and ξ_D is order preserving.

(E) δ^* *is injective*. Let $U, V \in [S, B]$ be such that $\delta_U = \delta_V$. Then, for $\tau \in \Theta$,

$$U_{\tau} \in \mathbb{N}_3 \iff \tau \delta_U \in \{L, T, R\} \iff \tau \delta_V \in \{L, T, R\} \iff V_{\tau} \in \mathbb{N}_3.$$

Also,

$$U_{\tau} = LNB \iff \tau \delta_U = L \iff \tau \delta_V = L \iff V_{\tau} = LNB$$

and similar implications hold for $U_{\tau} = S$ or $U_{\tau} = RNB$. Thus

$$U_{\tau} \in \mathbb{N}_3 \iff V_{\tau} \in \mathbb{N}_3 \text{ and, if } U_{\tau}, V_{\tau} \in \mathbb{N}_3, \text{ then } U_{\tau} = V_{\tau}.$$

Hence with $\pi \in \Pi$, $\tau = \pi^*$, by Lemma 3.1.9, we have

$$U_{\tau} = LNB \implies V_{\tau} = LNB \implies \zeta_U^{\pi} = \zeta_V^{\pi}.$$

Similarly

$$U_{\tau} = RNB \implies \zeta_U^{\pi} = \zeta_V^{\pi},$$
$$U_{\tau} = S \qquad \implies \zeta_U^{\pi} = \zeta_V^{\pi}.$$

In all cases, $U_{\tau} \in \mathbb{N}_3 \implies \zeta_U^{\pi} = \zeta_V^{\pi}$. Consequently, for $u, v \in U$ by Corollary 3.1.8, we obtain

$$u \, \zeta_U \, v \iff \begin{cases} \text{(i) } u\pi \, \eta \, v\pi & \text{for } \pi = \emptyset \text{ and } \pi \in \Pi \text{ with } U_{\pi^*} \notin \mathbb{N}_3, \\ \text{(ii) } u\pi \, \zeta_U^{\pi} \, v\pi & \text{for all } \pi \in \Pi \text{ with } U_{\pi^*} \in \mathbb{N}_3 \end{cases}$$

$$\iff \begin{cases} \text{(i) } u\pi \, \eta \, v\pi & \text{for } \pi = \emptyset \text{ and } \pi \in \Pi \text{ with } V_{\pi^*} \notin \mathbb{N}_3 \\ \text{(ii) } u\pi \, \zeta_V^{\pi} \, v\pi & \text{for all } \pi \in \Pi \text{ with } V_{\pi^*} \in \mathbb{N}_3. \end{cases}$$

$$\iff u \, \zeta_V \, v.$$

Therefore $U = V$.

(F) $\xi_D \delta^* = \iota_{\Phi_D}$. Let $\psi \in \Phi_D$. Our goal is to show that $\psi = \psi \xi_D \delta^* = \delta^*_{\psi \xi_D}$.

By the definition of Φ_D, we have $\emptyset \psi = 1$. Hence the A component in $\psi \xi_D$ is nonempty, that is, includes at least one B. Therefore $\psi \xi_D \in \mathcal{L}(B)$. In addition, the components A, B, C and D, when they come into play, all contain S. Consequently ξ_D maps Φ_D into $[S, B]$.

Since ψ *and* $\delta^*_{\psi\xi_D}$ are both mappings from Θ^1 to D, we need to show that $\sigma\psi = \sigma\delta^*_{\psi\xi_D} = \sigma\delta_{\psi\xi_D}$ for all $\sigma \in \Theta^1$. We consider the possible cases. Since $\emptyset\psi = 1 = \emptyset\delta^*_{\psi\xi_D}$ we need only consider $\sigma \in \Theta$.

Case: $\sigma\psi = 1$. Without loss of generality we may assume that $t(\sigma) = T_r$. We have

$$\sigma\delta^*_{\psi\xi_D} = 1 \iff (\psi\xi_D)_\sigma \notin \mathbb{N}_3.$$

So we consider the contribution to $(\psi\xi_D)_\sigma$ from each $\tau \in \Theta^1$. Since

$$(\psi\xi_D)_\sigma = (A \cap B \cap C \cap D)_\sigma = A_\sigma \cap B_\sigma \cap C_\sigma \cap D_\sigma$$

we must track the components to which $\tau\psi$ contributes for each $\tau \in \Theta$.

Subcase: $\sigma\psi = 1$, $|\tau| > |\sigma|$. Since $\sigma \in \Theta$ and $|\tau| > |\sigma|$ it follows that $|\tau| \geq 2$. We treat the different possible values of $\tau\psi$ and the consequent contributions of $\psi\xi_D$ to the components A, B, C, D.

Case: $\tau\psi = 1$. Then there is a contribution of $\mathcal{B}$ to component A and of $\mathcal{B}_\sigma = \mathcal{B}$ to $(\psi\xi_D)_\sigma$.

$\tau\psi = L$: Then $t(\tau) = T_r$ and there is a contribution of $(\mathcal{LNB})^{\overline{\tau}}$ to B and of $((\mathcal{LNB})^{\overline{\tau}})_\sigma$ to $(\psi\xi_D)_\sigma$. Since $t(\sigma) = t(\tau) = T_r$ and $|\tau| > |\sigma|$, it follows that $|\tau| \geq |\sigma| + 2$ and, from Lemma 3.1.3, that $((\mathcal{LNB})^{\overline{\tau}})_\sigma \supseteq \mathcal{LNB}^{T_r T_\ell} \supseteq \mathcal{LRB}$.

$\tau\psi = R$: Then $t(\tau) = T_\ell$ and in a manner similar to the previous case:

$$\left((\mathcal{RNB})^{\overline{\tau}}\right)_\sigma \supseteq (\mathcal{RNB})^{T_\ell} \supseteq \mathcal{LRB}.$$

$\tau\psi = T$: If $t(\overline{\tau}) = h(\tau) \neq h(\sigma)$, then

$$\left(\mathcal{S}^{\overline{\tau}}\right)_\sigma \supseteq \mathcal{S}^{\tau_1} \qquad \text{where } \tau_1 \in \Theta \text{ and } |\tau_1| \geq 2$$
$$\supseteq \mathcal{LRB}.$$

If $t(\overline{\tau}) = h(\tau) = h(\sigma)$, then $\tau = \sigma T_\ell \tau_2$ for some $\tau_2 \in \Theta^1$ so that

$$\left(\mathcal{S}^{\overline{\tau}}\right)_\sigma = \mathcal{S}^{T_\ell \tau_2} \supseteq \mathcal{S}^{T_\ell} \supseteq \mathcal{LRB}.$$

Subcase: $\sigma\psi = 1$, $|\tau| = |\sigma|$, $\tau = \sigma$. Then $\tau\psi = \sigma\psi = 1$ so that τ can only contribute to A and $\mathcal{B}_\sigma = \mathcal{B} \supseteq \mathcal{LRB}$.

Subcase: $\sigma\psi = 1$, $|\tau| = |\sigma|$, $\tau = \sigma^d$. Since $t(\sigma) = T_r$, it follows that $h(\overline{\tau}) = t(\tau) = T_\ell$ and $t(\overline{\tau}) = h(\tau) \neq h(\sigma)$. Hence $(\mathcal{S}^{\overline{\tau}})_\sigma \supseteq \mathcal{S}^{T_\ell} \supseteq \mathcal{LRB}$.

Subcase: $\sigma\psi = 1$, $|\tau| < |\sigma|$. Since ψ is order preserving, it follows that $\tau\psi = 1$. Hence τ can contribute only to the component A. Also $\mathcal{B}_\tau = \mathcal{B} \supseteq \mathcal{LRB}$.

Consequently, for all possible values of τ, the contribution of τ to the A, B, C or D component in this case contains $\mathcal{LRB}$. Hence $(\psi\xi_D)_\sigma \supseteq \mathcal{LRB}$, $(\psi\xi_D)_\sigma \notin \mathbb{N}_3$ and therefore $\sigma\delta^*_{\psi\xi_D} = 1 = \sigma\psi$.

Case: $\sigma\psi = L$. By (B3), $t(\sigma) = T_r$. Let $\tau \in \Theta^1$.

Subcase: $|\tau| < |\sigma|$, $\tau\psi = 1$. This contributes $\mathcal{B}$ to component A in $\psi\xi_D$ and $\mathcal{B}_\sigma = \mathcal{B}$.

Subcase: $|\tau| < |\sigma|$, $\tau\psi = L$. This contributes to component B in $\psi\xi_D$. By Lemma 3.2.2(ii), $|\tau| = |\sigma| - 1$ and by (B3) $t(\tau) = T_r$. Hence $\sigma = T_x\tau$ where $T_x \in \{T_\ell, T_r\}$ and $T_x \neq h(\tau) = t(\overline{\tau})$. We have

$$(\mathcal{LNB}^{\overline{\tau}})_\sigma = (\mathcal{LNB}^{\overline{\tau}})_{T_x\tau} = \mathcal{LNB}_{T_r} = \mathcal{LNB}.$$

Subcase: $|\tau| = |\sigma|$. Necessarily we have $\tau = \sigma$ or $\tau = \sigma^d$. If $\tau = \sigma$ then, since $\sigma\psi = L$ we have a contribution to component B:

$$(\mathcal{LNB}^{\overline{\tau}})_\sigma = (\mathcal{LNB}^{\overline{\sigma}})_\sigma = \mathcal{LNB}_{T_r} = \mathcal{LNB}.$$

Now assume that $\tau = \sigma^d$. Since $t(\tau) \neq t(\sigma) = T_r$, it follows that $t(\tau) = T_\ell$ and $\tau\psi \in \{1, R, T\}$. If $\tau\psi = 1$, then we have a contribution to component A.

With $\tau\psi = R$ we have a contribution to C and $(\mathcal{RNB}^{\overline{\tau}})_\sigma = \mathcal{RNB}^{T_\ell} \supseteq \mathcal{LNB}$. Now assume that $\tau\psi = T$. This gives a contribution to component D and $(S^{\overline{\tau}})_\sigma = S^{T_\ell} \supseteq \mathcal{LNB}$.

Subcase: $|\tau| > |\sigma|$. Then $\tau\psi \leq \sigma\psi$ so that $\tau\psi \in \{L, T\}$.

Suppose $\tau\psi = L$. This would contribute to component B. Also $t(\tau) = T_r$ so that $\tau = \tau_1\sigma$ for some $\tau_1 \in \Theta$. Consequently

$$(\mathcal{LNB}^{\overline{\tau}})_\sigma = (\mathcal{LNB}^{\overline{\sigma}\,\overline{\tau_1}})_\sigma \supseteq \mathcal{LNB}^{h(\overline{\sigma})} = \mathcal{LNB}^{T_r} \supseteq \mathcal{LNB}.$$

Now suppose that $\tau\psi = T$. This contributes to component D. There are two possibilities: Either $t(\tau) = T_r$ so that $\tau = \tau_1\sigma$ for some $\tau_1 \in \Theta$ where $t(\tau_1) \neq h(\sigma)$. Hence

$$(S^{\overline{\tau}})_\sigma = (S^{\overline{\sigma}\,\overline{\tau_1}})_\sigma \supseteq (S^{\overline{\sigma} t(\tau_1)})_\sigma = S^{T_r T_\ell} \supseteq \mathcal{LNB}.$$

Or, alternatively, $t(\tau) = T_\ell$ so that $\tau = \tau_1\sigma T_\ell$ for some $\tau_1 \in \Theta^1$. Then

$$(S^{\overline{\tau}})_\sigma = (S^{T_\ell \overline{\sigma}\,\overline{\tau_1}})_\sigma \supseteq S^{T_\ell} \supseteq \mathcal{LNB}.$$

Consequently $(\psi\xi_D)_\sigma$ is an intersection of components, the smallest of which is $\mathcal{LNB}$. Thus $(\psi\xi_D)_\sigma = \mathcal{LNB}$ and $\sigma\delta^*_{\psi\xi_D} = L = \sigma\psi$.

Case: $\sigma\psi = R$. This is the dual of the preceding case.

Case: $\sigma\psi = T$. Then there is a contribution of $S^{\overline{\sigma}}$ to component D so that $(\psi\xi_D)_\sigma \subseteq (S^{\overline{\sigma}})_\sigma = S$. Therefore $(\psi\xi_D)_\sigma = S$ and $\sigma\delta^*_{\psi\xi_D} = T = \sigma\psi$.

Thus $\sigma\psi = \sigma\delta^*_{\psi\xi_D}$ for all $\sigma \in \Theta^1$. Therefore $\psi = \delta^*_{\psi\xi_D} = \psi\xi_D\delta^*$ for all $\psi \in \Phi_D$ and so $\xi_D\delta^* = \iota_{\Phi_D}$ as required.

(i) By step (C), δ is a complete homomorphism of $[S, \mathcal{CR}]$ into Φ_D. By step (F), δ^* is surjective and therefore so also is δ.

(ii) By steps (A), (B) and (E) we know that δ^* is a one-to-one order preserving mapping of $[S, \mathcal{B}]$ into Φ_D. By the preamble in step (F), ξ_D maps Φ_D into $[S, \mathcal{B}]$ while by steps (D) and (F), ξ_D is one-to-one and order preserving. Also, from step (F) we have $\xi_D \delta^* = \iota_{\Phi_D}$. By Lemma 3.2.7, δ^* and ξ_D are mutually inverse isomorphisms.

(iii) Let $\mathcal{U} \in [\mathcal{S}, \mathcal{CR}]$, $\tau \in \Theta$ and suppose that $\mathcal{U}_\tau = \mathcal{LNB}$. Then

$$(\mathcal{U} \cap \mathcal{B})_\tau = \mathcal{U}_\tau \cap \mathcal{B}_\tau = \mathcal{LNB} \cap \mathcal{B} = \mathcal{LNB}.$$

Conversely, suppose that $(\mathcal{U} \cap \mathcal{B})_\tau = \mathcal{LNB}$. Then we must have $t(\tau) = T_r$ while $\mathcal{U}_\tau \cap \mathcal{B} = (\mathcal{U} \cap \mathcal{B})_\tau = \mathcal{LNB}$ which implies that $\mathcal{LRB} \nsubseteq \mathcal{U}_\tau$. By [PR1, Corollary IX.5.3(ii)], this implies that $\mathcal{U}_\tau \in [\mathcal{LNB}, \mathcal{LRRO}]$ which, in turn implies that $\mathcal{U}_\tau = \mathcal{LNB}$ since $\mathcal{U}_\tau$ is the least element in its T_r-class. A dual argument applies to the case of $\mathcal{RNB}$. In a similar fashion,

$$(\mathcal{U} \cap \mathcal{B})_\tau = \mathcal{S} \Longrightarrow \mathcal{U}_\tau \cap \mathcal{B} = \mathcal{S} \Longrightarrow \mathcal{LZ}, \mathcal{RZ} \nsubseteq \mathcal{U}_\tau \Longrightarrow \mathcal{U}_\tau \subseteq \mathcal{SG} \Longrightarrow \mathcal{U}_\tau = \mathcal{S}$$

while

$$\mathcal{U}_\tau = \mathcal{S} \Longrightarrow (\mathcal{U} \cap \mathcal{B})_\tau = \mathcal{U}_\tau \cap \mathcal{B} = \mathcal{S}.$$

Consequently we have $\mathcal{U}_\tau \in \mathbb{N}_3 \Leftrightarrow (\mathcal{U} \cap \mathcal{B})_\tau \in \mathbb{N}_3$ and, whenever $\mathcal{U}_\tau \in \mathbb{N}_3$, we have $\mathcal{U}_\tau = (\mathcal{U} \cap \mathcal{B})_\tau$. Hence $\mathcal{U}\delta = (\mathcal{U} \cap \mathcal{B})\delta$. Therefore

$$\mathcal{U}\delta\xi_D = (\mathcal{U} \cap \mathcal{B})\delta\xi_D = (\mathcal{U} \cap \mathcal{B})\delta^*\xi_D = \mathcal{U} \cap \mathcal{B}$$

by part (ii).

Thus $\delta\xi_D$ is precisely the retraction mapping $\mathcal{U} \to \mathcal{U} \cap \mathcal{B}$. Since δ is a complete homomorphism and ξ_D is an isomorphism from Φ_D to $[\mathcal{S}, \mathcal{B}]$, it follows that the retraction mapping is a complete homomorphism of $[\mathcal{S}, \mathcal{CR}]$ onto $[\mathcal{S}, \mathcal{B}]$.

(iv) Let $\psi = \delta\gamma$. For $\tau \in \Theta^1$,

$$\tau\psi = 1 \Longleftrightarrow \tau\delta\gamma = 1 \Longleftrightarrow \tau = 1 \text{ or } \mathcal{V}_\tau \notin \mathbb{N}_3,$$
$$\tau\psi = L \Longleftrightarrow \mathcal{V}_\tau = \mathcal{LNB},$$
$$\tau\psi = R \Longleftrightarrow \mathcal{V}_\tau = \mathcal{RNB},$$
$$\tau\psi = T \Longleftrightarrow \mathcal{V}_\tau = \mathcal{S}.$$

Then, from part (iii),

$$\mathcal{V} \cap \mathcal{B} = \mathcal{V}\delta\xi_D = \psi\xi_D$$

$$= \left(\bigcap_{\tau=1,\, \mathcal{V}_\tau \notin \mathbb{N}_3} \mathcal{B} \right) \cap \left(\bigcap_{\mathcal{V}_\tau = \mathcal{LNB}} \mathcal{LNB}^{\overline{\tau}} \right) \cap \left(\bigcap_{\mathcal{V}_\tau = \mathcal{RNB}} \mathcal{RNB}^{\overline{\tau}} \right) \cap \left(\bigcap_{\mathcal{V}_\tau = \mathcal{S}} \mathcal{S}^{\overline{\tau}} \right)$$

and the first equality in (iv) holds. The second equality holds by Lemma 3.2.5.

(v) For any family $\{\mathcal{U}_\alpha \mid \alpha \in A\}, \mathcal{U}_\alpha \in \mathcal{L}(\mathcal{CR})$, we have, with some help from [PR1, Theorem IX.9.1], $(\bigvee_{\alpha \in A} \mathcal{U}_\alpha) \cap \mathcal{B} = \bigvee_{\alpha \in A} (\mathcal{U}_\alpha \cap \mathcal{B})$ trivially if $\mathcal{U}_\alpha \in \mathcal{L}(\mathcal{CS})$ for all $\alpha \in A$, and from the above if $\mathcal{U}_\alpha \in [\mathcal{S}, \mathcal{CR}]$, for all $\alpha \in A$. So suppose that there exist $\beta, \gamma \in A$ with $\mathcal{U}_\beta \in \mathcal{L}(\mathcal{CS})$ and $\mathcal{U}_\gamma \in [\mathcal{S}, \mathcal{CR}]$. Then

$$\left(\bigvee_{\alpha \in A} \mathcal{U}_\alpha\right) \cap \mathcal{B} = \left(\bigvee_{\alpha \in A} (\mathcal{U}_\alpha \vee \mathcal{S})\right) \cap \mathcal{B}$$

$$= \bigvee_{\alpha \in A} (\mathcal{U}_\alpha \vee \mathcal{S}) \cap \mathcal{B}$$

$$= \bigvee_{\alpha \in A} (\mathcal{U}_\alpha \vee \mathcal{S}) \cap (\mathcal{B} \vee \mathcal{S})$$

$$= \bigvee_{\alpha \in A} ((\mathcal{U}_\alpha \cap \mathcal{B}) \vee \mathcal{S}) \qquad \text{by [PR1, Proposition IX.9.2]}$$

$$= \left(\bigvee_{\alpha \in A} (\mathcal{U}_\alpha \cap \mathcal{B})\right) \vee \mathcal{S}$$

$$= \bigvee_{\alpha \in A} (\mathcal{U}_\alpha \cap \mathcal{B}).$$

Thus (v) holds. $\qquad\qquad\square$

We conclude this section with a conceptually valuable presentation of the elements of Φ_D. Let $\psi \in \Phi_D$. Since $\Phi_D \subseteq D^{\Theta^1}$, the values $\tau\psi$, $\tau \in \Theta^1$, of ψ can be displayed in a diagram that simply adds labels from D to the vertices of the diagram for Θ^1 in Sect. 3.1:

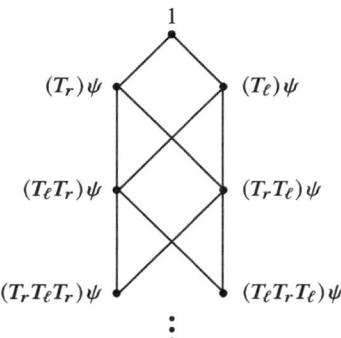

Definition 3.2.9 We refer to the above diagram as the *ladder* of ψ. By Theorem 3.2.8(ii), each $\psi \in \Phi_D$ is of the form $\delta_\mathcal{V}$ for some unique $\mathcal{V} \in [\mathcal{S}, \mathcal{B}]$. So we also refer to the ladder $\delta_\mathcal{V}$ as the *ladder* of $\mathcal{V}$.

For example, the ladders for $\mathcal{S}, \mathcal{B}, \mathcal{LRB}, \mathcal{NB}$ are in the nearby diagram.

Note that, because of condition (B3), L (respectively, R) can only appear on the left side (respectively, right side) of the ladder of $\psi \in \Phi_D$. Also all vertices in a ladder below a vertex labelled T must also be labelled T since ψ is order preserving. Thus each ladder has three regions.

Definition 3.2.10 (i) The elements of a ladder are arranged in *rows* or *layers* which we number from the topmost layer, numbered 0, containing just the single entry 1. All other layers contain two elements.

(ii) The set of layers below layer 0 with all entries equal to 1 is the *upper part* of the ladder.

(iii) The layers containing some, but possibly not all, entries labelled in $\{L, R\}$ are the *boundary layers* and constitute the *middle part*.

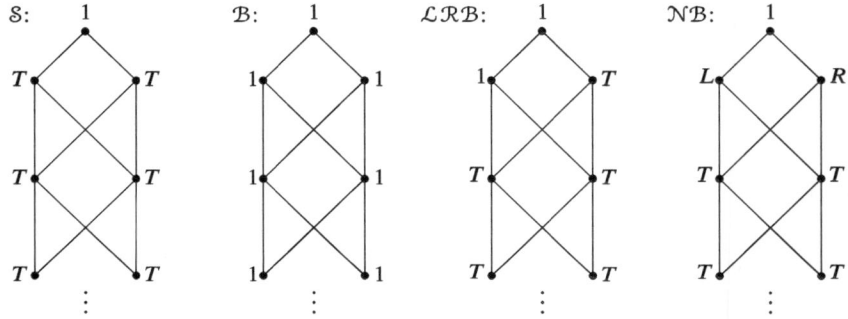

(iv) The remaining part, labelled entirely with T, is the *lower part*.

In the ladders displayed above, that for $\mathcal{S}$ has no upper or middle part, that for $\mathcal{B}$ has no middle or lower part, while that for $\mathcal{LRB}$ and that for $\mathcal{NB}$ have a middle and a lower part, but no upper part.

The material in this section is mostly a specialization to bands from Polák [1] to put the treatment of bands on a more independent footing. Theorem 3.2.8(v) is due to Trotter [1]. See also Reilly and Zhang [2, 4]. Sizer [1] characterizes bands that have a faithful representation over fields.

3.3   The Lattice $\mathcal{L}(\mathcal{B})$

In this section we describe the lattice $\mathcal{L}(\mathcal{B})$ and develop some of its basic properties.

Since δ^* in Theorem 3.2.8 is an isomorphism, it follows that for any proper subvariety $\mathcal{V}$ of $\mathcal{B}$ there must be some $\tau \in \Theta$ such that $\tau \delta_{\mathcal{V}} < \tau \delta_{\mathcal{B}} = 1$. Therefore we must have $\tau \delta_{\mathcal{V}} \in \{L, T, R\}$ for some $\tau \in \Theta$. Consider what happens as soon as one of the values in

$\{L, T, R\}$, say L, appears in the ladder. By Lemma 3.2.2 we must then have $(\tau T_\ell)\delta\gamma = T$ and therefore $\sigma\delta\gamma = T$ for all $\sigma \in \Theta$ with $|\sigma| \geq |\tau| + 2$.

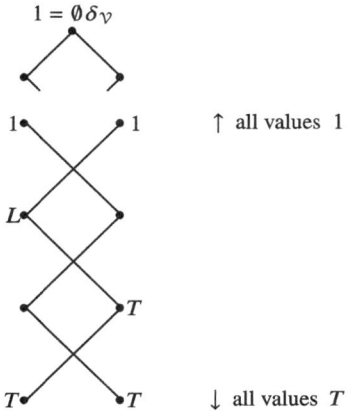

Thus in studying the subvarieties of $\mathcal{B}$, it is critical to consider the level at which the elements L, T and R might start to appear and the ensuing patterns for that and the next level, that is, the *boundary* layers . It is easy to see, by Lemma 3.2.2, that below these boundary layers all values are T. There are exactly ten possible patterns for these boundary layers that satisfy conditions (B1)–(B3) and these are displayed with the labels $E_1, \ldots E_{10}$.

$$E_9 = \begin{pmatrix} 1 & R \\ T & R \end{pmatrix} \qquad\qquad E_{10} = \begin{pmatrix} L & 1 \\ L & T \end{pmatrix}$$

$$E_7 = \begin{pmatrix} 1 & R \\ T & T \end{pmatrix} \qquad E_8 = \begin{pmatrix} L & 1 \\ T & T \end{pmatrix}$$

$$E_4 = \begin{pmatrix} 1 & T \\ T & T \end{pmatrix} \qquad E_5 = \begin{pmatrix} L & R \\ T & T \end{pmatrix} \qquad E_6 = \begin{pmatrix} T & 1 \\ T & T \end{pmatrix}$$

$$E_2 = \begin{pmatrix} L & T \\ T & T \end{pmatrix} \qquad E_3 = \begin{pmatrix} T & R \\ T & T \end{pmatrix}$$

$$E_1 = \begin{pmatrix} T & T \\ T & T \end{pmatrix}$$

Note that in these boundary layers there are always at most two positions to fill with L or R.

Let $E_{n,k}$ denote the variety that has the pattern E_k at the nth level (the left diagram). For a fixed $n \geq 1$, the $E_{n,k}$ relate to each other as in the right diagram D_n.

It is natural to expect that these diagrams can be assembled like the pieces of a jig-saw puzzle to represent the interval $[\mathcal{S}, \mathcal{B})$. The correct way to proceed will become evident if we start with the nine varieties in the interval $[\mathcal{S}, \mathcal{R}e\mathcal{B}]$. This was completely described in

[PR1, Chapter V] and displayed in [PR1, Diagram V.1.8]. From Theorem 1.5.5 and its dual, it is straightforward to see that the varieties in each of the sets

$$\{\mathcal{S}, \mathcal{LNB}, \mathcal{LRB}\}, \quad \{\mathcal{RNB}, \mathcal{NB}, \mathcal{LQNB}\}, \quad \{\mathcal{RRB}, \mathcal{RQNB}, \mathcal{ReB}\}$$

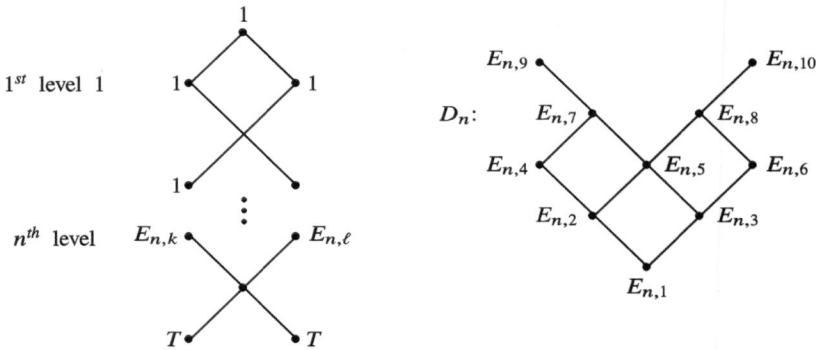

consists of T_ℓ-related varieties, while each of the sets

$$\{\mathcal{S}, \mathcal{RNB}, \mathcal{RRB}\}, \quad \{\mathcal{LNB}, \mathcal{NB}, \mathcal{RQNB}\}, \quad \{\mathcal{LRB}, \mathcal{LQNB}, \mathcal{ReB}\}$$

consists of T_r-related varieties. From this we obtain the diagram of their ladders as displayed, where only the top two or three labelled steps are displayed, all other labels being T.

$$
\begin{array}{ccccccccc}
 & & & & 1 & & & & \\
 & & & 1 & & 1 & & & \\
 & & & T & & T & & & \\
 & & 1 & & & & 1 & & \\
 & & 1 & R & & L & & 1 & \\
 & & T & T & & T & & T & \\
 & 1 & & & 1 & & & 1 & \\
 & 1 & T & & L & R & & T & 1 \\
 & T & T & & T & T & & T & T \\
 & & 1 & & & & 1 & & \\
 & & L & T & & T & R & & \\
 & & T & T & & T & T & & \\
 & & & & 1 & & & & \\
 & & & T & & T & & & \\
\end{array}
$$

Thus the ladders of the lower eight of these varieties coincide with the Polák ladders for the lower eight vertices in D_1. Consequently Diagram D_1 must represent the base of the interval $[\mathcal{S}, \mathcal{B})$. Close inspection of the ladders in D_2 now reveals that we also need to consider the left/right dual D_2' of D_2 (Diagram 3.1).

The diagrams D_1 and D_2' then fit together as in Diagram 3.2.

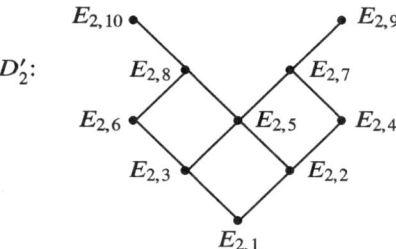

Diagram 3.1 Dual of D_2

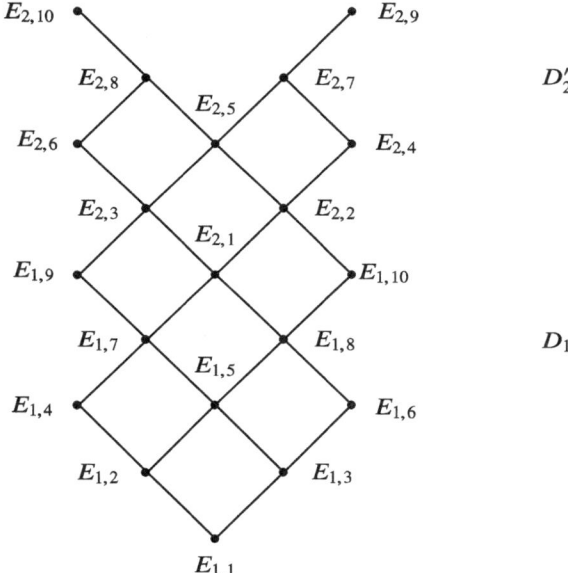

Diagram 3.2 D_1 and D'_2 combined

Continuing in this way, by stacking the diagrams $D_1, D'_2, D_3, D'_4, \ldots$, we obtain the lattice $[\mathcal{S}, \mathcal{B})$, to which we might add the variety $\mathcal{B}$ at the top thereby completing the interval $[\mathcal{S}, \mathcal{B}]$.

Note that all these varieties corresponding to ladders of the form $E_{n,k}$ are distinct for distinct pairs (n, k) by Theorem 3.2.8 because they have distinct ladders.

The only varieties of bands not containing $\mathcal{S}$ are the varieties $\mathcal{T}, \mathcal{LZ}, \mathcal{RZ}$ and $\mathcal{RB}$ (see [PR1, Chapter V]). Consequently blending the lattice $[\mathcal{S}, \mathcal{B}]$ with the lattice of subvarieties of $\mathcal{ReB}$ from [PR1, Diagram V.1.8], we obtain the full lattice $\mathcal{L}(\mathcal{B})$.

One feature of $\mathcal{L}(\mathcal{B})$ deserves highlighting immediately.

Definition 3.3.1 An element a in a lattice L is $\vee$-*irreducible* (respectively, $\wedge$-*irreducible* if there are no elements $b, c \in L$ such that $b, c < a$ and $a = b \vee c$ (respectively, $a < b, c$ and $a = b \wedge c$).

Note that since $\mathcal{L}(\mathcal{B})$ is an ideal in the lattice $\mathcal{L}(\mathcal{CR})$, a variety $\mathcal{V} \in \mathcal{L}(\mathcal{B})$ is join irreducible in $\mathcal{L}(\mathcal{B})$ if and only if it is join irreducible in $\mathcal{L}(\mathcal{CR})$.

Theorem 3.3.2 (i) *The lattice $\mathcal{L}(\mathcal{B})$ is as depicted in Diagram 3.3.*

(ii) *The varieties of bands that are $\vee$-irreducible varieties in $\mathcal{L}(\mathcal{CR})$ are $\mathcal{T}$, $\mathcal{LZ}$, $\mathcal{RZ}$, $\mathcal{S}$, $\mathcal{B}$ and all varieties in the extreme left and extreme right hand columns in Diagram 3.3.*

(iii) *The varieties of bands that are $\wedge$-irreducible in $\mathcal{L}(\mathcal{B})$ are $\mathcal{RB}$ and all the varieties in the extreme left and extreme right hand columns in Diagram 3.3.*

Proof (i) This follows from the discussion prior to the theorem.

(ii) This follows easily from direct inspection of Diagram 3.3 and the fact that $\mathcal{L}(\mathcal{B})$ is an ideal in the lattice $\mathcal{L}(\mathcal{CR})$.

(iii) This follows by inspection of Diagram 3.3.   □

Descriptions of the lattice of varieties of bands were provided independently and almost simultaneously by Birjukov [1], Gerhard [1] and Fennemore [1, 2, 3]. Their approach was subsequently refined by Gerhard and Petrich [8] who used a combination of systems of identities and invariants. The derivation in this section of the lattice of varieties of bands is based on Polák [1]. Koryakov [1] characterizes the basis ranks for varieties of bands. O. Sapir [1] shows that the variety $\mathcal{B}$ is a minimal inherently non-finitely generated variety, that is, it is locally finite and no variety containing it can be generated by a finite set of finite algebras. Reilly and Zhang [4] and Liu, Yan and Wang [1] study varieties of bands that are generated by bands for which the $\mathcal{D}$-classes form a chain. The variety of regular bands plays an important role in the next chapter. Pastijn and Albert [1, 2], establish that every regular band can be embedded in a regular band with a semilattice transversal of the $\mathcal{D}$-classes. Ćirić and Bogdanović [1] characterize $\mathcal{U} \vee \mathcal{V}$, where $\mathcal{U}$, $\mathcal{V}$ are band varieties defined by homotypical identities, as consisting of subdirect or spined products of elements from $\mathcal{U} \cup \mathcal{V}$. Gerhard [2] describes the sequences that are idempotent representable, in [3, 4] he characterizes all subdirectly irreducible bands and shows that every proper subvariety of $\mathcal{B}$ is generated by either one or two finite subdirectly irreducible bands and in [5] discusses injectives. See also Wang, Leng and Yu [1]. Aldhamri [1] shows that a quasivariety of bands allows a natural duality if and only if it is contained in the variety of normal bands and deduces that every finite normal band is of finite degree. Demlova and Koubek [1, 2, 3] study the endomorphism monoids of bands and determine which are categorically universal. Jones [3] showed

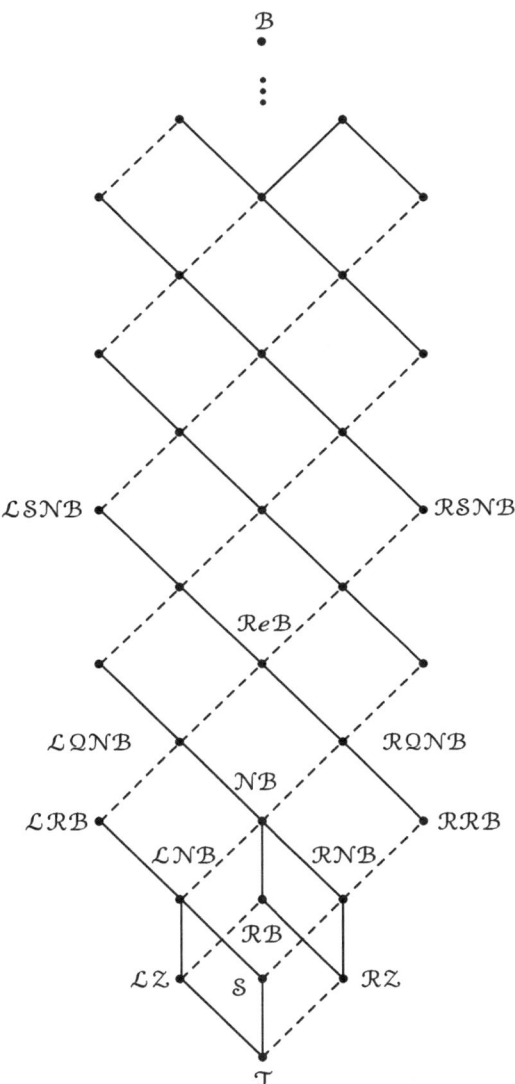

Diagram 3.3 $\mathcal{L}(\mathcal{B})$

that any band whose congruence lattice has ACC or DCC is finite. Fountain and Lockley [1] consider bands with distributive congruence lattices, Dean and Oehmke [1] consider idempotent semigroups with distributive right congruence lattices, Freese and Nation [1] and Hall [2] consider the lattice of congruences on semilattices and Gabovich [1] considers bands in which every finitely generated subband is an endomorphic image. Scheiblich [1] studies epimorphisms and dominions in band varieties. Zybina [1,2,3] studies orders on bands. Several variations of the concept of join (meet) irreducibility of the lattice of pseudovarieties of finite semigroups are considered by Lee, Rhodes and Steinberg [1]. For some early studies of varieties of bands, see Yamada [1,2,3,4].

3.4   Relations K_ℓ and K_r on $\mathcal{L}(\mathcal{B})$

Diagram 3.3 suggests that there exist relationships of some sort between certain varieties of bands. In particular, one might conjecture that varieties lying on the same line of positive slope (or negative slope) might be related in some way. It turns out that the relationship between varieties lying on the same line in Diagram 3.3 is extremely strong and that is what we will be demonstrating in this section.

In general, for $\mathcal{U}, \mathcal{V} \in \mathcal{L}(\mathcal{CR})$, we have $\mathcal{U}\ T_\ell\ \mathcal{V}$ if and only if $\mathcal{U}_{T_\ell} = \mathcal{V}_{T_\ell}$ (and dually for T_r). Consequently for $\mathcal{U}, \mathcal{V} \in [\mathcal{S}, \mathcal{B}]$,

$$\mathcal{U}\ T_\ell\ \mathcal{V} \Longleftrightarrow \delta_{\mathcal{U}_{T_\ell}} = \delta_{\mathcal{V}_{T_\ell}} \quad \text{Theorem 3.2.8(ii)}$$

$$\Longleftrightarrow \mathcal{U}_{T_\ell} \text{ and } \mathcal{V}_{T_\ell} \text{ have the same ladder.}$$

Since, for all $\tau \in \Theta$, $\delta_{(\mathcal{U}_{T_\ell})_\tau} = \delta_{\mathcal{U}_{T_\ell \tau}}$ the ladder of $\delta_{\mathcal{U}_{T_\ell}}$ can be read directly from the ladder of $\mathcal{U}$. To see this, let the ladder of $\mathcal{U}$ be as displayed on the left in Diagram 3.4, where $A_i, B_i \in D$.

For $\tau \in \Theta$, the value of $\tau\delta_{\mathcal{V}}$ is uniquely determined by the value of $\mathcal{V}_\tau$. Now let us focus on $\mathcal{V} = \mathcal{U}_{T_\ell}$. By definition, $\emptyset\delta_{\mathcal{V}} = 1$. Also, $\mathcal{V}_{T_\ell} = (\mathcal{U}_{T_\ell})_{T_\ell} = \mathcal{U}_{T_\ell}$ so that $T_\ell\delta_{\mathcal{V}} = T_\ell\delta_{\mathcal{U}} = B_1$. Furthermore $\mathcal{V}_{T_\ell T_r} = \mathcal{U}_{T_\ell T_\ell T_r} = \mathcal{U}_{T_\ell T_r}$ which implies that $(T_\ell T_r)\delta_{\mathcal{V}} = (T_\ell T_r)\delta_{\mathcal{U}} = A_2$. Now let $\sigma \in \Theta$ and $h(\sigma) \neq T_\ell$. Then $\mathcal{V}_{T_\ell\sigma} = (\mathcal{U}_{T_\ell})_{T_\ell\sigma} = \mathcal{U}_{T_\ell\sigma}$ so that $(T_\ell\sigma)\delta_{\mathcal{V}} = (T_\ell\sigma)\delta_{\mathcal{U}}$.

We now consider what happens to $\mathcal{V}$ under the lower operators associated with the sequence $T_r, T_r T_\ell, T_r T_\ell T_r, \ldots$. We have

$$\mathcal{V}_{T_r} = \mathcal{U}_{T_\ell T_r}, \quad \mathcal{V}_{T_r T_\ell} = \mathcal{U}_{T_\ell T_r T_\ell}, \quad \mathcal{V}_{T_r T_\ell T_r} = \mathcal{U}_{T_\ell T_r T_\ell T_r} \quad \text{and so on}$$

so that

$$T_r\delta_{\mathcal{V}} = T_\ell T_r\delta_{\mathcal{U}} = A_2, \quad T_r T_\ell\delta_{\mathcal{V}} = T_\ell T_r T_\ell\delta_{\mathcal{U}} = B_3, \quad T_r T_\ell T_r\delta_{\mathcal{V}} = T_\ell T_r T_\ell T_r\delta_{\mathcal{U}} = A_4.$$

Thus, if $\mathcal{U} \in [\mathcal{S}, \mathcal{CR}]$ has the ladder $\delta_{\mathcal{U}}$ displayed on the left in Diagram 3.4, then $\mathcal{U}_{T_\ell}$ has the ladder $\delta_{\mathcal{U}_{T_\ell}}$ displayed in the middle of Diagram 3.4, and the ladder for any $\mathcal{W} \in [\mathcal{S}, \mathcal{CR}]$ that is T_ℓ-related to $\mathcal{U}$ has the form of the ladder on the right in Diagram 3.4.

Note that the ladder $\delta_{\mathcal{U}_{T_\ell}}$ of $\mathcal{U}_{T_\ell}$ is completely determined by the even numbered labels on the left and the odd numbered labels on the right of $\delta_{\mathcal{U}}$.

In the following proposition, we consider only band varieties.

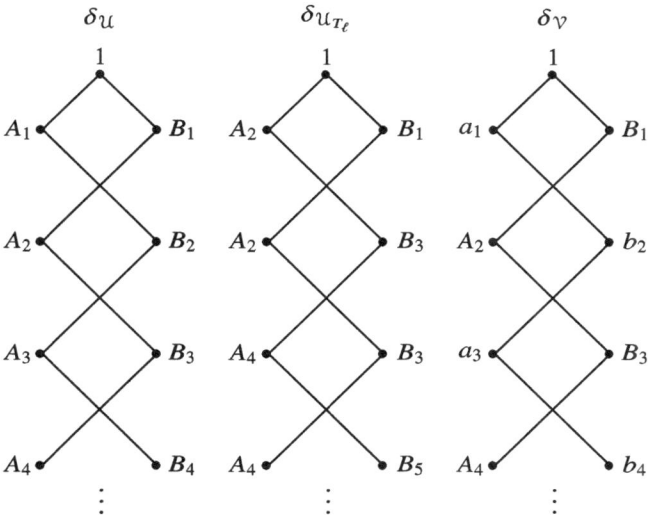

Diagram 3.4 Ladders of Three T_ℓ-equivalent varieties

Proposition 3.4.1 *Let* $\mathcal{U}, \mathcal{V} \in [\mathcal{S}, \mathcal{B}]$ *and the ladder of* $\mathcal{U}$ *be as in the ladder on the left in Diagram 3.4,*

(i) $\delta_{\mathcal{U}_{T_\ell}}$ *is as in the middle ladder in Diagram 3.4 and any variety in* $[\mathcal{S}, \mathcal{B}]$ *with a ladder of this form is the smallest variety in its* T_ℓ-*class.*

(ii) $\mathcal{U} \, T_\ell \, \mathcal{V}$ *if and only if the ladder for* $\mathcal{V}$ *is as displayed on the right in Diagram 3.4, for some* $a_i \in \{1, L, T\}$, $b_i \in \{1, R, T\}$.

(iii) *The variety* $\mathcal{U}$ *is the least element in its* T_ℓ-*class if and only if* $A_{2n-1} = A_{2n}$ *and* $B_{2n} = B_{2n+1}$ *for all* $n \geq 1$. *Moreover, any ladder of this form defines the least variety in a* T_ℓ-*class.*

(iv) *The variety* $\mathcal{U}$ *is the greatest element in its* T_ℓ-*class within* $\mathcal{B}$ *if and only if* $A_1 = 1$, $A_{2n+1} = A_{2n}$, *and* $B_{2n} = B_{2n+1}$, *for all* $n \geq 1$. *Moreover, any ladder of this form defines the greatest variety in* T_ℓ-*class.*

Proof (i) The first claim follows from the discussion prior to the proposition. Now suppose that $\mathcal{V}$ has the ladder as displayed on the left below. Then, from the discussion prior to the proposition, $\mathcal{V}_{T_\ell}$ has the ladder on the right below.

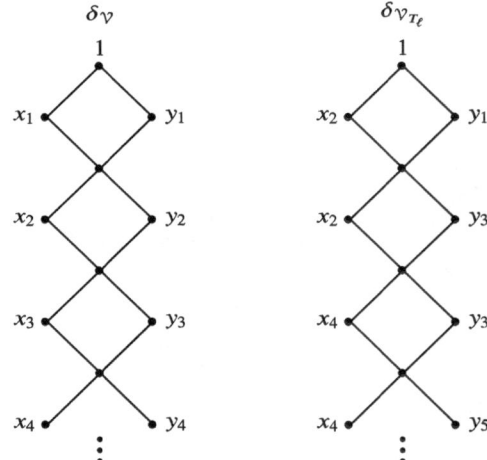

Hence

$$\mathcal{U} \, T_\ell \, \mathcal{V} \Longrightarrow \mathcal{U}_{T_\ell} = \mathcal{V}_{T_\ell}$$
$$\Longrightarrow \delta_{\mathcal{U}_{T_\ell}} = \delta_{\mathcal{V}_{T_\ell}}$$
$$\Longrightarrow x_{2n} = A_{2n}, \quad y_{2n-1} = B_{2n-1} \quad \text{for } n \geq 1$$
$$\Longrightarrow \delta_{\mathcal{V}} \text{ has the form shown in Diagram 3.4.}$$

It then follows that if the ladder of $\mathcal{V} \in [\mathcal{S}, \mathcal{B}]$ has the form of that in the middle of Diagram 3.4, then it. must be the smallest in its T_ℓ-class.

We now consider the converse. Let $\mathcal{U}, \mathcal{V} \in [\mathcal{S}, \mathcal{B}]$. Let the ladders of $\mathcal{U}$ and $\mathcal{V}$ be as in Diagram 3.4. We can divide Θ into the union of two sequences of words:

$$\Theta_\ell = \{T_\ell, \, T_\ell T_r, \, T_\ell T_r T_\ell, \, T_\ell T_r T_\ell T_r, \, \ldots\},$$
$$\Theta_r = \{T_r, \, T_r T_\ell, \, T_r T_\ell T_r, \, T_r T_\ell T_r T_\ell, \, \ldots\}$$

consisting of the words beginning with T_ℓ and the words beginning with T_r and we refer to Θ_ℓ as the T_ℓ-*chain* and to Θ_r as the T_r-*chain*, respectively. We will also write

$$\Theta_\ell^1 = \{\emptyset\} \cup \Theta_\ell, \quad \Theta_r^1 = \{\emptyset\} \cup \Theta_r.$$

Then the hypothesis for the converse is equivalent to the assumption that $\delta_{\mathcal{U}}$ and $\delta_{\mathcal{V}}$ agree on the T_ℓ-chain. Our goal is to show that it follows from this that $\mathcal{U}$ and $\mathcal{V}$ are T_ℓ-equivalent.

We will use the description of $\mathcal{U} = \mathcal{U}\delta\xi_D = \delta_{\mathcal{U}}\xi_D$ from Theorem 3.2.8(ii). Any $\tau \in \Theta$ for which $\tau\delta_{\mathcal{U}} = 1$ contributes a term $\mathcal{B}$ to $\delta_{\mathcal{U}}\xi_D$. Also, any $\tau \in \Theta$ for which $\tau\delta_{\mathcal{U}} = T$ contributes a term $\mathcal{S}^\tau$ to $\delta_{\mathcal{U}}\xi_D$. If $\sigma \in \Theta$ and $|\sigma| > |\tau|$ where $\tau\delta_{\mathcal{U}} = T$, then σ contributes a term $\mathcal{S}^{\overline{\sigma}}$ but this term can be ignored since $\mathcal{S}^\tau \subseteq \mathcal{S}^{\overline{\sigma}}$. Thus it suffices to focus on the region of the boundary layers. We consider the general situation first, by which we mean that we assume that $\mathcal{R}e\mathcal{B} \subseteq \mathcal{U}$.

From Sect. 3.3, we know that there are ten distinct boundary regions: $E_1, \ldots, E_{10}$. We must also take into account *how* the values of $\lambda\delta_\mathcal{U}$, for λ in the T_ℓ-chain transition through the boundary region. Since each boundary region consists of two layers, there will be two elements of the T_ℓ-chain λ_1, λ_2, say, for which the values $\lambda_i\delta_\mathcal{U}$, $i = 1, 2$ are in the boundary region. Let us consider one of the possible patterns for the boundary layers that is quite illustrative, say $E_{n,9}$. Let λ be the (shorter) element in the T_ℓ-chain for which $\lambda\delta_\mathcal{U}$ is in the first boundary layer of $\delta_\mathcal{U}$. Let $|\lambda| = n$. Since λ belongs to the T_ℓ-chain, $h(\lambda) = T_\ell$. Since the first (or upper) layer in E_9 consists of $\{1, R\}$, it follows that $\lambda\delta_\mathcal{U} \in \{1, R\}$. The arguments in the two cases are very similar, so considering one case will suffice to illustrate the calculation of $\delta_\mathcal{U}\xi_D$. So let us consider the case where $\lambda\delta_\mathcal{U} = 1$. Since the 1 in the first boundary layer occurs on the left hand side of the ladder, it follows that $t(\lambda)$ must be T_r. This is equivalent to assuming that n is even. Since all the entries in the layers $1, 2, \ldots, n - 1$ in the ladder $\delta_\mathcal{U}$ are equal to 1 and all the entries in layers $n + 3, n + 4, \ldots$ are equal to T, we can describe all the ladders that agree with $\delta_\mathcal{U}$ on the T_ℓ-chain by simply displaying the entries in the $n, n + 1, n + 2$ layers, which we do in (3.4.1):

$$
\begin{array}{ccccc}
\begin{pmatrix} 1 & 1 \\ 1 & R \\ T & R \end{pmatrix} & & & & \\
\delta_Z & \begin{pmatrix} 1 & 1 \\ 1 & R \\ T & T \end{pmatrix} & & & \\
& \delta_Y & \begin{pmatrix} 1 & 1 \\ L & R \\ T & T \end{pmatrix} & & \\
& & \delta_X & \begin{pmatrix} 1 & 1 \\ T & R \\ T & T \end{pmatrix} & \\
& & & \delta_W & \begin{pmatrix} 1 & R \\ T & R \\ T & T \end{pmatrix} \\
& & & & \delta_\mathcal{U}
\end{array}
\tag{3.4.1}
$$

Because of the constraints placed on $\mathcal{V}$ by the requirement that $\delta_\mathcal{U}$ and $\delta_\mathcal{V}$ agree on the T_ℓ-chain, these patterns for the two boundary layers and the layer below are the only possibilities. Let these be the boundary layers for the varieties $\mathcal{U}, \mathcal{W}, \mathcal{X}, \mathcal{Y}, \mathcal{Z}$ in $[\mathcal{S}, \mathcal{B})$, from smallest to largest. Then, necessarily, $\mathcal{V} \in \{\mathcal{U}, \mathcal{W}, \mathcal{X}, \mathcal{Y}, \mathcal{Z}\}$. All their corresponding ladders agree with $\lambda\delta_\mathcal{U} = 1$, $\lambda T_\ell\delta_\mathcal{U} = R$, $\lambda T_\ell T_r = T$. We will now show that these varieties are T_ℓ-related to each other. The first step will be to employ Theorem 3.2.8(ii) to generate specific descriptions of $\mathcal{U}, \mathcal{W}, \mathcal{X}, \mathcal{Y}, \mathcal{Z}$ based on their ladders.

Note that $\mathcal{U} \subset \mathcal{W} \subset \mathcal{X} \subset \mathcal{Y} \subset \mathcal{Z}$. Let $\rho = \lambda^d$, that is the element of Θ obtained from λ by swapping the occurrences of T_ℓ and T_r in λ. Then $|\rho| = |\lambda| = n$ and $\rho\delta_\mathcal{U} = R$. We also have $\lambda T_\ell = T_\ell\rho$, $T_r\lambda = \rho T_\ell$.

In the application of Theorem 3.2.8(ii) to $\mathcal{U}$, the critical values are:

$$\lambda\delta_\mathcal{U} = 1, \quad \lambda T_\ell\delta_\mathcal{U} = R, \quad \lambda T_\ell T_r\delta_\mathcal{U} = T, \quad \rho\delta_\mathcal{U} = R, \quad \rho T_r\delta_\mathcal{U} = T.$$

Thus

$$\mathcal{U} = \delta_\mathcal{U}\xi_D = \mathcal{B} \cap \mathcal{R}N\mathcal{B}^{\overline{\rho}} \cap \mathcal{R}N\mathcal{B}^{\overline{\lambda T_\ell}} \cap \mathcal{S}^{\overline{\rho T_r}} \cap \mathcal{S}^{\overline{\lambda T_\ell T_r}} = \mathcal{B} \cap \mathcal{R}N\mathcal{B}^{\overline{\rho}}$$

from which we can deduce that

$$\mathcal{U}_{T_\ell} = \mathcal{B}_{T_\ell} \cap ((\mathcal{RNB}^{\bar{\rho}})_{T_\ell} = \mathcal{B} \cap \mathcal{RNB}^{\bar{\rho}}.$$

and we see that $\mathcal{U}$ is the least element in its T_ℓ and K_ℓ-class.

Now we give the same treatment to $\mathcal{Z}$. The important values of $\delta_{\mathcal{Z}}$ are:

$$\lambda\delta_{\mathcal{Z}} = 1, \quad \lambda T_\ell \delta_{\mathcal{Z}} = R, \quad \lambda T_\ell T_r \delta_{\mathcal{Z}} = T, \quad \rho\delta_{\mathcal{Z}} = 1, \quad \rho T_r\delta_{\mathcal{Z}} = 1, \quad \rho T_r T_\ell \delta_{\mathcal{Z}} = R.$$

Note that the values on the T_ℓ-chain for $\delta_{\mathcal{Z}}$ are the same as for $\delta_{\mathcal{U}}$. From Theorem 3.2.8(ii), we obtain

$$\mathcal{Z} = \delta_{\mathcal{Z}}\xi_D = \mathcal{B} \cap \mathcal{RNB}^{\overline{\lambda T_\ell}} \cap \mathcal{S}^{\overline{\lambda T_\ell T_r}} \cap \mathcal{RNB}^{\overline{\rho T_r T_\ell}} = \mathcal{B} \cap (\mathcal{RNB})^{\overline{T_\ell \rho}}.$$

from which we obtain

$$\mathcal{Z}_{T_\ell} = \mathcal{B} \cap ((\mathcal{RNB})^{\overline{T_\ell \rho}})_{T_\ell} = \mathcal{B} \cap \mathcal{RNB}^{\bar{\rho}} = \mathcal{U}_{T_\ell}.$$

Thus, $\mathcal{U}$ and $\mathcal{Z}$ are T_ℓ-related and therefore the five varieties $\mathcal{U}, \mathcal{W}, \mathcal{X}, \mathcal{Y}, \mathcal{Z}$ are all T_ℓ (and therefore K_ℓ) related. The least variety in the K_ℓ-class of $\mathcal{U}$ is $\mathcal{U}$ itself. The greatest element in the K_ℓ-class of $\mathcal{U}$ is,

$$\mathcal{U}^{K_\ell} = \mathcal{B} \cap \mathcal{U}^{T_\ell} = \mathcal{B} \cap (\mathcal{B} \cap \mathcal{RNB}^{\bar{\rho}})^{T_\ell} = \mathcal{B} \cap \mathcal{RNB}^{\overline{T_\ell \rho}} = \mathcal{Z}.$$

In other words, the interval $[\mathcal{U}, \mathcal{Z}]$ is the complete K_ℓ-class of $\mathcal{U}$ and this derives from the set of all ladders which have the same values on the T_ℓ chain.

It should be noted that there were three important assumptions regarding the element λ in the above discussions. First it belongs to the T_ℓ-chain so that $h(\lambda) = T_\ell$. Then λ is assumed to be the first word in the T_ℓ chain for which $\delta_{\mathcal{U}}$ has a value in the boundary layers for $\delta_{\mathcal{U}}$. The assumption that $\lambda\delta_{\mathcal{U}}$ lies on the left hand side is equivalent to the assumption that the length of λ must be even. All this leads to the conclusion that $\mathcal{U}$ is the least element in its K_ℓ. Change any of these factors and the outcome changes. For instance, if n is odd, then $\mathcal{U}$ will be the greatest element in its K_ℓ-class.

This conclusion was reached by starting with one particular boundary layer and the assumption that the first element of the T_ℓ-chain with a value in the boundary region was on the left side of the ladder, in other words, it ended in T_r. There are ten patterns for boundary regions and the first element from the T_ℓ-chain to have a value in a boundary region can appear on the left or the right, depending on the level of the first layer of the boundary region. So there are $10 \times 2 = 20$ combinations to be considered. But in the above discussion, we actually covered five different combinations already with the five different elements in the K_ℓ class. We leave the verification of the remaining cases to the reader.

For varieties not containing $\mathcal{ReB}$, the argument is essentially the same. The only difference is that for the ladder of such a variety there are fewer choices of ladders with the same values on the T_ℓ-chain or T_r-chain. Instead of five varieties in the general K_ℓ-class or K_r-class, there will be just three or four. But the argument is just the same. This will be further clarified in Theorem 3.4.2.

(ii) This follows immediately from the discussion prior to the proposition.

(iii) We have that U is the least element in its T_ℓ-class if and only if $\delta_{\mathcal{U}}$ and $\delta_{\mathcal{U}T_\ell}$ agree on the T_ℓ-chain. These are exactly the conditions stated on the values of A_i, B_i. A simple induction argument shows that any ladder of this form is a member of Φ_D and therefore is the ladder of some band variety which must be the least element in its T_ℓ-class.

(iv). The varieties of bands that are T_ℓ-equivalent to $\mathcal{U}$ are those with ladders that agree with $\delta_{\mathcal{U}}$ on the T_ℓ-chain. The conditions listed give the maximum possibilities for the values on the T_r-chain for any such variety. It is a simple induction argument to show that these choices of values do produce an element of Φ_D and therefore define a variety that must be the greatest element in the T_ℓ-class of U.   □

Recall that, by definition, $K_\ell = K \cap T_\ell$ and $K_r = K \cap T_\ell$. Since the K-class of $\mathcal{B}$ is $\mathcal{L}(\mathcal{B})$, it follows that $\mathcal{L}(\mathcal{B})$ is a disjoint union of K_ℓ (respectively, K_r) classes and, moreover, that the restrictions of the congruences T_ℓ and T_r to $\mathcal{L}(\mathcal{B})$ are precisely the congruences K_ℓ and K_r on $\mathcal{L}(\mathcal{B})$.

We are now in a position to determine the K_ℓ- and K_r-classes on $\mathcal{L}(\mathcal{B})$.

Theorem 3.4.2 (i) *The relations K_ℓ and K_r on $\mathcal{L}(\mathcal{B})$ are represented by the solid lines and broken lines, respectively, in Diagram 3.3.*

(ii) *The varieties of the form $\mathcal{S}^{u(K_\ell,K_r)}$, $\mathcal{LNB}^{v(K_\ell,K_r)}$ and $\mathcal{RNB}^{w(K_\ell,K_r)}$ where u, v, $w \in \Theta$, $h(v) = T_r$, $h(w) = T_\ell$ are both $\vee$-irreducible in $\mathcal{L}(\mathcal{CR})$ and $\wedge$-irreducible in $\mathcal{L}(\mathcal{B})$ and are the only varieties in $[\mathcal{S}, \mathcal{B}]$ that are $\wedge$-irreducible in $\mathcal{L}(\mathcal{B})$.*

(iii) *The set $I = \{\mathcal{V}T_r \mid \mathcal{V} \in \mathcal{L}(\mathcal{B})\}$ is an ideal and $[\mathcal{S}T_r, \mathcal{B}T_r]$ is a chain in $\mathcal{L}(\mathcal{CR})/T_r$.*

(iv) *$\mathcal{L}(\mathcal{B})$ is a distributive lattice.*

(v) *For any $\mathcal{P} \in \mathcal{L}(\mathcal{B})$, the mapping $\mu_\mathcal{P}\colon \mathcal{V} \mapsto \mathcal{V} \cap \mathcal{P}$ $(\mathcal{V} \in \mathcal{L}(\mathcal{CR}))$ is a complete retraction of $\mathcal{L}(\mathcal{CR})$ onto $\mathcal{L}(\mathcal{P})$.*

Proof (i) Since the relations K_ℓ and K_r in $\mathcal{L}(\mathcal{B})$ are simply the T_ℓ and T_r relations restricted to $\mathcal{L}(\mathcal{B})$, we will cast the argument in terms of T_ℓ and T_r classes. From Theorem 1.5.5, we see that the following sets consist of T_r-equivalent varieties: $\{\mathcal{T}, \mathcal{RZ}\}$, $\{\mathcal{LZ}, \mathcal{RB}\}$. We also see that $\mathcal{S}$, $\mathcal{LNB}$ and $\mathcal{LRB}$ are the least elements in their T_r-classes. By the dual of Proposition 3.4.1 any $\mathcal{V} \in [\mathcal{S}, \mathcal{B}]$ that is T_r-equivalent to $\mathcal{S}$ must have a ladder that has the form (a) in (3.4.2).

$$
\begin{array}{llll}
 & \quad 1 & \qquad\qquad & \quad 1 \\
 & T \quad - & & L \quad - \\
\text{(a)} & T \quad T & \text{(b)} & - \quad T \qquad\qquad (3.4.2) \\
 & T \quad T & & T \quad T \\
 & \;\vdots & & \;\vdots
\end{array}
$$

There are only three possible ways to fill in the blank in order to satisfy condition (B2): T, R, 1. These give the ladders $E_{1,1}$, $E_{1,3}$, and $E_{1,6}$ of $\mathcal{S}$, $\mathcal{RNB}$ and $\mathcal{RRB}$, respectively.

Again by the dual of Proposition 3.4.1, any $\mathcal{V} \in [\mathcal{S}, \mathcal{B}]$ that is T_r-related to $\mathcal{LNB} = E_{1,2}$ must have ladder that looks as (b) in (3.4.2). This time we have two blank spaces to fill in. The one on the left may be T or L and the one on the right may be T, R or 1. However, if the blank on the right is filled with either T or R, then the blank on the left must be filled with T. This results in four varieties with ladders that have boundary regions $E_{1,2}$, $E_{1,5}$, $E_{1,8}$ and $E_{1,10}$. Thus so far, the restriction of T_r to $\mathcal{L}(\mathcal{B})$ yields classes corresponding to the broken lines in Diagram 3.3 for the classes $\mathcal{T}$, $\mathcal{LZ}$, $\mathcal{S}$ and $\mathcal{LNB}$.

Now we may restrict our attention to varieties of bands containing $\mathcal{LRB}$.

We start by identifying the varieties that are the least in their T_r-classes. Let $\mathcal{V} \in [\mathcal{LRB}, \mathcal{B}]$. By the dual of Proposition 3.4.1 we see that $\mathcal{V}$ will be least in its T_r-class if and only if its ladder has the form

$$
\begin{array}{cc}
 & 1 \\
1 & A \\
B & A \\
B & C \\
D & C \\
D & \\
& \vdots
\end{array}
$$

where $A, C \in \{T, R, 1\}$ and $B, D \in \{T, L, 1\}$.

The top element in the left hand column is necessarily 1 since we are assuming that $\mathcal{LRB} \subseteq \mathcal{V}$ and, therefore, $\mathcal{LRB} \subseteq \mathcal{V}_{T_r}$.

Consider the following ladders:

$$
E_{1,4} : \begin{array}{cc} & 1 \\ 1 & T \\ T & T \\ & \vdots \end{array}
\qquad
E_{1,9} : \begin{array}{cc} & 1 \\ 1 & R \\ T & R \\ T & T \\ & \vdots \end{array}
\qquad
E_{2,6} : \begin{array}{cc} & 1 \\ 1 & 1 \\ T & 1 \\ T & T \\ & \vdots \end{array}
\qquad
E_{2,10} : \begin{array}{cc} & 1 \\ 1 & 1 \\ L & 1 \\ L & T \\ T & T \\ & \vdots \end{array}.
$$

It is evident that the ladders $E_{1,4}$, $E_{1,9}$, $E_{2,6}$, $E_{2,10}$ all have the above pattern. It is also a simple exercise to show that if $E_{n,k}$ has this property then so does $E_{n+2,k}$. Consequently all the varieties on the extreme left of Diagrams 3.2 and 3.3 are the least varieties in their T_r-classes.

Consider the T_r-class of $\mathcal{LRB} = E_{1,4}$. This has ladder

$$
\begin{array}{cc}
1 & \\
1 & T \\
T & T \\
\vdots &
\end{array}
$$

Any variety of bands T_r-equivalent to $\mathcal{LRB}$ must have ladder of the form

$$
\begin{array}{cc}
1 & \\
1 & - \\
- & T \\
T & T \\
\vdots &
\end{array}
$$

There are two spaces to fill in subject to the constraints that the one on the left is either T, L or 1 and the one on the right is T, R or 1. However, if the right hand blank is filled with T or R, then the left hand blank must be filled with T. Thus there are exactly five possibilities. By checking, we see that these are exactly the varieties with ladders $E_{1,4}$ (corresponding to $\mathcal{LRB}$), $E_{1,7}$, $E_{2,1}$, $E_{2,2}$ and $E_{2,4}$. Thus the broken line from $\mathcal{LRB}$ in Diagram 3.3 again describes the T_r-class.

A similar analysis can then be performed on the ladders $E_{1,9}$, $E_{2,6}$ and $E_{2,10}$ establishing that the broken lines through the corresponding varieties of bands identify the varieties of bands in the corresponding T_r-classes.

Now let $\mathcal{V}$ be a variety of bands that is least in its T_r-class and contains $\mathcal{LRB}$ that is least in its T_r-class. By the dual of Proposition 3.4.1, its ladder is of the form

$$
E_{n,k} = \quad
\begin{array}{cc}
1 & \\
1 & A \\
B & A \\
B & C \\
 & C \\
\vdots &
\end{array}
$$

for suitable n, k. Then, by inserting two rows of ones, we obtain

$$
E_{n+2,k} = \quad
\begin{array}{cc}
1 & \\
1 & 1 \\
1 & 1 \\
1 & A \\
B & A \\
B & C \\
 & C \\
\vdots &
\end{array}
$$

which also represents a variety of bands that is least in its T_r-class. We know that the ladders of $E_{1,4}$, $E_{1,9}$, $E_{2,6}$, $E_{2,10}$ represent varieties of bands that are least in their T_r-classes. Hence all the ladders of the form

$$E_{2n+1,4}, \quad E_{2n+1,9}, \quad E_{2n+2,6}, \quad E_{2n+2,10} \quad \text{for} \quad n \geq 0$$

also represent varieties of bands that are least in their T_r-classes. These are precisely the varieties in Diagram 3.3 on the extreme left (and containing $\mathcal{LRB}$).

This completely determines the T_r-related varieties within the proper subvarieties of $\mathcal{B}$. Of course, $\mathcal{B}$ is not T_r-equivalent to any other variety of bands. The claim concerning the T_ℓ-classes within $\mathcal{L}(\mathcal{B})$ follows by duality.

(ii) By part (i) and inspection of Diagram 3.3, it is evident that each variety on the extreme left or extreme right is of one of the forms listed in the statement. By Theorem 3.3.2, each of these varieties is both $\vee$-irreducible in $\mathcal{L}(\mathcal{CR})$ and $\wedge$-irreducible in $\mathcal{L}(\mathcal{B})$ and these are the only varieties in $[\mathcal{S}, \mathcal{B})$ that are $\wedge$-irreducible in $\mathcal{L}(\mathcal{B})$

(iii) Let $\mathcal{V} \in \mathcal{L}(\mathcal{CR})$ be such that $\mathcal{V} = \mathcal{V}T_r$ and $\mathcal{V}T_r < \mathcal{B}T_r$. Then $\mathcal{V} = \mathcal{V}T_r \subseteq \mathcal{B}T_r = \mathcal{B}$. Therefore $\mathcal{V} \in \mathcal{L}(\mathcal{B})$ and $\mathcal{V}T_r \in I$. Consequently, I is an ideal in $\mathcal{L}(\mathcal{CR})/T_r$.

That the classes of the form $\mathcal{V}T_r$, $\mathcal{V} \in [\mathcal{S}, \mathcal{B}]$ form a chain follows immediately from part (i).

(iv) It is evident from Diagram 3.3 that $\mathcal{L}(\mathcal{B})$ contains no sublattice that is isomorphic to either of the forbidden lattices listed in [PR1, Lemma I.2.7]. Therefore, $\mathcal{L}(\mathcal{B})$ is distributive.

(v) From Theorem 3.2.8(v) we know that the claim holds for $\mathcal{P} = \mathcal{B}$.

Now consider $\mathcal{P} \in [\mathcal{T}, \mathcal{B})$. For any family $\{\mathcal{V}_\alpha \mid \alpha \in A\}$ with $\mathcal{V}_\alpha \in \mathcal{L}(\mathcal{B})$ either $\bigvee_{\alpha \in A} \mathcal{V}_\alpha = \mathcal{B}$ or, since $\mathcal{L}(\mathcal{B})$ has finite width, there exists a finite subset A_1 of A such that $\bigvee_{\alpha \in A} \mathcal{V}_\alpha = \bigvee_{\alpha \in A_1} \mathcal{V}_\alpha$. In the former case, there must exist $\beta \in A$ such that $\mathcal{P} \subseteq \mathcal{V}_\beta$ so that

$$\left(\bigvee_{\alpha \in A} \mathcal{V}_\alpha \right) \cap \mathcal{P} = \mathcal{B} \cap \mathcal{P} = \mathcal{P} = \mathcal{V}_\beta \cap \mathcal{P} = \bigvee_{\alpha \in A} (\mathcal{V}_\alpha \cap \mathcal{P}).$$

In the latter case we have, since $\mathcal{L}(\mathcal{B})$ is distributive,

$$\left(\bigvee_{\alpha \in A} \mathcal{V}_\alpha \right) \cap \mathcal{P} = \left(\bigvee_{\alpha \in A_1} \mathcal{V}_\alpha \right) \cap \mathcal{P} = \bigvee_{\alpha \in A_1} (\mathcal{V}_\alpha \cap \mathcal{P}).$$

In both cases we find that

$$\left(\bigvee_{\alpha \in A} \mathcal{V}_\alpha \right) \cap \mathcal{P} = \bigvee_{\alpha \in A} (\mathcal{V}_\alpha \cap \mathcal{P}).$$

Hence, for any family $\{\mathcal{V}_\alpha \mid \alpha \in A\}$ with $\mathcal{V}_\alpha \in \mathcal{L}(\mathcal{CR})$, and not necessarily just in $\mathcal{L}(\mathcal{B})$, we obtain

$$\left(\bigvee_{\alpha \in A} \mathcal{V}_\alpha\right) \cap \mathcal{P} = \left(\bigvee_{\alpha \in A} \mathcal{V}_\alpha\right) \cap \mathcal{B} \cap \mathcal{P}$$

$$= \left(\bigvee_{\alpha \in A} (\mathcal{V}_\alpha \cap \mathcal{B})\right) \cap \mathcal{P}$$

$$= \bigvee_{\alpha \in A} (\mathcal{V}_\alpha \cap \mathcal{B} \cap \mathcal{P}) \qquad \text{since } \mathcal{V}_\alpha \cap \mathcal{B} \in \mathcal{L}(\mathcal{B}) \text{ for all } \alpha \in A$$

$$= \bigvee_{\alpha \in A} (\mathcal{V}_\alpha \cap \mathcal{P}).$$

Thus $\mu_\mathcal{P}$ respects arbitrary joins. Clearly $\mu_\mathcal{P}$ respects arbitrary intersections. Therefore $\mu_\mathcal{P}$ is a complete retraction. □

It is important to appreciate that the K_ℓ and K_r relations on $\mathcal{L}(\mathcal{B})$ are, by their very definition, the *restrictions* of the relations T_ℓ and T_r to the K-class of $\mathcal{T}$, that is to $\mathcal{L}(\mathcal{B})$. However, the K_ℓ (respectively K_r) class of any variety of bands is distinct from its T_ℓ (respectively T_r) class. For example, for any $\mathcal{U} \in \mathcal{L}(\mathcal{B})$, we have $\mathcal{G}$ belongs to the T_ℓ (respectively T_r) class of $\mathcal{U}$ but not to its K_ℓ (respectively K_r) class.

Theorem 3.4.3 *Let $\mathcal{V} \in [\mathcal{S}, \mathcal{B})$. Then for suitable $\lambda, \rho \in \Theta$, $\mathcal{V}$ has one of the following forms*

$\mathcal{B} \cap \mathcal{LNB}^\lambda$	*where* $h(\lambda) = T_r$,
$\mathcal{B} \cap \mathcal{RNB}^\rho$	*where* $h(\rho) = T_\ell$,
$\mathcal{B} \cap \mathcal{LNB}^\lambda \cap \mathcal{RNB}^\rho$	*where* $h(\lambda) = T_r$, $h(\rho) = T_\ell$,
$\mathcal{B} \cap \mathcal{LNB}^\lambda \cap \mathcal{S}^\rho$	*where* $h(\lambda) = T_r$,
$\mathcal{B} \cap \mathcal{S}^\lambda \cap \mathcal{RNB}^\rho$	*where* $h(\rho) = T_\ell$,
$\mathcal{B} \cap \mathcal{S}^\lambda$ *or* $\mathcal{B} \cap \mathcal{S}^\lambda \cap \mathcal{S}^\rho$.	

Proof Equivalently, we will show that the claim holds for $\psi \xi_D$ where $\psi \in \Phi_D$, and ξ_D is the mapping from Theorem 3.2.8. If $\tau\psi = 1$ for all $\tau \in \Theta^1$, then $\psi\xi_D = \mathcal{B}$. So we assume that $\tau\psi \ne 1$ for some $\tau \in \Theta$. Once again we will work with the T_ℓ- and T_r-chains.

Since $\emptyset\psi = 1$, by condition (B1), there will always be a component $\mathcal{B}$ in $\psi\xi_D$. We will consider the contributions to $\psi\xi_D$ (see Theorem 3.2.8) from Θ_ℓ and Θ_r in all possible circumstances. From conditions (B2) and (B3) (see Definition 3.2.1), it is straightforward to show that $\Theta_\ell\psi$ and $\Theta_r\psi$ are such that each contains at most one occurrence of L and R and cannot contain both.

Case: $\Theta\psi$ *contains* 1. Since $\emptyset\psi = 1$, this is always the case. The contribution to $\psi\xi_D$ from any $\tau \in \Theta^1$ such that $\tau\psi = 1$ is $\mathcal{B}$.

Case: $\Theta_\ell\psi$ *contains neither L nor R.* Since $\psi \ne 1$ it follows that T appears in $\Theta\psi$ and therefore in both $\Theta_\ell\psi$ and $\Theta_r\psi$. Let σ be the greatest element in Θ_ℓ such that $\sigma\psi = T$.

Then the contribution to $\psi\xi_D$ from σ is $\mathcal{S}^{\overline{\sigma}}$. Let $\tau \in \Theta_\ell$. If $\tau > \sigma$ then $\tau\psi = 1$, since $\Theta_\ell\psi$ contains neither L nor R, so that the contribution from τ to $\psi\xi_D$ is $\mathcal{B}$. If $\sigma > \tau$, then $\tau\psi = T$ and $\mathcal{S}^{\overline{\tau}} \supseteq \mathcal{S}^{\overline{\sigma}}$. Hence the contribution to $\psi\xi_D$ from Θ_ℓ^1 is $\mathcal{B} \cap \mathcal{S}^\lambda$ where $\lambda = \overline{\sigma}$.

Case: Θ_r *contains neither L nor R.* An argument identical to the preceding case shows that the contribution to $\psi\xi_D$ from Θ_r^1 is of the form $\mathcal{B} \cap \mathcal{S}^\rho$.

Case: $\Theta_\ell\psi$ *contains L.* Let $\sigma \in \Theta_\ell$ be the unique element in Θ_ℓ such that $\sigma\psi = L$. Note that $t(\sigma) = T_r$. Let $\tau \in \Theta_\ell$. If $\tau > \sigma$, then $\tau\psi = 1$ and the contribution to the formula for $\psi\xi_D$ from τ is $\mathcal{B}$. If $\tau < \sigma$, then $\tau\psi = T$ and there exists $\sigma_1 \in \Theta^1$ such that $\tau = \sigma T_\ell \sigma_1$. The contribution of τ to $\psi\xi_D$ is $\mathcal{S}^{\overline{\tau}}$ where

$$\mathcal{S}^{\overline{\tau}} = \mathcal{S}^{\overline{\sigma}_1 T_\ell \overline{\sigma}} \supseteq \mathcal{S}^{T_\ell \overline{\sigma}} = \mathcal{LRB}^{\overline{\sigma}} \supseteq \mathcal{LNB}^{\overline{\sigma}}.$$

The contribution to $\psi\xi_D$ from σ is $\mathcal{LNB}^{\overline{\sigma}}$. Hence, in this case, the contribution to $\psi\xi_D$ from Θ_ℓ^1 is $\mathcal{B} \cap \mathcal{LNB}^\lambda$ where $\lambda = \overline{\sigma}$ and $h(\lambda) = t(\sigma) = T_r$.

Case: $\Theta_\ell\psi$ *contains R.* The argument is entirely similar to the preceding case leading to the conclusion that the contribution to $\psi\xi_D$ from Θ_ℓ^1 is of the form $\mathcal{B} \cap \mathcal{RNB}^\rho$ where $h(\rho) = T_\ell$.

Case: Θ_r *contains L or R.* The arguments and conclusions are similar to the preceding cases.

We now consider the different possible combinations.

Case: $\Theta_\ell\psi$ *and* $\Theta_r\psi$ *both contain L, say* $\lambda \in \Theta_\ell$, $\rho \in \Theta_r$ *are such that* $\lambda\psi = \rho\psi = L$. Then the formula for $\psi\xi_D$ in Theorem 3.2.8 reduces to

$$\mathcal{B} \cap \mathcal{LNB}^\lambda \cap \mathcal{LNB}^\rho \quad \text{where} \quad h(\lambda) = h(\rho) = T_r.$$

However, we must have either $|\lambda| > |\rho|$ or $|\rho| > |\lambda|$. Without loss of generality, we can assume that $|\rho| > |\lambda|$ and we have $\psi\xi_D = \mathcal{B} \cap \mathcal{LNB}^\lambda$ where $h(\lambda) = T_r$.

Case: $\Theta_\ell\psi$ *and* $\Theta_r\psi$ *both contain R.* The argument is similar to the preceding case but leads to $\psi\xi_D = \mathcal{B} \cap \mathcal{RNB}^\rho$ where $h(\rho) = T_\ell$.

Case: *One of* $\Theta_\ell\psi$, $\Theta_r\psi$ *contains L and the other contains R, say* $\lambda \in \Theta_\ell$, $\rho \in \Theta_r$ *are such that* $\lambda\psi = L$, $\rho\psi = R$. Then the different contributions to $\psi\xi_D$ are $\mathcal{B}$, $\mathcal{LNB}^\lambda$ with $h(\lambda) = T_r$ and $\mathcal{RNB}^\rho$ with $h(\rho) = T_\ell$ yielding

$$\psi\xi_D = \mathcal{B} \cap \mathcal{LNB}^\lambda \cap \mathcal{RNB}^\rho \quad \text{where} \quad h(\lambda) = T_r, \ h(\rho) = T_\ell.$$

Case: *One of* $\Theta_\ell\psi$ *and* $\Theta_r\psi$ *contains L and the other contains neither L nor R.* The one that contains L contributes $\mathcal{LNB}^\lambda$ for some $\lambda \in \Theta$ with $h(\lambda) = T_r$ while the one that contains neither L nor R contributes $\mathcal{S}^\rho$ for some $\rho \in \Theta$. Thus

$$\psi\xi_D = \mathcal{B} \cap \mathcal{LNB}^\lambda \cap \mathcal{S}^\rho \quad \text{where} \quad h(\lambda) = T_r.$$

Case: One of $\Theta_\ell \psi$ and $\Theta_r \psi$ contains R and the other contains neither L nor R. This is similar to the preceding case and yields, for suitable $\lambda, \rho \in \Theta$,

$$\psi \xi_D = \mathcal{B} \cap \mathcal{S}^\lambda \cap \mathcal{RNB}^\rho \quad \text{where} \quad h(\rho) = T_\ell.$$

Case: Neither Θ_ℓ nor Θ_r contains L or R. Then each Θ_ℓ and Θ_r contributes a term of the for $\mathcal{S}^\lambda$ and we have $\psi \xi_D = \mathcal{B} \cap \mathcal{S}^\lambda \cap \mathcal{S}^\rho$ for some $\lambda, \rho \in \Theta$ where $\mathcal{S}$ and $\mathcal{S}^\rho$ may be comparable, in which case only the smaller (with the smaller of $|\lambda|, |\rho|$) is retained.
   This exhausts all the possibilities. □

Corollary 3.4.4 *Let $\mathcal{U} \in [\mathcal{S}, \mathcal{B})$. Then there exist $\lambda, \rho \in \Theta^1$ such that $\mathcal{U}$ has one of the following forms:*

(i) $\mathcal{S}^{\lambda(K_\ell, K_r)}$.
(ii) $\mathcal{LNB}^{\lambda(K_\ell, K_r)}$ or $\mathcal{RNB}^{\rho(K_\ell, K_r)}$ where $h(\lambda) = T_r$ and $h(\rho) = T_\ell$.
(iii) $\mathcal{LNB}^{\lambda(K_\ell, K_r)} \cap \mathcal{RNB}^{\rho(K_\ell, K_r)}$ where $h(\lambda) = T_r$ and $h(\rho) = T_\ell$.
(iv) $\mathcal{LNB}^{\lambda(K_\ell, K_r)} \cap \mathcal{S}^\rho$ where $h(\lambda) = T_r$ or
   $\mathcal{RNB}^{\rho(K_\ell, K_r)} \cap \mathcal{S}^\lambda$ where $h(\rho) = T_\ell$.
(v) $\mathcal{S}^{\lambda(K_\ell, K_r)} \cap \mathcal{S}^{\rho(K_\ell, K_r)}$ where $h(\lambda) = T_\ell, h(\rho) = T_r$ and $|\lambda| = |\rho| \geq 2$.

Proof This follows immediately from Theorem 3.4.3 and Lemma 3.2.5. □

   To this point we have been considering the generation of varieties of bands by the application of the upper K_ℓ and K_r operators. We conclude this section by showing how to generate all proper varieties of bands with the help of an elementary construction. The following concept captures the nature of the objects that we will construct.

Notation 3.4.5 Let

$$\mathcal{DCH}_{LR} = \{S \in \mathcal{B} \mid S/\mathcal{D} \text{ is a chain and } D_a \in \mathcal{LZ} \cup \mathcal{RZ} \text{ for all } a \in S\}.$$

Theorem 3.4.6 *Let $\mathcal{V} \in [\mathcal{S}, \mathcal{B})$. Then either $\mathcal{V} = \langle S \rangle$ or $\mathcal{V} = \langle S_1, S_2 \rangle$ for some $S, S_1, S_2 \in \mathcal{CR}$ are finite and belong to $\mathcal{DCH}_{LR}$.*

Proof We take advantage of the fact that, from Theorem 3.3.2, $\mathcal{V}$ is either one of the varieties

$$\mathcal{S}, \mathcal{LNB}, \mathcal{RNB},$$
$$\mathcal{S}^{u(K_\ell, K_r)}, \quad u \in \Theta$$
$$\mathcal{LNB}^{v(K_\ell, K_r)}, \quad v \in \Theta \text{ and } h(v) = T_r$$
$$\mathcal{RNB}^{w(K_\ell, K_r)}, \quad w \in \Theta \text{ and } h(w) = T_\ell$$

or is the *join* of two of these varieties. It therefore suffices to show that each of the listed varieties is generated by a single element in $\mathcal{DCH}_{LR}$. By [PR1, Theorem V.1.9], we know that the claim holds for $\mathcal{S}$, $\mathcal{LNB}$ and $\mathcal{RNB}$.

We consider just one typical case. Let $\mathcal{V} \in [\mathcal{S}, \mathcal{B})$ be such that $\mathcal{V} = \mathcal{V}^{K_\ell}$ and let $T \in \mathcal{V}$ be such that $\mathcal{V} = \langle T \rangle$, T is finite and $T \in \mathcal{DCH}_{LR}$. Let

$$[T^1] = \{[t] \mid t \in T^1\}, \quad S = T \cup [T^1].$$

Thus each $[t]$ is an *element* of the set $[T^1]$. We define a binary operation $\cdot$ on S by

$$
\begin{aligned}
t_1 \cdot t_2 &= t_1 t_2 & &\text{for } t_1, t_2 \in T, \\
[t_1] \cdot [t_2] &= [t_2] & &\text{for } t_1, t_2 \in T, \\
t_1 \cdot [t_2] &= [t_2] & &\text{for } t_1 \in T,\ t_2 \in T^1, \\
[t_1] \cdot t_2 &= [t_1 t_2] & &\text{for } t_1 \in T^1,\ t_2 \in T.
\end{aligned}
$$

It is straightforward to check that the associative law is satisfied by the operation $\cdot$ so that S is a semigroup with respect to this operation. Clearly $[T^1]$ is a right zero ideal in S so that S is a finite band and a member of $\mathcal{DCH}_{LR}$. Moreover, $S/[T^1] \cong T^0 \in \mathcal{V}$ so that $S \in \mathcal{V}^{K_r}$ and $\langle S \rangle \supseteq \langle T \rangle = \mathcal{V}$. Therefore $\langle S \rangle K_r \mathcal{V}$.

Now let $a, b \in T$ and $a \mathcal{L}^0 b$. Then we have $[a] = [1] \cdot a \mathcal{L}^0 [1] \cdot b = [b]$ which, since $[T^1]$ is a right zero subsemigroup of S, implies that $a = b$. Thus $\mathcal{L}^0 = \varepsilon_S$ and $S \cong S/\mathcal{L}^0 \in \langle S \rangle_{T_\ell}$. Consequently $\langle S \rangle = \langle S \rangle_{K_\ell}$.

But it is evident from Theorems 3.3.2 and 3.4.2(i) that the only variety in $\mathcal{V} K_r$ with the property that $\mathcal{V}_{K_\ell} = \mathcal{V}$ is $\mathcal{V}^{K_r}$. Hence $\mathcal{V}^{K_r} = \langle S \rangle$. Thus the claim holds for $\mathcal{V}^{K_r}$. Dually, we also have that $\mathcal{V}^{K_\ell}$ is of the form $\langle S \rangle$ for some finite $S \in \mathcal{DCH}_{LR}$ whenever $\mathcal{V} \in [\mathcal{S}, \mathcal{B})$, $\mathcal{V} = \mathcal{V}^{K_r}$ and $\mathcal{V} = \langle T \rangle$ for some finite $T \in \mathcal{DCH}_n$. The proof can then be completed with a simple induction argument. $\qquad\square$

The results in this section concerning the relations K_ℓ and K_r can be found in Reilly and Zhang [2].

3.5   Bases of Identities for Varieties in $\mathcal{L}(\mathcal{B})$

In Sect. 3.3 we determined the lattice of subvarieties of $\mathcal{B}$ and in Sect. 3.4 the relations K_ℓ, K_r on $\mathcal{L}(\mathcal{B})$. Bases of identities can be found in [PR1, Theorem V.1.9] for the following varieties of bands: $\mathcal{T}$, $\mathcal{LZ}$, $\mathcal{RZ}$, $\mathcal{RB}$, $\mathcal{S}$, $\mathcal{LNB}$, $\mathcal{RNB}$, $\mathcal{LRB}$, $\mathcal{RRB}$ and $\mathcal{NB}$. In this section we will build on these results to obtain a basis of identities for all proper subvarieties of $\mathcal{B}$. We will accomplish this by combining the characterization of K_ℓ- and K_r-related varieties of bands from Sect. 3.4 with the basis of identities provided for $\mathcal{V}^{T_\ell}$ and $\mathcal{V}^{T_r}$ in Theorem 2.5.2(vi).

In [PR1, Section I.9], we defined the reverse w^ρ of a word $w = x_1 \cdots x_n \in X^+$ to be $x_n \cdots x_1$. However, in order to align our notation with the notation that prevails in the literature on the subjects to follow, we now define the *mirror image* $\overline{w}$ of a word $w = x_1 \cdots x_n \in X^+$ to be the word $x_n \cdots x_1$. In the spirit of what is to follow, we can also define the mirror image of a word inductively as follows:

$$\overline{x} = x, \quad \overline{uv} = \overline{v}\,\overline{u} \quad (x \in X, \; u, v \in X^+).$$

Definition 3.5.1 We define inductively three systems of words in X^+ as follows:

$$G_2 = x_2x_1, \quad H_2 = x_2, \quad I_2 = x_2x_1x_2,$$
$$G_n = x_n\overline{G}_{n-1}, \quad P_n = G_nx_n\overline{P}_{n-1} \quad \text{for} \quad P \in \{H, I\} \text{ and } n \geq 3.$$

For example, $G_3 = x_3x_1x_2$, $H_3 = x_3x_1x_2x_3x_2$, $I_3 = x_3x_1x_2x_3x_2x_1x_2$.

The identity $x = x^2$ can be omitted when writing bases of identities for varieties of bands in many circumstances.

Lemma 3.5.2 *Let $\mathcal{V} \in \mathcal{L}(\mathcal{CR})$. If there exists $n \geq 2$ such that $\mathcal{V}$ satisfies the identity $G_n = P_n$ or $\overline{G}_n = \overline{P}_n$ for $P \in \{H, I\}$, then $\mathcal{V}$ satisfies the identity $x^2 = x$.*

Proof If $\mathcal{V}$ satisfies either of the identities $G_2 = H_2$ or $G_2 = I_2$, then it clearly follows that $\mathcal{V}$ satisfies the identity $x^2 = x$. If $\mathcal{V}$ satisfies an identity of the form $G_n = P_n$, for $P \in \{H, I\}$, then the substitution $x_i \rightarrow x_n^0$ for all $i < n$, yields the identity $x_n^2 = x_n$. □

It follows that in any basis of identities of the form $[G_n = P_n, \, x = x^2]$ or $[\overline{G}_n = \overline{P}_n, \, x = x^2]$, the identity $x = x^2$ is superfluous and we may simply write $[G_n = P_n]$ or $[\overline{G}_n = \overline{P}_n]$. However, in the bases

$$[axy = ayx, \, x = x^2], \; [xy = yx, \, x = x^2],$$
$$[xya = yxa, \, x = x^2], \; [axya = ayxa, \, x = x^2]$$

the identity $x = x^2$ cannot be dropped since in each case any commutative completely regular semigroup will satisfy the remaining identity.

Lemma 3.5.3 *We have the following.*

(i) $\mathcal{S}^{K\ell} = \mathcal{LNB}^{K\ell} = \mathcal{LRB} = [G_2 = I_2]$.
(ii) $\mathcal{S}^{K_r} = \mathcal{RNB}^{K_r} = \mathcal{RRB} = [\overline{G}_2 = \overline{I}_2]$.
(iii) $\mathcal{RNB}^{K\ell} = \mathcal{NB}^{K\ell} = \mathcal{LQNB}^{K\ell} = [G_3 = H_3]$.
(iv) $\mathcal{LNB}^{K_r} = \mathcal{NB}^{K_r} = \mathcal{RQNB}^{K_r} = [\overline{G}_3 = \overline{H}_3]$.
(v) $\mathcal{RRB}^{K\ell} = [G_3 = I_3]$.

(vi) $\mathcal{LRB}^{K_r} = [\overline{G}_3 = \overline{I}_3]$.

Proof By duality, it suffices to prove (ii), (iv) and (vi).

(ii) The first two equalities follow from Theorem 3.4.2(i). By Theorem 2.5.2(vi) and the fact that $\mathcal{S}^{K_r} = \mathcal{S}^K \cap \mathcal{S}^{T_r} = \mathcal{B} \cap \mathcal{S}^{T_r}$, we have

$$\mathcal{S}^{K_r} = [xyz = (yxz)^0 xyz, x^2 = x] = [xyz = yxzxyz, x^2 = x].$$

Substituting y for z, this implies that

$$\mathcal{S}^{K_r} \subseteq [xy = yxyxy, x^2 = x] = [xy = yxy, x^2 = x] = [\overline{G}_2 = \overline{I}_2].$$

Now, assuming the identity $\overline{G}_2 = \overline{I}_2$, we have $xyz = z(xy)z = xzxyz = yxzxyz$ so that $[\overline{G}_2 = \overline{I}_2] \subseteq \mathcal{S}^{K_r}$ and equality prevails.

(iv) The first two equalities follow from Theorem 3.4.2(i). From [PR1, Theorem V.1.7], we have $\mathcal{LNB} = [axy = ayx, x^2 = x]$. Then, by Theorem 2.5.2(vi)

$$\mathcal{V}^{K_r} = \mathcal{V}^{T_r} \cap \mathcal{B} = [axyz = ayxzaxyz, x^2 = x].$$

On the other hand $[\overline{G}_3 = \overline{H}_3] = [axy = ayaxy]$ Assuming the identity $\overline{G}_3 = \overline{H}_3$, we obtain

$$axyz = (ay)axyz = a(yx)yaxyz$$
$$= ay(xz)xyaxyz = (ayxz)xy(axyz)$$
$$= ayxzaxyz$$

so that $[\overline{G}_3 = \overline{H}_3] \subseteq \mathcal{V}^{K_r}$. Conversely, if we assume the identities for $\mathcal{V}^{K_r}$ and substitute ax for x and y for z in $axyz = ayxzaxyz$ we obtain

$$axy = a(ax)yy = ay(ax)ya(ax)yy = a(yax)(yax)y = ayaxy.$$

Thus $\mathcal{V}^{K_r} \subseteq [\overline{G}_3 = \overline{H}_3]$ and equality prevails.

(vi) We have

$$\mathcal{LRB}^{K_r} = (\mathcal{S}^{K_\ell})^{K_r} = [G_2 = I_2]^{K_r}$$
$$= [x_2 x_1 = x_2 x_1 x_2]^{K_r} = [x_2 x_1 = x_2 x_1 x_2]^{T_\ell} \cap [x_2 x_1 = x_2 x_1 x_2]^K$$
$$= [x_2 x_1 x_3 = (x_2 x_1 x_2 x_3)^0 x_2 x_1 x_3, x^2 = x]$$
$$= [x_2 x_1 x_3 = x_2 x_1 x_2 x_3 x_2 x_1 x_3]$$
$$= [\overline{G}_3 = \overline{I}_3]. \qquad \qquad \square$$

Lemma 3.5.3 deals with all varieties of bands in the extreme left- or right-hand columns involving G_2, H_2 or I_2 in their bases of identities. We now consider the general case.

Theorem 3.5.4 *Let $P \in \{H, I\}$ and $n \geq 3$.*

(i) $[G_n = P_n]^{K_r} = [\overline{G}_{n+1} = \overline{P}_{n+1}]$.   (ii) $[\overline{G}_n = \overline{P}_n]^{K_\ell} = [G_{n+1} = P_{n+1}]$.

Proof (i) We have

$$[G_n = P_n]^{K_r} = [G_n = P_n]^{T_r} \cap [G_n = P_n]^K = [G_n = P_n]^{T_r} \cap \mathcal{B}$$
$$= [G_n x_{n+1} = (P_n x_{n+1})^0 G_n x_{n+1}] \cap \mathcal{B} \quad \text{by Theorem 2.6.2(vi)}$$
$$= [\overline{G}_{n+1} = P_n x_{n+1} \overline{G}_{n+1}] \cap \mathcal{B}$$
$$= [\overline{G}_{n+1} = \overline{P}_{n+1}] \cap \mathcal{B}$$
$$= [\overline{G}_{n+1} = \overline{P}_{n+1}] \quad \text{by Lemma 3.5.2.}$$

(ii) The argument here is similar to that of part (i).   □

In the light of Theorem 3.4.2, it follows that the varieties on the extreme left of Diagram 3.3 are precisely those that are greatest in their K_ℓ-classes while those on the extreme right of Diagram 3.3 are precisely those that are greatest in their K_r-classes.

Recall from Theorem 2.6.2(x) that, for any $\mathcal{V} \in \mathcal{L}(\mathcal{B})$, $\mathcal{V} = \mathcal{V}^{K_\ell} \cap \mathcal{V}^{K_r} = \mathcal{V}_{K_\ell} \vee \mathcal{V}_{K_r}$.

Corollary 3.5.5 *Let $\mathcal{V} \in [\mathcal{S}, \mathcal{B})$. Then for some integer n and some $P \in \{H, I\}$, $\mathcal{V}$ has a basis of identities of the form*

$$[G_n = P_n] \text{ if } \mathcal{V} = \mathcal{V}^{K_\ell} \text{ and } [\overline{G}_n = \overline{P}_n] \text{ if } \mathcal{V} = \mathcal{V}^{K_r},$$

and otherwise one of the forms

$$[G_n = H_n, \overline{G}_n = \overline{H}_n], \ [G_n = H_n, \overline{G}_n = \overline{I}_n], \ [G_n = H_n, \overline{G}_{n-1} = \overline{I}_{n-1}] \text{ if } n \geq 3,$$
$$\text{or } [G_n = I_n, \overline{G}_{n+1} = \overline{H}_{n+1}], \ [G_n = I_n, \overline{G}_n = \overline{I}_n], \text{ if } n \geq 2,$$
$$\text{or } [G_n = I_n, \overline{G}_n = \overline{H}_n], \text{ if } n \geq 3.$$

Proof This follows from Theorem 3.5.4 and Corollary 3.4.4.   □

The results of Corollary 3.5.5 are displayed in Diagram 3.5 (while remembering that the identity $x^2 = x$ must be included for $\mathcal{LNB}$ and $\mathcal{RNB}$). It might be tempting to amalgamate certain pairs of identities used for the bases of varieties in the middle of the diagram into a single identity. For example, would $[G_n \overline{G}_n = I_n \overline{I}_n]$ define the same variety as $[G_n = I_n, \overline{G}_n = \overline{I}_n]$? The answer is 'no', unless you change some of the variables—see examples below—so that the variables in $G_n = I_n$ are completely distinct from the variables in $\overline{G}_n = \overline{I}_n$.

In contrast to the identities considered in Lemma 3.5.2, in general the identities of the form $\overline{G}_n G_n = \overline{T}_n T_n$ and $\overline{G}_n x_{n+1} G_n = T_n x_{n+1} T_n$ do not imply the identity $x^2 = x$.

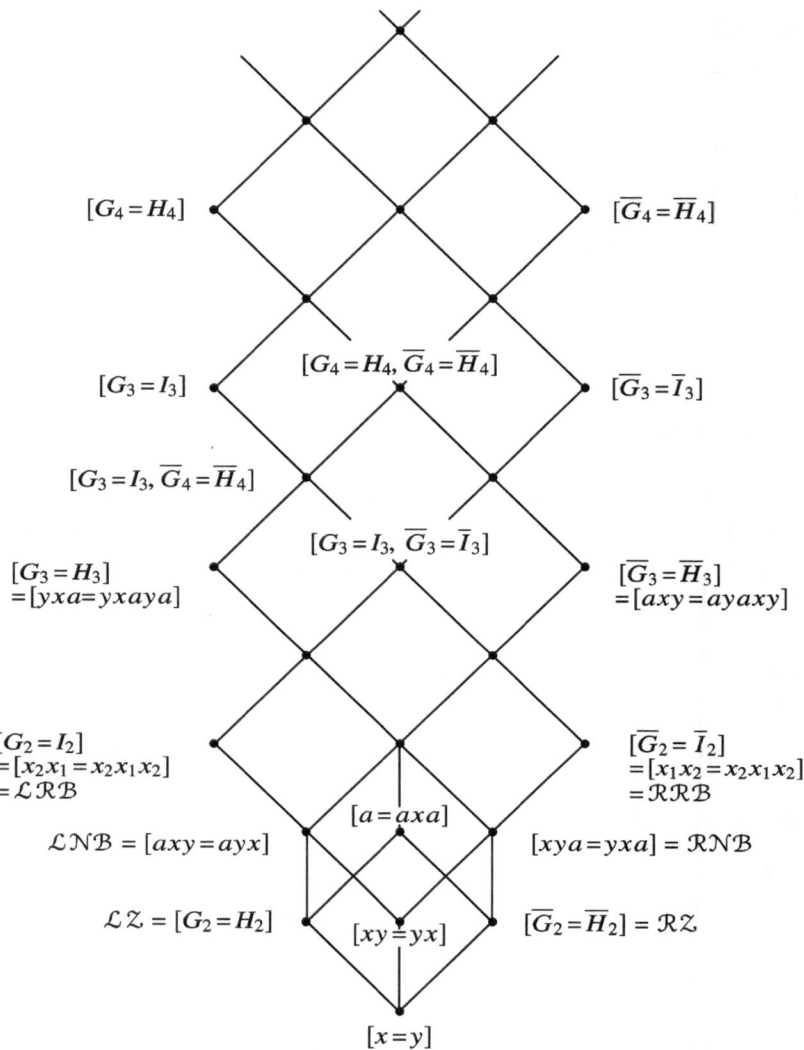

Diagram 3.5 Bases of identities for proper varieties of bands

Example 3.5.6 We have

$$\overline{G}_3 G_3 = x_2 x_1 x_3 x_3 x_1 x_2$$

$$\overline{I}_3 I_3 = x_2 x_1 x_2 x_3 x_2 x_1 x_3 \cdot x_3 x_1 x_2 x_3 x_2 x_1 x_2$$

so that the identity $\overline{G}_3 G_3 = \overline{I}_3 I_3$ is satisfied by $\mathbb{Z}_2$ and does not define a variety of bands.

The following is relevant to the above discussion. The property of being defined by a single identity is not unusual.

Lemma 3.5.7 *Let* $\mathcal{V} = [u_\alpha = v_\alpha]_{\alpha \in A} \in \mathcal{L}(\mathcal{CR})$ *where* A *is finite and* $\mathcal{RB} \subseteq \mathcal{V}$. *Then* $\mathcal{V} = [u = v]$ *for some* $u, v \in CR_X$.

Proof Let $A = \{1, \ldots, m\}$ and $c(u_\alpha v_\alpha) = \{x_{\alpha 1}, x_{\alpha 2}, \ldots, x_{\alpha n_\alpha}\}$, $\alpha \in A$. Since X is infinite while A and each n_α is finite, we may assume that $c(u_\alpha v_\alpha) \cap c(u_\beta v_\beta) = \emptyset$ for all $\alpha \neq \beta$. $u = u_1 \cdots u_m$, $v = v_1 \cdots v_m$. Clearly, $\mathcal{V} \subseteq [u = v]$. Since $\mathcal{RB} \subseteq \mathcal{V}$, for each $\alpha \in A$, u_α and v_α must begin with the same variable, $x_{\alpha 1}$ say, and end with the same variable, $x_{\alpha n_\alpha}$ say. If we now replace each $x_{\beta i}$ ($\beta < \alpha$) by $x_{\alpha 1}^0$ and each $x_{\gamma 1}$ ($\gamma > \alpha$) by $x_{\alpha n_\alpha}^0$, then the identity $u = v$ reduces to the identity $u_\alpha = v_\alpha$. Thus $[u = v] \subseteq \mathcal{V}$ and equality prevails.   □

The following system of words can also be useful in the study of varieties of bands.

Notation 3.5.8 We introduce another three systems of words in X^+ inductively as follows. Let

$$P_2 = x_2, \quad Q_2 = x_2 x_1, \quad R_2 = x_2 x_1 x_2,$$
$$T_n = R_{n-1} x_n \overline{T_{n-1}} \quad (T \in \{P, Q, R\}, \ n > 2).$$

It follows easily from the definition that $\overline{R_n} = R_n$ and $R_n = R_{n-1} x_n R_{n-1}$. We now use these words to create an alternative system of identities for defining varieties of bands.

Lemma 3.5.9 *For* $n > 2$, *the identity* $T_n = R_n$ *for* $T \in \{P, Q\}$ *implies the identity* $x^2 = x$.

Proof Clearly the number of occurrences of x_2 in R_n is 2^{n-1} while a simple induction argument will show that the number of occurrences of x_2 in T_n ($T \in \{P, Q\}$) is $2^{n-1} - 1$. Hence any group satisfying the identity $T_n = R_n$ must be trivial. Therefore any completely regular semigroup satisfying $T_n = R_n$ must be a band.   □

Lemma 3.5.10 *For* $T = P$, $n > 2$ *and* $T = Q$, $n \geq 2$, *we have*

$$[T_n = R_n]^{K_r} = [\overline{T_{n+1}} = R_n], \quad [\overline{T_n} = R_n]^{K_\ell} = [T_{n+1} = R_{n+1}].$$

Proof By Lemma 3.2.5 and Theorem 2.5.2(vi),

$$[T_n = R_n]^{K_r} = [T_n = R_n]^{T_r} \cap \mathcal{B} = [R_n x_{n+1} = (T_n x_{n+1})^0 R_n x_{n+1}] \cap \mathcal{B}$$
$$= [R_n x_{n+1} = T_n x_{n+1} R_n x_{n+1}].$$

On the other hand, in the context of bands, $\overline{T}_{n+1} = T_n x_{n+1} R_n = T_n x_{n+1} R_n x_{n+1} R_n$ so that

$$[\overline{T_{n+1}} = R_{n+1}] = [T_n x_{n+1} R_n x_{n+1} R_n = R_n x_{n+1} R_n]$$
$$\supseteq [T_n x_{n+1} R_n x_{n+1} = R_n x_{n+1}] = [T_n = R_n]^{K_r}.$$

Conversely,

$$R_{n+1} = \overline{T}_{n+1} \implies R_n x_{n+1} R_n = T_n x_{n+1} R_n$$
$$\implies R_n x_{n+1} R_n x_{n+1} = T_n x_{n+1} R_n x_{n+1}$$
$$\implies R_n x_{n+1} = T_n x_{n+1} R_n x_{n+1}.$$

Therefore $[\overline{T_{n+1}} = R_{n+1}] \subseteq [T_n = R_n]^{K_r}$. This proves the first equality and the second follows by a dual argument. $\qquad\square$

In both the systems of words $\{G_n, H_n, I_n\}$ and $\{P_n, Q_n, R_n\}$ we see that the inductive step amounts to applying the upper K_ℓ or K_r operators on varieties of bands. This makes it possible to compare the four sets of varieties which have arisen from these systems of identities. By duality, it suffices to consider the unbarred versions.

Lemma 3.5.11 *The following statements hold.*

(i) $[G_n = I_n] = [Q_n = R_n]$, $n \geq 2$.   (ii) $[G_n = H_n] = [P_n = R_n]$, $n \geq 3$.

Proof (i) The claim follows immediately from the definition for $n = 2$. In general, for $n \geq 3$

$$[G_n = I_n] = [Q_n = R_n] \implies [\overline{G}_n = \overline{I}_n] = [\overline{Q}_n = \overline{R}_n]$$
$$\implies [G_{n+1} = I_{n+1}] = [\overline{G}_n = \overline{I}_n]^{K_\ell} = [\overline{Q}_n = \overline{R}_n]^{K_\ell}$$
$$\implies [G_{n+1} = I_{n+1}] = [Q_{n+1} = R_{n+1}].$$

(ii) We begin by showing that $[G_3 = H_3] = [P_3 = R_3]$. We write these identities explicitly.

$$G_3 = H_3 : \ x_3 x_1 x_2 = x_3 x_1 x_2 x_3 x_2,$$
$$R_3 = P_3 : \ x_2 x_1 x_2 x_3 x_2 x_1 x_2 = x_2 x_1 x_2 x_3 x_2.$$

For easier comparison, we write the first identity as

$$x_1 x_2 x_3 = x_1 x_2 x_3 x_1 x_3. \tag{3.5.1}$$

Premultiplying the second by x_1 and swapping sides, we get the identity

$$x_1 x_2 x_3 x_2 = x_1 x_2 x_3 x_2 x_1 x_2. \tag{3.5.2}$$

Premultiplying this identity by x_2 we recover the identity $R_3 = P_3$ so that these identities are actually equivalent.

The substitution $x_1 \to x_1, x_2 \to x_2 x_3, x_3 \to x_2$ in (3.5.1) yields (3.5.2). If (3.5.2) holds, then

$$x_1 x_2 x_3 = (x_1 x_2 x_3 x_1) x_2 (x_3 x_1 x_2 x_3) = x_1 x_2 x_3 [x_1 x_3 (x_1 x_2) x_3]$$
$$= x_1 x_2 x_3 (x_1 x_3 x_1 x_2 x_3) x_1 x_3 = (x_1 x_2 x_3) x_1 x_3 x_1 (x_2 x_3 x_1 x_3)$$
$$= x_1 x_2 x_3 (x_2 x_3) x_1 x_3 = x_1 x_2 x_3 x_1 x_3$$

giving (3.5.1). The argument for the induction step in this part is entirely analogous to the argument in part (i) and is omitted. □

The lattice $\mathcal{L}(\mathcal{B})$ was described by Birjukov [1], Gerhard [1] and Fennemore [1]. While Fennemore obtained his result by deriving bases of identities for all varieties of bands, Gerhard used fully invariant congruences. More systematic bases of identities were found by Gerhard and Petrich [3, 4] where they also explored the left trace relation on the free band. Lemma 3.5.11 appears in Petrich [3], but the proof here is new and more direct. See also Petrich [4]. Fennemore [4, 5] characterized those bands that satisfy no nontrivial identity, see also Hall [1]. Varieties of normal bands with involution are considered in Dolinka [1] and Dolinka [4] proved that every finite band is finitely related. Koryakov [1] studied basis ranks for bands. Siekman and Szabó considered confluent rewrite systems for bands. For information on bands with involution, see Petrich [3], Dolinka [2, 3] and for word problems in certain varieties of bands, see Petrich [5]. Sapir [1] shows that for any proper variety $\mathcal{V}$ of bands containing the variety $[G_3 = H_3]$, there exists a continuum of quasivarieties of bands each of which generates the variety $\mathcal{V}$. Tamura [1] considers the attainability of identities on bands. Wismath [2] considers hyperidentities for *-bands.

3.6   The B-Relation

We shall see in this section that the variety $\mathcal{B}$ of all bands plays an important role in the context of the lattice of subvarieties of the variety $\mathcal{CR}$ of all completely regular semigroups. Clearly $\mathcal{L}(\mathcal{B})$ constitutes an important ideal of $\mathcal{L}(\mathcal{CR})$ but its structure has important ramifications across the whole of $\mathcal{L}(\mathcal{CR})$. Earlier, in our studies of the kernel and trace relations, we have seen how helpful complete congruences can be to our efforts to understand the structure of the lattice $\mathcal{L}(\mathcal{CR})$. In this section we will introduce one more complete congruence that will help us in the analysis of $\mathcal{L}(\mathcal{CR})$ globally and also play an important role in our efforts to understand the detailed structure of kernel classes.

Definition 3.6.1 The relation **B** defined by

$$\mathcal{U} \, \mathbf{B} \, \mathcal{V} \text{ if } \mathcal{U} \cap \mathcal{B} = \mathcal{V} \cap \mathcal{B} \quad (\mathcal{U}, \mathcal{V} \in \mathcal{L}(\mathcal{CR}))$$

is the **B**-*relation* on $\mathcal{L}(\mathcal{CR})$.

We have seen in Theorem 3.2.8(v) that the mapping

$$\mu_{\mathcal{B}} : \mathcal{V} \mapsto \mathcal{V} \cap \mathcal{B} \quad (\mathcal{V} \in \mathcal{L}(\mathcal{CR}))$$

is a complete retraction of $\mathcal{L}(\mathcal{CR})$ onto $\mathcal{L}(\mathcal{B})$. This yields that $\mathbf{B} = \overline{\mu_{\mathcal{B}}}$ is a complete congruence on $\mathcal{L}(\mathcal{CR})$. As a consequence, since $\mathcal{L}(\mathcal{CR})$ is a complete lattice, the **B**-classes are intervals, so for any $\mathcal{V} \in \mathcal{L}(\mathcal{CR})$, we may use the standard notation

$$\mathcal{V}\mathbf{B} = [\mathcal{V}_{\mathbf{B}}, \mathcal{V}^{\mathbf{B}}]$$

to denote the interval that is the **B**-class of $\mathcal{V}$. Clearly, $\mathcal{V}_B = \mathcal{V} \cap \mathcal{B}$.
In this way, the relation **B** induces two operators on $\mathcal{L}(\mathcal{CR})$ as follows.

Definition 3.6.2 The operators $\mathcal{V} \mapsto \mathcal{V}_B$, $\mathcal{V} \mapsto \mathcal{V}^B$, $(\mathcal{V} \in \mathcal{L}(\mathcal{CR}))$ are the *lower* and *upper* **B**-*operators* · on $\mathcal{L}(\mathcal{CR})$. The varieties of the form $\mathcal{V}^B$, $\mathcal{V} \in \mathcal{L}(\mathcal{CR})$ are the *uppermost* elements in their **B**-classes and the set of all such varieties will be denoted

$$\Upsilon = \{\mathcal{V}^B \mid \mathcal{V} \in \mathcal{L}(\mathcal{CR})\} = \{\mathcal{V}^B \mid \mathcal{V} \in \mathcal{L}(\mathcal{B})\}.$$

For these operators, we have the following basic properties.

Proposition 3.6.3 *The following statements hold.*

(i)   *The lower* **B**-*operator is a complete retraction of* $\mathcal{L}(\mathcal{CR})$ *onto* $\mathcal{L}(\mathcal{B})$.
(ii) *The upper* **B**-*operator is a complete* $\cap$-*endomorphism.*
(iii) Υ *is an* $\cap$-*subsemilattice of* $\mathcal{L}(\mathcal{CR})$. *Neither* $\Upsilon \cap [\mathcal{T}, \mathcal{CS}]$ *nor* $\Upsilon \cap [\mathcal{S}, \mathcal{CR}]$ *is a sublattice of* $\mathcal{L}(\mathcal{CR})$.
(iv) $\bigvee_{\mathcal{V} \in \Upsilon \setminus \{\mathcal{CR}\}} \mathcal{V} = \mathcal{CR}$.
(v)   *Let* $\mathcal{V} \in \mathcal{L}(\mathcal{CR})$. *Then* $\mathcal{V}^B = (\mathcal{V} \cap \mathcal{B})^B$.
(vi) *We have* $T \subseteq \mathbf{B}$ *and, for any* $\mathcal{V} \in \mathcal{L}(\mathcal{CR})$, *we have* $\mathcal{V}^B = (\mathcal{V}^B)^T$.
(vii) *The restriction of* T *to* Υ *is the identity relation.*

Proof (i) This is just a restatement of Theorem 3.2.8(v).

(ii) and first claim in (iii) These follow from the fact that B is a complete congruence on $\mathcal{L}(\mathcal{CR})$ and the dual of [PR1, Lemma I.2.2].

For the remaining claims in part (iii), first we have $\mathcal{LG} = \mathcal{LZ}^B$ and $\mathcal{RG} = \mathcal{LZ}^B$ are elements of Υ, however, $\mathcal{LG} \vee \mathcal{RG} = \mathcal{ReG} \neq \mathcal{CS} = \mathcal{RB}^B$ so that $\Upsilon \cap [\mathcal{T}, \mathcal{CS}]$ is not a sublattice of $\mathcal{L}(\mathcal{CR})$.

For the final claim in part (iii), let $\mathcal{V} = \mathcal{SG}$, the variety of all semilattices of groups, then

$$\mathcal{V}^{T_\ell} = \mathcal{LRO} \quad - \quad \text{the variety of all left regular orthogroups}$$
$$\mathcal{V}^{T_r} = \mathcal{RRO} \quad - \quad \text{the variety of all right regular orthogroups}$$
$$\mathcal{V}^{T_\ell} \vee \mathcal{V}^{T_r} = \mathcal{RO} \quad - \quad \text{the variety of all regular orthogroups}$$
$$\mathcal{V}^{T_\ell T_r} = \mathcal{R}^* \quad - \quad \text{the variety of all completely regular semigroups}$$
$$\qquad\qquad\qquad\qquad\qquad \text{on which } \mathcal{R} \text{ is a congruence}$$
$$\mathcal{V}^{T_r T_\ell} = \mathcal{L}^* \quad - \quad \text{the dual of } \mathcal{R}^*$$
$$\mathcal{V}^{T_\ell T_r} \cap \mathcal{V}^{T_r T_\ell} = \mathcal{RBG} \quad - \quad \text{the variety of all regular bands of groups.}$$

For the equality $\mathcal{LRO} \vee \mathcal{RRO} = \mathcal{RO}$ see [PR1, Theorem V.3.3].

Let $\mathcal{U} \in \mathcal{L}(\mathcal{CS})$, be a non-orthodox variety. Then we have $\mathcal{RO} \vee \mathcal{U} \in \{\mathcal{RO}, \mathcal{RBG}\}$. Thus

$$\mathcal{SG}^{T_\ell} \vee \mathcal{SG}^{T_r} = \mathcal{LRO} \vee \mathcal{RRO} \neq \mathcal{RBG} = \mathcal{SG}^{T_\ell T_r} \cap \mathcal{SG}^{T_r T_\ell}.$$

The situation is illustrated below.

(iv) A stronger result will be proved in [PR3, Theorem V.2.3(viii)].

(v) This is obvious, but is important to remember.

(vi) By Theorem 2.4.3, we have

$$\mathcal{U} \, T \, \mathcal{V} \Longrightarrow \mathcal{U} \cap \mathcal{F} = \mathcal{V} \cap \mathcal{F} \Longrightarrow \mathcal{U} \cap \mathcal{B} = \mathcal{V} \cap \mathcal{B} \Longrightarrow \mathcal{U} \, \mathbf{B} \, \mathcal{V}.$$

The second claim then holds since a variety that is maximal in its B-class must also be maximal in its T-class.

(vii) This follows from part (vi) and the obvious fact that the restriction of $\mathbf{B}$ to Υ is the identity relation. $\qquad\qquad\qquad\qquad\qquad\qquad\qquad\qquad\qquad\qquad\qquad\qquad\qquad$ □

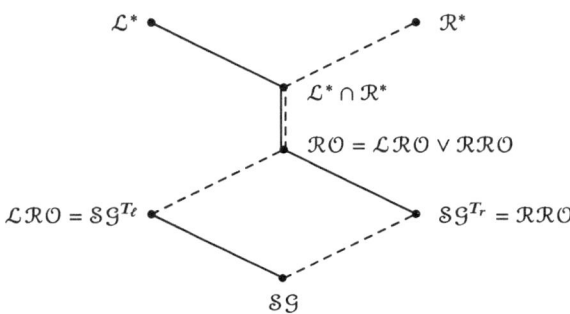

A very important and surprising fact about the elements of Υ is that they can all be generated or constructed by successively applying the upper operators associated with T_ℓ, T_r to just a handful of elements of $\mathcal{L}(\mathcal{B})$ and then taking the intersection of the varieties so obtained in pairs.

Theorem 3.6.4 *Let* $\lambda, \rho, \sigma, \tau \in \Theta$, $h(\lambda) = T_r$ *and* $h(\rho) = T_\ell$.

(i) $\mathcal{T}^B = \mathcal{G} = [x^0 = y^0]$, $\mathcal{L}\mathcal{Z}^B = \mathcal{L}\mathcal{G} = [x^0 y^0 = x^0]$,
   $\mathcal{R}\mathcal{Z}^B = \mathcal{R}\mathcal{G} = [x^0 y^0 = y^0]$, $\mathcal{R}\mathcal{B}^B = \mathcal{C}\mathcal{S} = [(axa)^0 = a^0]$,
   $\mathcal{S}^B = \mathcal{S}\mathcal{G} = [x^0 y^0 = y^0 x^0]$, $\mathcal{B}^B = \mathcal{C}\mathcal{R}$.

(ii) $\mathcal{L}\mathcal{N}\mathcal{B}^B = \mathcal{L}\mathcal{N}\mathcal{O} = [axy^0 = ay^0 x]$,
   $\mathcal{R}\mathcal{N}\mathcal{B}^B = \mathcal{R}\mathcal{N}\mathcal{O} = [y^0 xa = xy^0 a]$.

(iii) $\mathcal{L}\mathcal{R}\mathcal{B}^B = \mathcal{L}\mathcal{R}\mathcal{O} = [ax = axa^0] = \mathcal{S}^{T_\ell}$.
   $\mathcal{R}\mathcal{R}\mathcal{B}^B = \mathcal{R}\mathcal{R}\mathcal{O} = [xa = a^0 xa] = \mathcal{S}^{T_r}$,

(iv) $(\mathcal{L}\mathcal{N}\mathcal{B}^{\lambda(K_\ell, K_r)})^B = \mathcal{L}\mathcal{N}\mathcal{B}^\lambda$, $(\mathcal{R}\mathcal{N}\mathcal{B}^{\rho(K_\ell, K_r)})^B = \mathcal{R}\mathcal{N}\mathcal{B}^\rho$,
   $(\mathcal{S}^{\sigma(K_\ell, K_r)})^B = \mathcal{S}^\sigma$.

(v) $(\mathcal{L}\mathcal{N}\mathcal{B}^{\lambda(K_\ell, K_r)} \cap \mathcal{R}\mathcal{N}\mathcal{B}^{\rho(K_\ell, K_r)})^B = \mathcal{L}\mathcal{N}\mathcal{B}^\lambda \cap \mathcal{R}\mathcal{N}\mathcal{B}^\rho$.

(vi) $(\mathcal{L}\mathcal{N}\mathcal{B}^{\lambda(K_\ell, K_r)} \cap \mathcal{S}^{\sigma(K_\ell, K_r)})^B = \mathcal{L}\mathcal{N}\mathcal{B}^\lambda \cap \mathcal{S}^\sigma$,
   $(\mathcal{R}\mathcal{N}\mathcal{B}^{\rho(K_\ell, K_r)} \cap \mathcal{S}^{\sigma(K_\ell, K_r)})^B = \mathcal{R}\mathcal{N}\mathcal{B}^\rho \cap \mathcal{S}^\sigma$.

(vii) $(\mathcal{S}^{\sigma(K_\ell, K_r)} \cap \mathcal{S}^{\tau(K_\ell, K_r)})^B = \mathcal{S}^\sigma \cap \mathcal{S}^\tau$.

(viii) *Parts (i)–(vii) describe all elements of* Υ *and Diagram 3.6 is an accurate representation of* Υ *as an intersection subsemilattice of* $\mathcal{L}(\mathcal{C}\mathcal{R})$. *All the labels of the form* $\mathcal{U}^B$ *and* $\mathcal{P}^\lambda$, *for* $\mathcal{U} \in \mathcal{L}(\mathcal{B})$, $\mathcal{P} \in \{\mathcal{L}\mathcal{N}\mathcal{B}, \mathcal{S}, \mathcal{R}\mathcal{N}\mathcal{B}\}$, $\lambda \in \Theta$ *are accurately positioned.*

(ix) *Let* $\mathcal{V} \in [\mathcal{S}, \mathcal{C}\mathcal{R}]$. *Then*

$$\mathcal{V}^B = \mathcal{C}\mathcal{R} \cap \left(\bigcap_{\mathcal{V}_\tau = \mathcal{L}\mathcal{N}\mathcal{B}} \mathcal{L}\mathcal{N}\mathcal{B}^\tau \right) \cap \left(\bigcap_{\mathcal{V}_\tau = \mathcal{R}\mathcal{N}\mathcal{B}} \mathcal{R}\mathcal{N}\mathcal{B}^\tau \right) \cap \left(\bigcap_{\mathcal{V}_\tau = \mathcal{S}} \mathcal{S}^\tau \right)$$

which may be abbreviated to

$$\mathcal{V}^B = \mathcal{C}\mathcal{R} \cap \left(\bigcap_{\mathcal{V}_\tau \in \mathbb{N}_3} (\mathcal{V}_\tau)^\tau \right).$$

(x) *Let* $\mathcal{V} \in [\mathcal{S}, \mathcal{B}]$. *If* $\mathcal{V} = \mathcal{B}$ *or is* $\cap$-*irreducible in* $\mathcal{L}(\mathcal{B})$, *then* $\mathcal{V}^B$ *is also* $\cap$-*irreducible. Otherwise* $\mathcal{V}^B$ *is reducible.*

(xi) *Let* $\mathcal{U} \in \mathcal{L}(\mathcal{B})$, $\mathcal{V} \in \mathcal{L}(\mathcal{C}\mathcal{R})$ *and* $\mathcal{V} \cap \mathcal{B} \subseteq \mathcal{U}$. *Then* $\mathcal{V} \subseteq \mathcal{U}^B$.

Proof (i) Proofs of these claims are all straightforward.

(ii) It is clear from [PR1, Corollary IV.2.12] that $\mathcal{L}\mathcal{N}\mathcal{O} \cap \mathcal{B} = \mathcal{S} \vee \mathcal{L}\mathcal{Z} = \mathcal{L}\mathcal{N}\mathcal{B}$. Hence $\mathcal{L}\mathcal{N}\mathcal{O} \subseteq \mathcal{L}\mathcal{N}\mathcal{B}^B$. On the other hand, for any $S \in \mathcal{L}\mathcal{N}\mathcal{B}^B$, we know that S has no subsemigroup isomorphic to R_2 (the two element right zero semigroup) and so S must be a semilattice of left groups. By [PR1, Lemma V.1.4], S cannot contain a copy of L_2^1. Hence S satisfies $\mathcal{D}$-majorization for idempotents and therefore, by [PR1, Lemma II.4.14], also $\mathcal{D}$-majorization. From [PR1, Theorem IV.1.6], it now follows that S is a strong semilattice of left groups, that is, $S \in \mathcal{LNO}$. Thus $\mathcal{LNB}^B \subseteq \mathcal{LNO}$ and equality prevails. The second equality also follows from [PR1, Lemma V.3.1] while the claims for $\mathcal{RNB}^B$ follow by duality.

(iii) Since $\mathcal{LRB}$ is the greatest variety of bands not containing R_2 by [PR1, Lemma V.1.2], it follows from [PR1, Lemma V.3.1] that $\mathcal{LRO} \cap \mathcal{B} \subseteq \mathcal{LRB}$ and therefore that $\mathcal{LRO} \subseteq \mathcal{LRB}^B$. On the other hand, no element in $\mathcal{LRB}^B$ can contain a copy of R_2 which implies, by [PR1, Lemma V.3.1] that $\mathcal{LRB}^B \subseteq \mathcal{LRO}$ and equality prevails. The second equality follows from [PR1, Lemma I.5.5 while the claims for $\mathcal{RRB}^B$ follow by duality.

(iv) *Case*: $\mathcal{LNB}^{\lambda(K_\ell,K_r)}$ where $h(\lambda) = T_r$. By Lemma 3.2.5, we have

$$(\mathcal{LNB}^\lambda) \cap \mathcal{B} = \mathcal{LNB}^{\lambda(K_\ell,K_r)}$$

which implies that $\mathcal{LNB}^\lambda \subseteq (\mathcal{LNB}^{\lambda(K_\ell,K_r)})^B$. Now let $\mathcal{V} \in \mathcal{L}(\mathcal{CR})$ be such that $\mathcal{V} \cap \mathcal{B} \subseteq \mathcal{LNB}^{\lambda(K_\ell,K_r)}$. Then

$$\mathcal{V}_{\bar{\lambda}} \cap \mathcal{B} = (\mathcal{V} \cap \mathcal{B})_{\bar{\lambda}} \subseteq (\mathcal{LNB}^{\lambda(K_\ell,K_r)})_{\overline{\lambda(K_\ell,K_r)}}$$
$$= \mathcal{LNB}_{K_r} = \mathcal{LNB}.$$

Therefore, by [PR1, Corollary IV.2.12], $\mathcal{V}_{\bar{\lambda}} \subseteq \mathcal{LNO} \subseteq \mathcal{LNB}^T$. Consequently,

$$\mathcal{V} \subseteq (\mathcal{V}_{\bar{\lambda}})^\lambda \subseteq (\mathcal{LNB}^T)^\lambda = \mathcal{LNB}^\lambda.$$

Hence $(\mathcal{LNB}^{\lambda(K_\ell,K_r)})^B = \mathcal{LNB}^\lambda$.

Case: $\mathcal{RNB}^{\rho(K_\ell,K_r)}$. This is the dual of the preceding case.

Case: $\mathcal{S}^{\sigma(K_\ell,K_r)}$. By Lemma 3.2.5, $\mathcal{S}^\sigma \cap \mathcal{B} = \mathcal{S}^{\sigma(K_\ell,K_r)}$ so that $\mathcal{S}^\sigma \subseteq (\mathcal{S}^{\sigma(K_\ell,K_r)})^B$. Now let $\mathcal{V} \in \mathcal{L}(\mathcal{CR})$ be such that $\mathcal{V} \cap \mathcal{B} \subseteq \mathcal{S}^{\sigma(K_\ell,K_r)}$. Then

$$\mathcal{V}_{\bar{\sigma}} \cap \mathcal{B} = (\mathcal{V} \cap \mathcal{B})_{\bar{\sigma}} \subseteq (\mathcal{S}^{\sigma(K_\ell,K_r)})_{\overline{\sigma(K_\ell,K_r)}} = \mathcal{S}$$

which implies that $\mathcal{V}_{\bar{\sigma}} \subseteq \mathcal{SG} = \mathcal{S}^T$. Hence $\mathcal{V} \subseteq (\mathcal{V}_{\bar{\sigma}})^\tau \subseteq (\mathcal{S}^T)^\sigma = \mathcal{S}^\sigma$. Consequently, $\mathcal{S}^\sigma = (\mathcal{S}^{\sigma(K_\ell,K_r)})^B$.

(v)–(vii) It suffices for all these cases to know that for $\mathcal{U}, \mathcal{V} \in \mathcal{L}(\mathcal{B})$, we have $(\mathcal{U} \cap \mathcal{V})^B = \mathcal{U}^B \cap \mathcal{V}^B$. But this follows immediately from Proposition 3.6.3.

(viii) Since the least element in every B-class is, necessarily, a variety of bands, it follows from Theorem 3.3.2 and Corollary 3.4.4 that every B-class is covered in the statement of the theorem.

Clearly it follows from Proposition 3.6.3(i) that the restriction of the mapping μ_B to Υ is an order preserving bijection of Υ onto $\mathcal{L}(\mathcal{B})$. Therefore Diagram 3.6 faithfully represents Υ as a partially ordered set. By Proposition 3.6.3(ii), the mapping $\mathcal{V} \to \mathcal{V}^B$ is an $\cap$-preserving

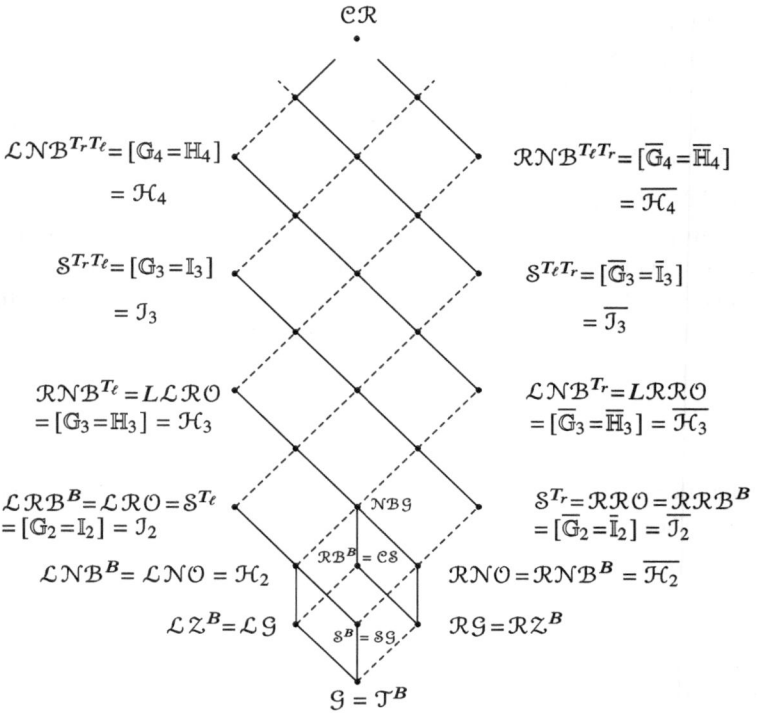

Diagram 3.6 The intersection subsemilattice Υ

homomorphism. Therefore Diagram 3.6 also faithfully represents Υ as an $\cap$-semilattice. The final claim concerning the positioning of labels follows from parts (i)–(vii).

(ix) If $\mathcal{B} \subseteq \mathcal{V}$, then the claim is trivial since for any $\tau \in \Theta$, we have $\mathcal{V}_\tau \supseteq \mathcal{B}_\tau = \mathcal{B}$. So assume that $\mathcal{B} \nsubseteq \mathcal{V}$. By Theorem 3.2.8(iv),

$$\mathcal{V} \cap \mathcal{B} = \left(\bigcap_{\mathcal{V}_\tau = \mathcal{LNB}} \mathcal{LNB}^{\overline{\tau}(K_\ell, K_r)} \right) \cap \left(\bigcap_{\mathcal{V}_\tau = \mathcal{RNB}} \mathcal{RNB}^{\overline{\tau}(K_\ell, K_r)} \right) \cap \left(\bigcap_{\mathcal{V}_\tau = \mathcal{S}} \mathcal{S}^{\overline{\tau}(K_\ell, K_r)} \right).$$

By Proposition 3.6.3(ii),

$$\mathcal{V}^B = (\mathcal{V} \cap \mathcal{B})^B$$

$$= \left(\bigcap_{\mathcal{V}_\tau = \mathcal{LNB}} \mathcal{LNB}^{\overline{\tau}(K_\ell, K_r)} \right)^B \cap \left(\bigcap_{\mathcal{V}_\tau = \mathcal{RNB}} \mathcal{RNB}^{\overline{\tau}(K_\ell, K_r)} \right)^B \cap \left(\bigcap_{\mathcal{V}_\tau = \mathcal{S}} \mathcal{S}^{\overline{\tau}(K_\ell, K_r)} \right)^B$$

$$= \left(\bigcap_{\mathcal{V}_\tau = \mathcal{LNB}} \mathcal{LNB}^{\overline{\tau}} \right) \cap \left(\bigcap_{\mathcal{V}_\tau = \mathcal{RNB}} \mathcal{RNB}^{\overline{\tau}} \right) \cap \left(\bigcap_{\mathcal{V}_\tau = \mathcal{S}} \mathcal{S}^{\overline{\tau}} \right)$$

where the last step follows from parts (iv)–(vii).

(x) If $\mathcal{V} = \mathcal{B}$, then $\mathcal{V}^B = \mathcal{B}^B = \mathcal{CR}$ so that the claim is trivially true in this case. So now assume that $\mathcal{V} \in [\mathcal{S}, \mathcal{B})$ and is $\cap$-irreducible. By Theorem 3.4.2(ii), this means that $\mathcal{V}$ has one of the following forms: $\mathcal{S}^{\lambda(K_\ell, K_r)}$, $\mathcal{LNB}^{\rho(K_\ell, K_r)}$ or $\mathcal{RNB}^{\sigma(K_\ell, K_r)}$, for some $\lambda, \rho, \sigma \in \Theta$. Without loss of generality, we may assume that $\mathcal{V} = \mathcal{S}^{\lambda(K_\ell, K_r)}$. By part (iv), it then follows that $\mathcal{V}^B = \mathcal{S}^{\lambda(T_\ell, T_r)} = \mathcal{S}^\lambda$.

Now let $\mathcal{U}, \mathcal{W} \in \mathcal{L}(\mathcal{CR})$ be such that $\mathcal{S}^\lambda \subset \mathcal{U}, \mathcal{W}$. Then $\mathcal{S}^\lambda \subset \mathcal{U}^B, \mathcal{W}^B$ and it follows that $(\mathcal{S}^\lambda)\mathbf{B} < \mathcal{U}\mathbf{B}, \mathcal{W}\mathbf{B}$ in the inherited order on $\mathcal{L}(\mathcal{CR})/\mathbf{B}$. On the other hand, $\mathcal{S}^{\lambda(K_\ell, K_r)}$, $\mathcal{U} \cap \mathcal{B}, \mathcal{W} \cap \mathcal{B}$ are all the least elements in their respective $\mathbf{B}$-classes and so we must have $\mathcal{S}^{\lambda(K_\ell, K_r)} \subset \mathcal{U} \cap \mathcal{B}, \mathcal{V} \cap \mathcal{B}$. But $\mathcal{V} = \mathcal{S}^{\lambda(K_\ell, K_r)}$ is $\cap$-irreducible in $\mathcal{L}(\mathcal{B})$. Hence

$$\mathcal{S}^{\lambda(K_\ell, K_r)} \subset (\mathcal{U} \cap \mathcal{B}) \cap (\mathcal{W} \cap \mathcal{B}) = (\mathcal{U} \cap \mathcal{W}) \cap \mathcal{B}$$

so that $\mathcal{S}^{\lambda(K_\ell, K_r)}\mathbf{B} < (\mathcal{U} \cap \mathcal{W})\mathbf{B}$. Therefore $\mathcal{V}^B = \mathcal{S}^\lambda = \mathcal{S}^B \neq \mathcal{U} \cap \mathcal{W}$, as required.

Now assume that $\mathcal{V} \neq \mathcal{B}$ and is reducible. That means that for some choice of $\mathcal{U}, \mathcal{V} \in \{\mathcal{LNB}, \mathcal{S}, \mathcal{RNB}\}$ and $\rho, \sigma \in \Theta$ the varieties $\mathcal{U}^{\rho(K_\ell, K_r)}, \mathcal{V}^{\sigma(K_\ell, K_r)}$ are distinct,

$$\mathcal{V} \subset \mathcal{U}^{\rho(K_\ell, K_r)}, \mathcal{V}^{\sigma(K_\ell, K_r)} \quad \text{and} \quad \mathcal{V} = \mathcal{U}^{\rho(K_\ell, K_r)} \cap \mathcal{V}^{\sigma(K_\ell, K_r)}.$$

Consequently,

$$\mathcal{V}^B = (\mathcal{U}^{\rho(K_\ell, K_r)} \cap \mathcal{V}^{\sigma(K_\ell, K_r)})^B = (\mathcal{U}^{\rho(K_\ell, K_r)})^B \cap (\mathcal{V}^{\sigma(K_\ell, K_r)})^B = \mathcal{U}^\rho \cap \mathcal{V}^\sigma$$

where $\mathcal{V} \subset \mathcal{U}^\rho, \mathcal{V}^\sigma$. Thus $\mathcal{V}^B$ is reducible.

(xi) We have

$$\mathcal{V} \cap \mathcal{B} \subseteq \mathcal{U} \Longrightarrow (\mathcal{V} \vee \mathcal{U}) \cap \mathcal{B} = (\mathcal{V} \cap \mathcal{B}) \vee (\mathcal{U} \cap \mathcal{B}) = (\mathcal{V} \cap \mathcal{B}) \vee \mathcal{U} = \mathcal{U}$$

$$\Longrightarrow (\mathcal{V} \vee \mathcal{U})^B = \mathcal{U}^B \Longrightarrow \mathcal{V} \subseteq \mathcal{U}^B. \qquad \square$$

Definition 3.6.5 The part of a ladder L labelled by elements of $\mathbb{N}_3$, if there are any, including their position (level) in L is the *socle* of L and will be denoted by $soc(L)$. If there are no such elements, then $soc(L) = \emptyset$.

The following lemma will be useful in recognizing **B**-related elements.

Lemma 3.6.6 (i) *Let* $\mathcal{V} \in [\mathcal{S}, \mathcal{CR}]$ *and* $\tau \in \Theta$. *Then* $\mathcal{V}_\tau \in \mathbb{N}_3 \Leftrightarrow (\mathcal{V} \cap \mathcal{B})_\tau \in \mathbb{N}_3$. *and* $\mathcal{V}_\tau = (\mathcal{V} \cap \mathcal{B})_\tau$ *whenever* $(\mathcal{V} \cap \mathcal{B})_\tau \in \mathbb{N}_3$.

(ii) *Let* $\mathcal{U}, \mathcal{V} \in [\mathcal{S}, \mathcal{CR}]$. *Then* $\mathcal{U} \mathbf{B} \mathcal{V} \Leftrightarrow soc(\delta_\mathcal{u}) = soc(\delta_v)$.

Proof (i) We have

$$V_\tau \in N_3 \implies (V \cap B)_\tau = V_\tau \cap B_\tau = V_\tau \cap B = V_\tau.$$

Conversely, suppose that $(V \cap B)_\tau \in N_3$. First assume that $(V \cap B)_\tau = \mathcal{LNB}$. Then $V_\tau \cap B = (V \cap B)_\tau = \mathcal{LNB} = \mathcal{LNB} \cap B$ so that $\mathcal{LNB} \subseteq V_\tau \subseteq \mathcal{LNO}$. By Theorem 1.5.5, the only variety in $[\mathcal{LNB}, \mathcal{LNO}]$ that is the least element in its T_ℓ or T_r-class is $\mathcal{LNB}$ and $\mathcal{LNB}$ is least in its T_r-class. Hence $t(\tau) = T_r$ and

$$\mathcal{LNB} = \mathcal{LNB}_{T_r} \subseteq V_{\tau T_r} = V_\tau \subseteq \mathcal{LNO}_{T_r} = \mathcal{LNB}.$$

Therefore $V_\tau = \mathcal{LNB}$. Dually, we have $(V \cap B)_\tau = \mathcal{RNB} \Rightarrow V_\tau = \mathcal{RNB}$.

Finally, suppose that $(V \cap B)_\tau = \mathcal{S}$. Then $V_\tau \cap B = \mathcal{S}$ which implies that $V_\tau \subseteq \mathcal{SG}$. But $(\mathcal{SG})_{T_\ell} = (\mathcal{SG})_{T_r} = \mathcal{S}$. Therefore $V_\tau = \mathcal{S}$ and the proof is complete.

(ii) First note that $\mathcal{U} \cap B = B$ if and only if $\mathrm{soc}(\mathcal{U}) = \emptyset$. The claim then follows from this observation and Theorem 3.2.8(ii). $\qquad\square$

The meet subsemilattice Υ is depicted in Diagram 3.6. Note that each vertex in the diagram represents a distinct variety since it is of the form $\mathcal{U}^B$ for a distinct unique variety $\mathcal{U}$ in $\mathcal{L}(B)$.

Note that the $\cap$-irreducible varieties in Υ are $\mathcal{LG}, \mathcal{RG}, \mathcal{CS}$ and those (containing $\mathcal{S}$) in the extreme left and right hand columns of Diagram 3.6. The intersections of these varieties generate all the remaining varieties in Υ.

Warning. Diagram 3.6 accurately represents the relative ordering of the *greatest elements in each B-class* and their intersections. It does *not* accurately reflect the relative position of joins to meets. A casual look at Diagram 3.6 might suggest, for instance, that $\mathcal{H}_{n+1} \cap \overline{\mathcal{H}}_{n+1} \subseteq \mathcal{I}_n \vee \overline{\mathcal{I}}_n$. But that is not the case. For example $\mathcal{H}_3 \cap \overline{\mathcal{H}}_3 \not\subseteq \mathcal{I}_2 \vee \overline{\mathcal{I}}_2$, since the latter variety is orthodox and the former variety is not.

Notation 3.6.7 Let $X = \{x_1, x_2, \ldots\}$. We now define sequences of words $\mathbb{G}_n, \overline{\mathbb{G}}_n, \mathbb{P}_n, \overline{\mathbb{P}}_n$ in U_X, for $\mathbb{P} \in \{\mathbb{H}_n, \mathbb{I}_n\}$, inductively and related varieties $\mathcal{H}_n, \mathcal{I}_n$ as follows:

$$\mathbb{G}_2 = x_2 x_1, \quad \mathbb{H}_2 = x_2, \quad \mathbb{I}_2 = x_2 x_1 x_2^0, \quad \overline{\mathbb{G}}_2 = x_1 x_2, \quad \overline{\mathbb{H}}_2 = x_2, \quad \overline{\mathbb{I}}_2 = x_2^0 x_1 x_2,$$

while, for $n > 2$,

$$\mathbb{G}_n = x_n \overline{\mathbb{G}}_{n-1}, \qquad \overline{\mathbb{G}}_n = \mathbb{G}_{n-1} x_n$$
$$\mathbb{P}_n = \mathbb{G}_n (x_n \overline{\mathbb{P}}_{n-1})^0, \quad \overline{\mathbb{P}}_n = (\mathbb{P}_{n-1} x_n)^0 \overline{\mathbb{G}}_n$$
$$\mathcal{H}_n = [\mathbb{G}_n = \mathbb{H}_n], \quad \mathcal{I}_n = [\mathbb{G}_n = \mathbb{I}_n].$$

Note that, within bands, the words $\mathbb{G}_n, \mathbb{H}_n$ and $\mathbb{I}_n$ simplify to exactly the words G_n, H_n and I_n used in Sect. 3.5 to generate bases of identities for varieties of bands. In Theorem 3.6.4(i)–

(iii), we provided bases of identities for the varieties of the form $\mathcal{U}^B$ where $\mathcal{U}$ is near the base of $\mathcal{L}(\mathcal{B})$. These were all familiar varieties. In the next theorem we provide bases that are defined inductively for the remaining proper varieties of the form $\mathcal{U}^B$.

Following a path similar to the one that we used when dealing with varieties of bands, we can now provide bases of identities for the varieties in Υ and also show how they lie in a neat pattern of T_ℓ and T_r-classes.

Theorem 3.6.8 (i) *For $\lambda \in \Theta$ with $h(\lambda) = T_r$, $n \in \mathbb{N}$,*

$$\mathcal{S}^\lambda = \begin{cases} [\overline{\mathbb{G}_{2n}} = \overline{\mathbb{I}_{2n}}] & \text{if } |\lambda| = 2n - 1 \\ [\mathbb{G}_{2n+1} = \mathbb{I}_{2n+1}] & \text{if } |\lambda| = 2n. \end{cases}$$

(ii) *For $\rho \in \Theta$ with $h(\rho) = T_\ell$, $n \in \mathbb{N}$,*

$$\mathcal{S}^\rho = \begin{cases} [\mathbb{G}_{2n} = \mathbb{I}_{2n}] & \text{if } |\rho| = 2n - 1 \\ [\overline{\mathbb{G}_{2n+1}} = \overline{\mathbb{I}_{2n+1}}] & \text{if } |\rho| = 2n. \end{cases}$$

(iii) *For $\lambda \in \Theta$ with $h(\lambda) = T_r$, $n \in \mathbb{N}$,*

$$\mathcal{LNB}^\lambda = \begin{cases} [\overline{\mathbb{G}_{2n+1}} = \overline{\mathbb{H}_{2n+1}}] & \text{if } |\rho| = 2n - 1 \\ [\mathbb{G}_{2n} = \mathbb{H}_{2n}] & \text{if } |\rho| = 2n. \end{cases}$$

(iv) *For $\rho \in \Theta$ with $h(\rho) = T_\ell$, $n \in \mathbb{N}$,*

$$\mathcal{RNB}^\rho = \begin{cases} [\mathbb{G}_{2n+1} = \mathbb{H}_{2n+1}] & \text{if } |\lambda| = 2n - 1 \\ [\overline{\mathbb{G}_{2n}} = \overline{\mathbb{H}_{2n}}] & \text{if } |\lambda| = 2n. \end{cases}$$

(v) *The following intervals are T_ℓ-classes (for $n \geq 2$):*

$$[\mathcal{J}, \mathcal{LG}], \ [\mathcal{RZ}, \mathcal{CS}], \ [\mathcal{S}, \mathcal{LRO}], \ [\mathcal{RNB}, \mathcal{LLRO}],$$

$$\left[[\overline{\mathbb{G}_n} = \overline{\mathbb{I}_n}], [\mathbb{G}_{n+1} = \mathbb{I}_{n+1}] \right], \ \left[[\overline{\mathbb{G}_{n+1}} = \overline{\mathbb{H}_{n+1}}], [\mathbb{G}_{n+2} = \mathbb{H}_{n+2}] \right].$$

(vi) *The following intervals are T_r-classes (for $n \geq 2$):*

$$[\mathcal{J}, \mathcal{RG}], \ [\mathcal{LZ}, \mathcal{CS}], \ [\mathcal{S}, \mathcal{RRO}], \ [\mathcal{LNB}, \mathcal{LRRO}],$$

$$\left[[\mathbb{G}_n = \mathbb{I}_n], [\overline{\mathbb{G}_{n+1}} = \overline{\mathbb{I}_{n+1}}] \right], \ \left[[\mathbb{G}_{n+1} = \mathbb{H}_{n+1}], [\overline{\mathbb{G}_{n+2}}, \overline{\mathbb{H}_{n+2}}] \right].$$

(vii) *The T_ℓ-relation (respectively, T_r-relation) between varieties in Υ is represented in Diagram 3.6 by solid (respectively, broken) lines.*

Proof (i) Let $\lambda \in \Theta$ and $h(\lambda) = T_r$. We proceed by induction on $|\lambda|$. For $|\lambda| = 1$, we have $\lambda = T_r$ so that $\mathcal{S}^\lambda = \mathcal{S}^{T_r} = \mathcal{RRO}$, by Theorem 1.5.5. By the dual of [PR1, Lemma V.3.1], this implies that $\mathcal{S}^{T_r} = [xa = a^0xa]$. On the other hand, $\left[\overline{\mathbb{G}_2} = \overline{\mathbb{I}}_2\right] = \left[x_1x_2 = x_2^0x_1x_2\right]$ which is the same variety.

Now assume that $|\lambda| = 2n - 1$ $(n \geq 1)$ and that $\mathcal{S}^\lambda = \left[\overline{\mathbb{G}_{2n}} = \overline{\mathbb{I}}_{2n}\right]$. Then $t(\lambda) = T_r$ and we have

$$\left[\overline{\mathbb{G}_{2n}} = \overline{\mathbb{I}}_{2n}\right]^{T_\ell} = \left[x_{2n+1}\overline{\mathbb{G}_{2n}} = x_{2n+1}\overline{\mathbb{G}_{2n}}(x_{2n+1}\overline{\mathbb{I}}_{2n})^0\right]$$

by the dual of Theorem 2.5.2(vi).

$$= [\mathbb{G}_{2n+1} = \mathbb{I}_{2n+1}].$$

Now assume that $|\lambda| = 2n$ $(n \geq 1)$ and that $\mathcal{S}^\lambda = [\mathbb{G}_{2n+1} = \mathbb{I}_{2n+1}]$. Then $t(\lambda) = T_\ell$ and we have

$$[\mathbb{G}_{2n+1} = \mathbb{I}_{2n+1}]^{T_r} = [\mathbb{G}_{2n+1}x_{2n+2} = (\mathbb{I}_{2n+1}x_{2n+2})^0\mathbb{G}_{2n+1}x_{2n+2}]$$
$$= \left[\overline{\mathbb{G}_{2n+2}} = (\mathbb{I}_{2n+1}x_{2n+2})^0\overline{\mathbb{G}_{2n+2}}\right]$$
$$= \left[\overline{\mathbb{G}_{2n+2}} = \overline{\mathbb{I}}_{2n+2}\right].$$

This completes the proof of (i).

(ii), (iii) and (iv) The proofs of the claims in these parts follow the same pattern as that for part (i).

(v), (vi) The first row of classes in each of these parts was determined in Theorem 1.5.5 (and its dual). Consider $V = [\mathbb{G}_{2n} = \mathbb{I}_{2n}]$. By part (i), $V = \mathcal{S}^\rho$ where $h(\rho) = T_\ell$ and $|\rho| = 2n - 1$. Hence $t(\rho) = T_\ell$ and we can write $\rho = \rho_1 T_\ell$ for some $\rho_1 \in \Theta^1$. By Theorem 2.5.2(xiii), we then have $V_{T_r} = (\mathcal{S}^{\rho_1 T_\ell})_{T_r} = \mathcal{S}^{\rho_1 T_\ell} = \mathcal{S}^\rho = V$. Thus V is the least element in its T_r-class. Moreover, $V^{T_r} = \mathcal{S}^{\rho T_r}$ where $h(\rho T_r) = T_\ell$ and $|\rho T_r| = 2n$. By part (ii), $V^{T_r} = \left[\overline{\mathbb{G}_{2n+1}}, \overline{\mathbb{I}}_{2n+1}\right]$. Consequently $\left[[\mathbb{G}_{2n} = \mathbb{I}_{2n}], [\overline{\mathbb{G}_{2n+1}}, \overline{\mathbb{I}}_{2n+1}]\right]$ is a T_r-class. Similar arguments apply to the solid lines.

(vii) It suffices to consider the T_r-relation. By Theorem 1.5.5, the following pairs of varieties are T_r-related: $(\mathcal{G}, \mathcal{RG})$, $(\mathcal{LG}, \mathcal{CS})$, $(\mathcal{S}, \mathcal{RRO})$, $(\mathcal{LNO}, \mathcal{LRRO})$, confirming the corresponding broken lines in Diagram 3.6. The accuracy of the broken lines in Diagram 3.6, starting from $\mathcal{LRO}$ and above was established in part (v). Similar arguments apply to the solid lines.

Similar arguments establish all the remaining cases.   □

Parts (v) and (vi) of Theorem 3.6.8 take care of all the T_ℓ- and T_r-classes of members of Υ.

We can summarize the description of the bases for elements of Υ, as follows.

Corollary 3.6.9 *Let* $V \in \Upsilon \cap [\mathcal{S}, \mathcal{CR})$. *Then for some integer* n *and some* $P \in \{\mathbb{H}, \mathbb{I}\}$, V *has a basis of identities of the form*

$$[\mathbb{G}_n = P_n] \quad if \quad V = V^{T_\ell} \quad and \quad [\overline{\mathbb{G}_n} = \overline{P}_n] \quad if \quad V = V^{T_r},$$

and otherwise one of the forms

$$[\mathbb{G}_n = \mathbb{H}_n, \overline{\mathbb{G}_n} = \overline{\mathbb{H}_n}], \ [\mathbb{G}_n = \mathbb{H}_n, \overline{\mathbb{G}_n} = \overline{\mathbb{I}_n}], \ [\mathbb{G}_n = \mathbb{H}_n, \overline{\mathbb{G}_{n-1}} = \overline{\mathbb{I}_{n-1}}] \ (n \geq 3)$$
$$or \ [\mathbb{G}_n = \mathbb{I}_n, \overline{\mathbb{G}_{n+1}} = \overline{\mathbb{H}_{n+1}}], \ [\mathbb{G}_n = \mathbb{I}_n, \overline{\mathbb{G}_n} = \overline{\mathbb{I}_n}] \ (n \geq 2),$$
$$or \ [\mathbb{G}_n = \mathbb{I}_n, \overline{\mathbb{G}_n} = \overline{\mathbb{H}_n}] \ (n \geq 3).$$

Proof This follows from Theorems 3.6.4, 3.6.8(i)–(iv) and Corollary 3.6.9(i)–(iv). □

For the material in this section, see Reilly and Zhang [2] and Petrich [8]. The relation on $\mathcal{L}(\mathcal{CR})$ defined by forming joins with $\mathcal{B}$ is considered in Petrich [9]. See Reilly and Zhang [5, 6] and Trotter and Weil [1] for the treatment of similar topics in the lattice of pseudovarieties of finite semigroups.

3.7 Free Bands

In this section we describe the free band on a nonempty set X and use that to provide a solution to the word problem for the free band. In the context of bands, we have $x^0 = x^{-1} = x$ and so it is immaterial whether we work in the signature of $(2, 1)$ for completely regular semigroups or simply the signature for semigroups. For this section, we choose the latter.

We start by recalling some basic notation and terminology concerning the free semigroup X^+ on a nonempty set X that we require here from [PR1, Chapter I]. The solution to the word problem for the free band is based on X^+.

As usual the elements of X^+ are referred to as *words* and the elements of X are called *variables*. A word $w \in X^+$ is thought of as a finite product of variables. We sometimes use the free monoid X^* on X obtained by adjoining the empty word $\emptyset$ to X^+.

Recall that the content $c(w)$ of a word w in X^+ is the set of variables in w. Write $w = uxp$ where $c(w) = c(ux), c(w) \neq c(u)$ and $x \in X$. Let $s(w) = u$ and $\sigma(w) = x$. Note that $s(w) = \emptyset$ if $c(w) = \{x\}$. Thus $s(w)\sigma(w)$ is the shortest left cut of w containing all variables of w. Dually write $w = qyv$ where $c(w) = c(yv), c(w) \neq c(u)$ and $y \in X$. Let $\varepsilon(w) = y$ and $e(w) = v$. Thus $\varepsilon(w)e(w)$ is the shortest right cut of w containing all variables of w.

For convenience we treat c, s, σ etc. as operators and omit parentheses. For example $c\sigma s(w)$ means $c(\sigma(s(w)))$. The operators $s, \sigma, \varepsilon, e$ are combined to form $b : X^+ \to X^+$ which is defined by

$$b(w) = bs(w)\sigma(w)\varepsilon(w)be(w) \quad (= b(s(w))\sigma(w)\varepsilon(w)b(e(w))).$$

This is an inductive definition on $|c(w)|$. In particular $b(x) = xx$, since $s(x) = e(x) = \emptyset$. Note that $cb(w) = c(w)$.

Lemma 3.7.1 *The above mappings have the following properties.*

$$(i)\ bs = sb,\ be = eb.\quad (ii)\ \sigma b = \sigma,\ \varepsilon b = \varepsilon.\quad (iii)\ b^2 = b.$$

Proof Let $u \in X^+$, $x \in X$.

(i) If $u = x^k$ for some $k \geq 1$, then $bs(x^k) = \emptyset = s(xx) = sb(x^k)$. If $|c(u)| > 1$, then

$$sb(u) = s(bs(u)\sigma(u)\varepsilon(u)be(u)) = bs(u).$$

This proves that $bs = sb$; the equality $be = eb$ follows dually.

(ii) If $u = x^k$ for some $k \geq 1$, then $\sigma b(x^k) = \sigma(xx) = x = \sigma(x^k)$. If $|c(u)| > 1$, then

$$\sigma b(u) = \sigma(bs(u)\sigma(u)\varepsilon(u)be(u)) = \sigma(u).$$

This shows that $\sigma b = \sigma$; the equality $\varepsilon b = \varepsilon$ follows dually.

(iii) The argument is by induction on $|c(u)|$. If $u = x^k$ for some $k \geq 1$, then

$$b(x^k) = xx = b(xx) = bb(x^k).$$

If $|c(u)| > 1$, then

$$
\begin{aligned}
bb(u) &= bsb(u)\sigma b(u)\varepsilon b(u)beb(u) \\
&= b^2 s(u)\sigma(u)\varepsilon(u)b^2 e(u) \quad \text{by parts (i) and (ii)} \\
&= bs(u)\sigma(u)\varepsilon(u)be(u) \quad\ \ \text{by the induction hypothesis} \\
&= b(u)
\end{aligned}
$$

Therefore $b^2 = b$, as required.   □

Definition 3.7.2 We define the extensions of the definition of $s, \sigma, \varepsilon, e$ as follows:

$$
\begin{aligned}
s^0(w) &= w, & e^0(w) &= w. \\
s^{k+1}(w) &= s(s^k(w)), & e^{k+1}(w) &= e(e^k(w)) \quad \text{for all } k \geq 0. \\
\sigma^{k+1}(w) &= \sigma(s^k(w)), & \varepsilon^{k+1}(w) &= \varepsilon(e^k(w)) \quad \text{for all } k \geq 0.
\end{aligned}
$$

(For large enough k, $s^k(w)$, $\sigma^k(w)$, $\varepsilon^k(w)$, $e^k(w)$ are equal to $\emptyset$.)

Define $s^A(w)$ and $\sigma^A(w)$ so that $\sigma^A(w) \in X$ and $s^A(w)\sigma^A(w)$ is the shortest left cut of w containing all variables of $c(w) \setminus A$. Define dually $e^A(w)$ and $\varepsilon^A(w)$ so that $\varepsilon^A(w) \in X$ and $\varepsilon^A(w)e^A(w)$ is the shortest right cut of w containing all variables of $c(w) \setminus A$. (Again these may be $\emptyset$.)

Lemma 3.7.3 *Let $u \in X^+$ and A be a nonempty subset of X. Then*

$$s^A(u) = s^k(u), \quad \sigma^A(u) = \sigma^k(u),$$

where k is the least integer such that $\sigma^k(u) \notin A$.

Proof Note that if $c(u) \subseteq A$, then $\sigma^A(u) = \emptyset$ and if $|c(u) \setminus A| \leq 1$, then $s^A(u) = \emptyset$. If $t = |c(u)|$, then $(\sigma^t(u), \ldots, \sigma(u))$ is the sequence of variables of u in order of first occurrence. By definition of k, $s^k(u)\sigma^k(u)$ is the shortest left cut of u containing all variables of u not in A. Therefore $s^k(u)\sigma^k(u) = s^A(u)\sigma^A(u)$ and the lemma follows. $\qquad\square$

Lemma 3.7.4 *For all $k \geq 1$, $\sigma^k b = \sigma^k$.*

Proof The argument is by induction on k. For $k = 1$, the statement is just Lemma 3.7.1(ii). For $k \geq 1$ and $u \in X^+$,

$$\begin{aligned}
\sigma^{k+1}b(u) &= \sigma s^k b(u) \\
&= \sigma b s^k(u) \quad \text{by Lemma 3.7.1(i)} \\
&= \sigma s^k(u) \quad\;\; \text{by case } k = 1 \\
&= \sigma^{k+1}(u).
\end{aligned}$$

$\qquad\square$

Corollary 3.7.5 *For any nonempty set A of X, we have $s^A b = b s^A$, $\sigma^A b = \sigma^A$.*

Proof Lemmas 3.7.3 and 3.7.4 imply that $s^A(bu) = s^k(bu)$ and $s^A(u) = s^k(u)$ for the same k. The first formula therefore follows from Lemma 3.7.1(i). A similar argument yields the second formula using Lemma 3.7.1(ii). $\qquad\square$

Lemma 3.7.6 *Let τ be any congruence on X^+ such that X^+/τ is a band. Then, for any $u \in X^+$, $u \, \tau \, s(u)\,\sigma(u)\,\varepsilon(u)\,e(u)$.*

Proof We show first that $c(u) = c(v) \Rightarrow u\tau \mathrel{\mathcal{D}} v\tau$. Let $c(u) = c(v) = \{x_1, \ldots, x_n\}$. Then $u\tau \mathrel{\mathcal{D}} (x_1\tau) \cdots (x_n\tau) \mathrel{\mathcal{D}} v\tau$, since $(X^+/\tau)/\mathcal{D}$ is a semilattice. It then follows that $c(u) \subseteq c(v) \Rightarrow D_{v\tau} \leq D_{u\tau}$ since $c(y) \subseteq c(v)$ implies $c(uv) = c(v)$ so that $D_{v\tau} \leq D_{(uv)\tau} \leq D_{u\tau}$.

To prove the lemma note that since X^+/τ is a band,

$$u \, \tau \, uu = s(u)\sigma(u)w\varepsilon(u)e(u),$$

for some $w \in X^+ \cup \{\emptyset\}$ with $c(w) \subseteq c(u) = c(s(u)\sigma(u)) = c(\varepsilon(u)e(u))$. $\qquad\square$

Notation 3.7.7 Let $B = b(X^+)$ with the multiplication $u \cdot v = b(uv)$.

Note that in view of Lemma 3.7.1(iii), we have $B = \{w \in X^+ \mid b(w) = w\}$.
We now arrive at a representation of the free band.

Theorem 3.7.8 *The mapping b is a homomorphism of X^+ onto B and B is a free band on X.*

Proof According to the definition of multiplication in B, we must prove that for any u, $v \in X^+$,

$$b(u) \cdot b(v) = b(b(u)b(v)) = b(uv). \tag{3.7.1}$$

Applying the definition of b, we thus have to show that

$$[bs(b(u)b(v))][\sigma(b(u)b(v))][\varepsilon(b(u)b(v))][be(b(u)b(v))]$$
$$= [bs(uv)][\sigma(uv)][\varepsilon(uv)][be(uv)]. \tag{3.7.2}$$

The argument is by induction on $|c(u) \cup c(v)|$. For the first step, we have $u = x^m$ and $v = x^n$ for some $x \in X$ and $m, n \geq 1$. We now compute

$$b(b(x^m)b(x^n)) = b(xxxx) = xx = b(x^m x^n).$$

For the inductive step, we first prove that

$$bs(b(u)b(v)) = bs(uv). \tag{3.7.3}$$

On the one hand, we have

$$bs(b(u)b(v)) = \begin{cases} bsb(u) & \text{if } c(v) \subseteq c(u), \\ b(b(u)s^{c(u)}b(v)) & \text{otherwise,} \end{cases}$$

and on the other hand,

$$bs(uv) = \begin{cases} bs(u) & \text{if } c(v) \subseteq c(u), \\ b(us^{c(u)}(v)) & \text{otherwise.} \end{cases}$$

The desired equality in the case $c(v) \subseteq c(u)$ follows directly from Lemma 3.7.1(i)(ii). For the case $c(v) \not\subseteq c(u)$, $s^{c(u)}b(v) \neq s^0(b(v))$ so that

$$|c(b(u)s^{c(u)}b(v))| < |c(u) \cup c(v)|,$$

and analogously $|c(us^{c(u)}v)| < |c(u) \cup c(v)|$, and we may use the induction hypothesis. Indeed,

$$b(b(u)s^{c(u)}b(v)) = b^2(u)bs^{c(u)}b(v) \text{ by the induction hypothesis}$$
$$= b(u)bs^{c(u)}(v) \qquad \text{by Lemma 3.7.1(iii) and}$$
$$\text{Corollary 3.7.5(i)(ii)}$$
$$= b(us^{c(u)}(v)) \qquad \text{by the induction hypothesis.}$$

This proves (3.7.3).

Next we show that

$$\sigma(b(u)b(v)) = \sigma(uv). \tag{3.7.4}$$

Indeed,

$$\sigma(b(u)b(v)) = \begin{cases} \sigma b(u) & \text{if } c(v) \subseteq c(u) \\ \sigma^{c(u)}b(v) & \text{otherwise} \end{cases}$$

$$= \begin{cases} \sigma(u) & \text{if } c(v) \subseteq c(u) \text{ by Lemma 3.7.1(ii).} \\ \sigma^{c(u)}(v) & \text{otherwise} \qquad \text{by Corollary 3.7.5} \end{cases}$$

$$= \sigma(uv)$$

which proves (3.7.4).

Relations (3.7.3) and (3.7.4) imply the equality of the first and second brackets on the left and on the right in (3.7.2). The equality of the remaining two pairs of brackets follows by duality. This proves (3.7.2) and thus also (3.7.1), as required.

To prove that B is a free band, we show that if $\overline{b}$ is the congruence induced on X^+ by b, then $\overline{b} \subseteq \tau$ for any congruence τ on X^+ such that X^+/τ is a band.

Assume $b(u) = b(v)$. By Lemma 3.7.1(i)(ii) we have

$$bs(u) = sb(u) = sb(v) = bs(v),$$
$$\sigma(u) = \sigma b(u) = \sigma b(v) = \sigma(v).$$

The proof that $u \tau v$ is by induction of $|c(u) \cup c(u)|$. Note that $x^n \tau x^m$ for $x \in X$ since X^+/τ is a band. By induction we have $s(u) \tau s(v)$ since $bs(u) = bs(v)$ as we just proved and $c(s(u)) \not\subseteq c(u)$. Using this, the fact that $\sigma(u) = \sigma(v)$ (as just shown) and the dual result gives

$$s(u)\,\sigma(u)\,\varepsilon(u)\,e(u) \quad \tau \quad s(v)\,\sigma(v)\,\varepsilon(v)\,e(v).$$

Finally, an application of Lemma 3.7.6 shows that $u \tau v$.   □

Remark 3.7.9 A straightforward inductive argument can be used to show that the congruence induced on X^+ by b is the congruence β defined inductively as follows:

$$u \beta v \iff s(u) \beta s(v), \quad \sigma(u) = \sigma(v), \quad \varepsilon(u) = \varepsilon(v), \quad e(u) \beta e(v),$$

where we formally set $\emptyset \beta \emptyset$.

We now describe an algorithm for computing $b(w)$.

Step 1: Double the word w to ww.

Step 2: Write $ww = w_0 x_0 w' x_1 w_1$ where $x_0 \in X$ and $w_0 x_0$ is the shortest left cut of ww (or w) which contains all variables of w and $x_1 \in X$ and $x_1 w_1$ is the shortest right cut of ww (or w) which contains all variables of w. This factorization is possible because of the form of ww.

Step 3: Delete the word w' retaining the word $w_0 x_0 x_1 w_1$. Note that this amounts to deleting all letters in ww which occur earlier and later in the word ww.

Step 4: Apply Steps 1, 2 and 3 to w_0 thereby obtaining the words $w_{00} x_{00}$ and $x_{01} w_{01}$. Also apply the same steps to w_1 thereby obtaining the words $w_{10} x_{10}$ and $x_{11} w_{11}$.

Continue this procedure on w_{00}, w_{01}, w_{10} and w_{11} until the end. This procedure must finish since at each step, the content of each $w_{i_1 \ldots i_k}$ is one less than in the preceding step.

We thus arrive at a word of the form

$$x_{00\ldots0} \cdots x_{00} x_{01} x_0 x_1 x_{10} x_{11} \cdots x_{11\ldots1},$$

where the number of 0's in the first subscript equals the number of 1's in the last subscript which in its turn equals $|c(w)|$. This word is our $b(w)$. The length of $b(w)$ is $2(2^{|c(w)|} - 1)$.

From the above steps one can easily devise a test for a word to have the property that $b(w) = w$.

We illustrate the above procedure by the following example. Let $w = (xyxy^3xz)^2 y$. The algorithm consists, briefly, of two steps: doubling and taking certain subwords. In each of the following steps the underlined subword is doubled in the next line.

Step 1: $(xyxy^3xz)^2 y (xyxy^3xz)^2 y$ doubling w.

Step 2: $\underline{xyxy^3x}\ zx\ \underline{zy}$ deleting (where $s(w) = xyxy^3x$, $\sigma(w) = z$, $\varepsilon(w) = x$, $e(w) = zy$).

Step 3: $xyxy^3xxyxy^3xzxzyzy$ doubling.

Step 4: $\underline{x}\ yy\ \underline{x}\ zx\ \underline{z}\ yz\ \underline{y}$ deleting.

Step 5: $xxyyxxzxzzyzyy$ doubling.
$b(w) = xxyyxxzxzzyzyy$.

A different canonical form for words in the treatment of the free band was devised by Siekmann and Szabó [1]. The attraction of their approach is that the canonical word is the unique shortest word representing an element of the free band. They show that a word can be reduced to the shortest form by replacing uu by u and pqr by pr if $c(q) \subseteq c(p) = c(r)$. It may be of some interest that the canonical word discussed above can be reduced to theirs by making only the substitutions of the form uu by u.

The expanded canonical word used here may be much longer than that of Siekmann and Szabó (which is the word of least length). But the expanded canonical word reflects the intrinsic nature of the usual solution of the word problem. The principal advantage of this treatment is that at every step the procedure is unique and mechanical; there is no search for suitable reductions and so it is easier to apply. In particular, the route which leads from a word to its canonical form is unique.

It is evident from Remark 3.7.9 that if X is a finite set, then so also is B. Thus we have the following corollary.

Corollary 3.7.10 *If X is a finite set, then so also is the free band on X.*

The material in this section is drawn from Gerhard and Petrich [2]. Corollary 3.7.10 follows from Green and Rees [1], McLean [1] and Brown [1]. For a general construction of an arbitrary band and applications, see Petrich [P1, Theorem II.1.6 and Sects. II.3–II.7]. Pastijn and Albert [1] solved the word problem for free split bands. Petrich and Silva [1, 3] solve the word problem and provide a structure for relatively free bands in varieties of bands and study *-bands in [2, 4]. See also Gerhard [9]. Jones [4] studied the free product of bands. Araújo and Konieczny [1] describe the automorphisms of endomorphism monoids of relatively free bands. Nordahl [1] and Olin [1] consider free products of bands.

3.8 Lattice of Band Monoids

In this section we will calculate the lattice of band monoids. In order to be precise, for this section we make the following definition.

Definition 3.8.1 By a *band monoid* we mean a triple $(S, \cdot, 1)$ where $(S, \cdot)$ is band and $1 \in S$ is a nullary operation or constant such that $1s = s1 = s$ for all $s \in S$. The class $\mathcal{BM}$ of all band monoids is clearly a variety and we denote the lattice of subvarieties of $\mathcal{BM}$ by $\mathcal{L}(\mathcal{BM})$.

Recall from [PR1, Section IX.8] that $\mathcal{M}$ denotes the class of all completely regular semigroups with an identity. Clearly every completely regular semigroup with an identity can be considered as being endowed with a nullary operation or constant 1 and so can be viewed as, or has a reduct that is, an algebra of type $(2, 0)$. Thus it is fair to write $\mathcal{BM} = \mathcal{B} \cap \mathcal{M}$.

Recall the definitions of G_n, H_n, I_n and P_n in the context of bands from Definition 3.5.1. We adopt a similar notation for *band monoids*.

Notation 3.8.2 We will write

$$G_n = G_n(x_1, \ldots, x_n), \quad P_n = P_n(x_1, \ldots, x_n) \text{ for } n \geq 2, \ P \in \{H, I\}$$
$$\mathcal{BM}P_n = [G_n = P_n], \ n \geq 2, \ P \in \{H, I\}$$
$$\mathcal{BM}\overline{P_n} = [\overline{G}_n = \overline{P}_n], \ n \geq 2, \ P \in \{H, I\}$$

Here $\mathcal{BM}P_n$ and $\mathcal{BM}\overline{P_n}$ stand for $\mathcal{B} \cap \mathcal{M} \cap P_n$ and $\mathcal{B} \cap \mathcal{M} \cap \overline{P_n}$, respectively.

We indicate the deletion of a variable in $G_n(x_1, \ldots, x_n)$ or $P_n(x_1, \ldots, x_n)$ by placing 1 in the position of that variable. For example

$$G_3(x_1, 1, x_3) = x_3 x_1, \quad H_4(x_1, x_2, 1, x_4) = x_4 x_2 x_1 x_4 x_2 x_2 x_1.$$

Theorem 3.8.3 (i) *The mapping* $\theta_{\mathcal{BM}} : \mathcal{V} \to \mathcal{V} \cap \mathcal{BM}$ $(\mathcal{V} \in \mathcal{L}(\mathcal{B}))$ *is a complete homomorphism of* $\mathcal{L}(\mathcal{B})$ *onto* $\mathcal{L}(\mathcal{BM})$.

(ii) *For* $\mathcal{V} \in \mathcal{L}(\mathcal{B})$, *the smallest element in* $\mathcal{V}_{\overline{\mathcal{BM}}}$ *is* $\langle \mathcal{V} \cap \mathcal{M} \rangle$.

(iii) *Let* $\mathcal{V} = [u_\alpha = v_\alpha]_{\alpha \in A} \in \mathcal{L}(\mathcal{B})$. *Then*

$$\langle \mathcal{V} \cap \mathcal{BM} \rangle = [u'_\alpha = v'_\alpha \,|\, u'_\alpha, \, v'_\alpha \text{ can be obtained from } u_\alpha = v_\alpha \text{ by the simultaneous}$$
$$\text{deletion of a (possibly empty) subset of}$$
$$\text{variables from both } u_\alpha \text{ and } v_\alpha]_{\alpha \in A}.$$

(iv) *The classes of* $\overline{\theta_{\mathcal{BM}}}$ *are as follows*

$$\{\mathcal{J}, \mathcal{LZ}, \mathcal{RZ}, \mathcal{RB}\}, \quad \{\mathcal{S}, \mathcal{LNB}, \mathcal{RNB}, \mathcal{NB}\}, \quad \{\mathcal{B}\}$$
$$\{\mathcal{B} \cap \mathcal{J}_{n-1}, \overline{\mathcal{J}_n} \cap \mathcal{B} \cap \mathcal{H}_n, \mathcal{B} \cap \mathcal{H}_n\}, \quad n \geq 3$$
$$\{\mathcal{B} \cap \overline{\mathcal{J}_{n-1}}, \mathcal{J}_n \cap \mathcal{B} \cap \overline{\mathcal{H}_n}, \mathcal{B} \cap \overline{H_n}\}, \quad n \geq 3$$
$$\{\mathcal{J}_{n-1} \cap \mathcal{B} \cap \overline{\mathcal{J}_{n-1}}, \mathcal{J}_{n-1} \cap \mathcal{B} \cap \overline{\mathcal{H}_n},$$
$$\mathcal{H}_n \cap \mathcal{B} \cap \overline{\mathcal{J}_{n-1}}, \mathcal{H}_n \cap \mathcal{B} \cap \overline{\mathcal{H}_n}\} \quad n \geq 3.$$

(v) *The relation* $\overline{\theta_{\mathcal{BM}}}$ *is displayed in Diagram 3.7, where vertices corresponding to* $\overline{\theta_{\mathcal{BM}}}$-*related elements are decorated with similar circular patterns, and the lattice* $\mathcal{L}(\mathcal{BM})$ *is displayed in Diagram 3.8.*

Proof (i) The mapping $\theta_{\mathcal{BM}}$ is simply the restriction of the mapping $\theta_{\mathcal{M}}$ in [PR1, Sections IX.7 and IX.8] to $\mathcal{L}(\mathcal{B})$. For any $\mathcal{V} \in \mathcal{L}(\mathcal{B})$, it is evident that $\mathcal{V} \cap \mathcal{BM}$ is closed under taking homomorphisms, products and submonoids. Therefore $\mathcal{V} \cap \mathcal{BM}$ is a variety of band monoids, that is, $\mathcal{V} \cap \mathcal{BM} \in \mathcal{L}(\mathcal{BM})$. Thus $\theta_{\mathcal{BM}}$ is a complete homomorphism and is surjective by [PR1, Theorem IX.7.3] or by verifying directly that for any $\mathcal{W} \in \mathcal{L}(\mathcal{BM})$, we have $\langle \mathcal{W} \rangle \theta_{\mathcal{BM}} = \mathcal{W}$.

(ii) Let $\mathcal{U}, \mathcal{V} \in \mathcal{L}(\mathcal{B})$ and $\mathcal{U}\theta_{\mathcal{BM}} = \mathcal{V}\theta_{\mathcal{BM}}$. Then $\langle \mathcal{V} \cap \mathcal{BM} \rangle = \langle \mathcal{U} \cap \mathcal{BM} \rangle \subseteq \mathcal{U}$ while, by part (i), $\langle \mathcal{V} \cap \mathcal{BM} \rangle \theta_{\mathcal{BM}} = \mathcal{V} \cap \mathcal{BM}$.

(iii) Let $\mathcal{V} = [u_\alpha = v_\alpha]_{\alpha \in A} \in \mathcal{L}(\mathcal{B})$ and $\mathcal{W}$ denote the variety defined by the identities of the form $u'_\alpha = v'_\alpha$. It is evident that every element of $\mathcal{V} \cap \mathcal{BM}$ must satisfy all the identities defining $\mathcal{W}$ so that $\langle \mathcal{V} \cap \mathcal{BM} \rangle \subseteq \mathcal{W}$. On the other hand, if $S \in \mathcal{W}$, then S satisfies all the identities of the form $u'_\alpha = v'_\alpha$, including $u_\alpha = v_\alpha$ and therefore S^1 also satisfies all these identities, including $u_\alpha = v_\alpha$. Consequently $S^1 \in \mathcal{V} \cap \mathcal{M}$ so that $S \in \langle \mathcal{V} \cap \mathcal{BM} \rangle$. Thus $\mathcal{W} \subseteq \langle \mathcal{V} \cap \mathcal{BM} \rangle$ and equality prevails.

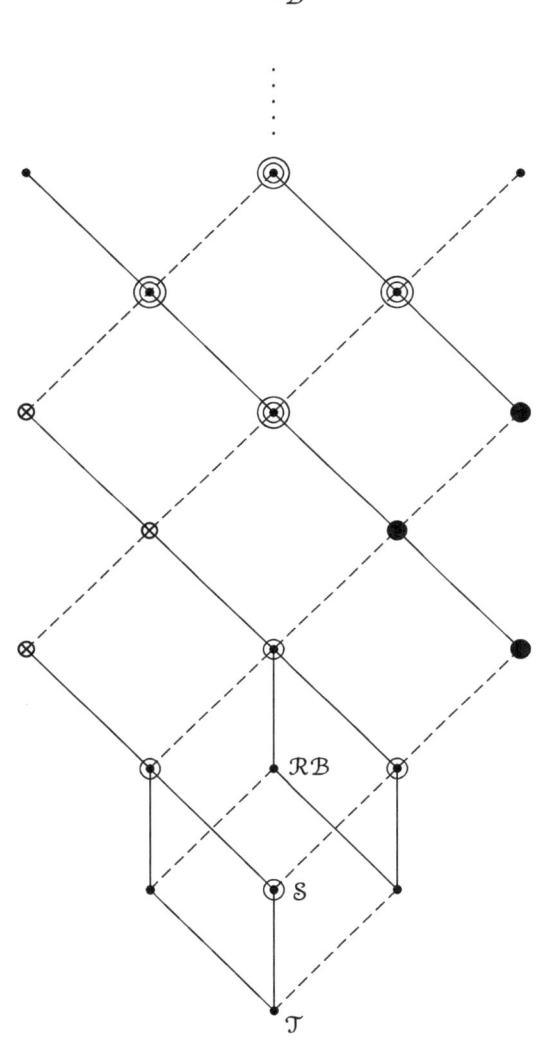

Diagram 3.7 The relation $\overline{\theta}_{\mathcal{BM}}$

(iv) The first three claims follow from the obvious fact that

$$\mathcal{T} = \mathcal{T} \cap \mathcal{BM} = \mathcal{LZ} \cap \mathcal{BM} = \mathcal{RZ} \cap \mathcal{BM} = \mathcal{RB} \cap \mathcal{BM},$$
$$\mathcal{S} \cap \mathcal{BM} = \mathcal{LNB} \cap \mathcal{BM} = \mathcal{RNB} \cap \mathcal{BM} = \mathcal{NB} \cap \mathcal{BM},$$

while $\mathcal{B} \cap \mathcal{BM} = \mathcal{BM}$.

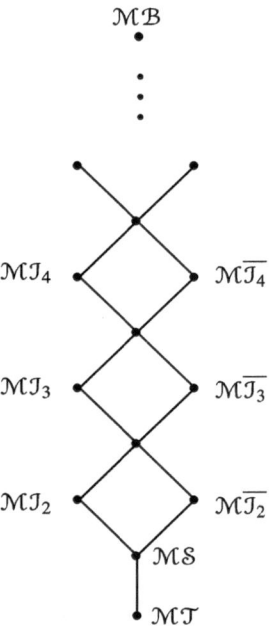

Diagram 3.8 The lattice of varieties of band monoids

In order to establish the fourth claim, we will argue by induction on the impact on the words G_n, H_n, I_n of deleting variables.

Claim: For $n \geq 3$ and in presence of the identity $x^2 = x$,

$$G_n(x_1, 1, x_3, \ldots, x_n) = G_{n-1}(x_1, x_3, \ldots, x_n)$$
$$H_n(x_1, 1, x_3, \ldots, x_n) = I_{n-1}(x_1, x_3, \ldots, x_n)$$

and if $Y = \{y_1, \ldots, y_n\}$ is such that, for each $i \in \{1, \ldots, n\}$, either $y_i = x_i$ or $y_i = 1$ *and*, for at least one value of i different from 2, $y_i = 1$, then $G_n(y_1, \ldots, y_n) = H_n(y_1, \ldots, y_n)$ is an identity that holds in all bands.

Case: $n = 3$. We have

$$G_3(x_1, x_2, x_3) = x_3 x_1 x_2, \quad H_3(x_1, x_2, x_3) = x_3 x_1 x_2 x_3 x_2, \quad I_2(x_1, x_2) = x_2 x_1 x_2$$

so that

$$G_3(1, x_2, x_3) = x_3 x_2, \quad H_3(1, x_2, x_3) = x_3 x_2,$$
$$G_3(x_1, 1, x_3) = x_3 x_1 = G_2(x_1, x_3), \quad H_3(x_1, 1, x_3) = x_3 x_1 x_3 = I_2(x_1, x_3),$$
$$G_3(x_1, x_2, 1) = x_1 x_2, \quad H_3(x_1, x_2, 1) = x_1 x_2 x_2 = x_1 x_2,$$

The deletion of any two variables reduces G_3 and H_3 to just the remaining variable. Thus the claim holds for $n = 3$. Now assume that the claim holds up to $n - 1$ and consider $n > 3$. We have

$$
\begin{aligned}
G_n(x_1, 1, x_3, \ldots, x_n) &= x_n \overline{G_{n-1}(x_1, 1, x_3, \ldots, x_{n-1})} \\
&= x_n \overline{G_{n-2}(x_1, x_3, \ldots, x_{n-1})} \text{ by induction hypothesis} \\
&= G_{n-1}(x_1, x_3, \ldots, x_n), \\
H_n(x_1, 1, x_3, \ldots, x_n) &= G_n(x_1, 1, x_3, \ldots, x_n) x_n \overline{H_{n-1}(x_1, 1, x_3, \ldots, x_{n-1})} \\
&= G_{n-1}(x_1, x_3, \ldots, x_n) x_n \overline{I_{n-2}(x_1, 1, x_3, \ldots, x_{n-1})} \\
&= I_{n-1}(x_1, x_3, \ldots, x_n).
\end{aligned}
$$

Now let $Y = \{y_1, \ldots, y_n\}$ be as described with at least one $i \neq 2$ such that $y_i = 1$. First consider the case where $y_n = 1$. Obviously $G_n(y_1, \ldots, y_{n-1}, 1) = \overline{G_{n-1}(y_1, \ldots, y_{n-1})}$ while

$$
\begin{aligned}
H_n(y_1, \ldots, y_{n-1}, y_n) &= G_n(y_1, \ldots, y_n) y_n \overline{H_{n-1}(y_1, \ldots, y_{n-1})} \\
&= y_n \overline{G_{n-1}(y_1, \ldots, y_{n-1})} y_n \overline{G_{n-1}(y_1, \ldots, y_{n-1}) y_{n-1} \overline{H_{n-2}(y_1, \ldots, y_{n-2})}} \\
&= \overline{G_{n-1}(y_1, \ldots, y_{n-1}) H_{n-2}(y_1, \ldots, y_{n-2}) \overline{G_{n-1}(y_1, \ldots, y_{n-1})}} \\
&= \overline{G_{n-1}(y_1, \ldots, y_{n-1})}
\end{aligned}
$$

where the final equality follows from the fact that, in a band, we always have $xax = x$ if $D_x \leq D_a$.

Now consider the case where $y_n \neq 1$ (and still $y_2 \neq 1$). Then there must exist i such that $1 \leq i \leq n - 1$ and $i \neq 2$ such that $y_i = 1$. Consequently,

$$
\begin{aligned}
H_n(y_1, \ldots, y_n) &= G_n(y_1, \ldots, y_n) y_n \overline{H_{n-1}(y_1, \ldots, y_{n-1})} \\
&= G_n(y_1, \ldots, y_n) y_n \overline{G_{n-1}(y_1, \ldots, y_{n-1})} \text{ induction hypothesis} \\
&= G_n(y_1, \ldots, y_n) G_n(y_1, \ldots, y_n) \\
&= G_n(y_1, \ldots, y_n)
\end{aligned}
$$

Therefore the claim holds.

Consequently, the only nontrivial identities obtained from the identity $H_n = G_n$ from the deletion of variables are $G_n = H_n$ (from deleting no variables) and $G_{n-1} = I_{n-1}$ (from deleting the variable x_2). However, the identity $G_n = H_n$ is a basis for $\mathcal{B} \cap \mathcal{H}_n$ and the identity $G_{n-1} = I_{n-1}$ is a basis for $\mathcal{B} \cap \mathcal{I}_{n-1}$. Since $\mathcal{B} \cap \mathcal{I}_{n-1} \subseteq \mathcal{B} \cap \mathcal{H}_n$, it follows that

$$
[G_n = H_n, \ G_{n-1} = I_{n-1}] = [G_{n-1} = I_{n-1}] = \mathcal{B} \cap \mathcal{I}_{n-1}.
$$

Furthermore

$$
\mathcal{B} \cap \mathcal{I}_{n-1} \subseteq \overline{\mathcal{I}_n} \cap \mathcal{B} \cap \mathcal{H}_n \subseteq \mathcal{B} \cap \mathcal{H}_n, \quad (n \geq 3)
$$

so that these three varieties lie in the same $\overline{\theta}_{\mathcal{B}\mathcal{M}}$ class in which, by the above and part (iii), $\mathcal{B} \cap \mathcal{I}_{n-1}$ is the smallest element. By duality, the varieties $\mathcal{B} \cap \overline{\mathcal{I}_{n-1}}, \mathcal{I}_n \cap \mathcal{B} \cap \overline{\mathcal{H}_n}, \mathcal{B} \cap \overline{\mathcal{H}_n}$ ($n \geq 3$) also lie in the same $\overline{\theta}_{\mathcal{B}\mathcal{M}}$-class with $\mathcal{B} \cap \overline{\mathcal{I}_{n-1}}$ as its smallest member.

Since $\overline{\theta}_{\mathcal{B}\mathcal{M}}$ is a complete congruence in $\mathcal{L}(\mathcal{B})$, we deduce from the above that the following varieties are $\overline{\theta}_{\mathcal{B}\mathcal{M}}$-equivalent:

$$\mathcal{I}_{n-1} \cap \mathcal{B} \cap \overline{\mathcal{I}_{n-1}}, \quad \mathcal{I}_{n-1} \cap \mathcal{B} \cap \overline{\mathcal{H}_n},$$
$$\mathcal{H}_n \cap \mathcal{B} \cap \overline{\mathcal{I}_{n-1}}, \quad \mathcal{H}_n \cap \mathcal{B} \cap \overline{\mathcal{H}_n}, \quad (n \geq 3).$$

In addition, combining the bases for $\mathcal{B} \cap \mathcal{H}_n$ and $\mathcal{B} \cap \overline{\mathcal{H}_n}$, we have

$$\mathcal{B} \cap \mathcal{H}_n \cap \mathcal{B} \cap \overline{\mathcal{H}_n} = \left[G_n = H_n, \ \overline{G_n} = \overline{H_n} \right].$$

Part (ii) and the claim above tell us that the least element in the $\overline{\theta}_{\mathcal{B}\mathcal{M}}$-class of $\mathcal{B} \cap \mathcal{H}_n \cap \mathcal{B} \cap \overline{\mathcal{H}_n}$ is

$$[G_{n-1} = I_{n-1}, \ \overline{G_{n-1}} = \overline{I_{n-1}}] = \mathcal{B} \cap \mathcal{I}_{n-1} \cap \mathcal{B} \cap \overline{I_{n-1}}.$$

Since the smallest elements in the sets of $\overline{\theta}_{\mathcal{B}\mathcal{M}}$-equivalent varieties listed in the statement are all distinct, there is no overlap between those sets. In addition every variety of bands is in one of the listed sets. Consequently the assertion is validated.

(v) These assertions follow from part (iv).   □

We can now describe the L-relation on $\mathcal{L}(\mathcal{B})$. Note that, if $\mathcal{U} \in \mathcal{L}(\mathcal{B})$, $\mathcal{V} \in \mathcal{L}(\mathcal{CR})$ are such that $\mathcal{U} \ L \ \mathcal{V}$, then $\mathcal{V} \in \mathcal{L}(\mathcal{B})$.

Theorem 3.8.4 *The classes of the L-relation on $\mathcal{L}(\mathcal{B})$ consist of the following intervals*

$$[\mathcal{I}, \mathcal{RB}], \quad [\mathcal{S}, \mathcal{NB}], \quad \{\mathcal{B}\},$$
$$[\mathcal{MI}_{n-1}, \mathcal{MH}_n], \quad [\mathcal{M\overline{I}}_{n-1}, \mathcal{M\overline{H}}_n], \quad (n \geq 3)$$
$$[\mathcal{MI}_{n-1} \cap \mathcal{M\overline{I}}_{n-1}, \mathcal{MH}_n \cap \mathcal{M\overline{H}}_n], \quad (n \geq 3)$$

Proof This follows immediately from Theorem 3.8.3.   □

Thus we can identify the lattice of varieties of band monoids with the lattice of $\overline{\theta}_{\mathcal{B}\mathcal{M}}$-classes. It makes sense to identify each of these classes with a distinguished member, such as the least member of the class.

Notation 3.8.5 We write

$$\mathcal{MI} = [\mathcal{I}, \mathcal{RB}], \quad \mathcal{MS} = [\mathcal{S}, \mathcal{NB}], \quad \mathcal{MB} = \{\mathcal{B}\},$$
$$\mathcal{MI}_{n-1} = [\mathcal{B} \cap \mathcal{I}_{n-1}, \mathcal{B} \cap \mathcal{H}_n], \quad \mathcal{M\overline{I}}_{(n-1)} = [\mathcal{B} \cap \overline{\mathcal{I}_{n-1}}, \mathcal{B} \cap \overline{\mathcal{H}_n}], \quad (n \geq 3).$$

Corollary 3.8.6 *In the context of band monoids, each variety of band monoids has a basis of identities consisting of one or two identities as follows:*

$$\mathcal{MJ} = [x = 1], \quad \mathcal{MS} = [xy = yx, x^2 = x], \quad \mathcal{MB} = [x^2 = x],$$
$$\mathcal{MJ}_{n-1} = [G_{n-1} = I_{n-1}], \quad \mathcal{M\overline{J}}_{(n-1)} = [\overline{G_{n-1}} = \overline{I_{n-1}}], \quad (n \geq 3).$$

Proof It is clear that the identities used to define the least element in each $\overline{\theta_{\mathcal{BM}}}$-class are sufficient to determine the appropriate variety of band monoids in each situation. □

The description of $\overline{\theta}_{\mathcal{BM}}$ in Theorem 3.8.3 is due to Wismath [1].

3.9 Review

The first two sections concern $\mathcal{L}(\mathcal{CR})$ and culminate in a non-faithful but important representation of on [$\mathcal{S}$, $\mathcal{CR}$]. This led to the determination of the lattice $\mathcal{L}(\mathcal{B})$, one of the highlights of the chapter. This, in turn, led to the expressions for the K_{ℓ}- and K_r-relations on $\mathcal{L}(\mathcal{B})$ which made it possible to construct bases of identities for all varieties of bands in terms of a simple inductive procedure. That is followed by the introduction of the B-relation and its basic properties. The B-relation is shown to be a complete congruence on $\mathcal{L}(\mathcal{CR})$ and thereby provides an important decomposition of $\mathcal{L}(\mathcal{CR})$. This section describes the upper ends of the B-classes and provides bases of identities for them. The penultimate section is devoted to the description of the free band on a nonempty set. The chapter concludes with a description of the lattice of varieties of band monoids as a homomorphic image of the lattice of varieties of bands. The relation induced by this homomorphism will appear in a different context in [PR3].

Polák Theorem

4

One of the principal results of the study of varieties of completely regular semigroups is the construction of an isomorphic copy of the lattice $[\mathcal{S}, \mathcal{CR}]$ in terms of functions between certain sets satisfying mostly natural conditions. Even though this construction is quite involved, as it necessarily must be if it is to describe such a complicated lattice, it fortunately turns out to be remarkably useful in its various applications. These include diverse global properties of the interval $[\mathcal{S}, \mathcal{CR}]$ as well as detailed descriptions of certain of its parts. The proof of this theorem involves somewhat lengthy considerations and repeated applications of many concepts introduced and results obtained heretofore especially in Chap. 2.

We considered in [PR1, Chapter VIII] the lattice $\mathcal{L}(\mathcal{CS})$ of varieties of completely simple semigroups. The theorem alluded to above complements this effort in the sense that the lattice $\mathcal{L}(\mathcal{CR})$ of varieties of completely regular semigroups is a disjoint union of $\mathcal{L}(\mathcal{CS})$ and $[\mathcal{S}, \mathcal{CR}]$.

4.1 Basic Characterization

Generally, in the study of varieties, it is both interesting and useful to be able to describe a variety in terms of smaller varieties. We have seen an important instance of this (Theorem 2.5.2(xi)) in the formula $\mathcal{V} = \mathcal{V}_{T_\ell} \vee \mathcal{V}_K \vee \mathcal{V}_{T_r}$ ($\mathcal{V} \in \mathcal{L}(\mathcal{CR})$). However, in some situations we may have $\mathcal{V} = \mathcal{V}_K$ or $\mathcal{V}_{T_\ell}$ or $\mathcal{V}_{T_r}$, in which case this formula provides no insight into the nature of $\mathcal{V}$. Here we consider the situation where $\mathcal{V} = \mathcal{V}_{T_r}$ and obtain an alternative description of $\mathcal{V}$ in terms of (possibly) smaller varieties. In the event that $\mathcal{V} = \mathcal{V}_{T_\ell} = \mathcal{V}_{T_r}$, we shall see later in this chapter that $\mathcal{V} = \mathcal{V}^K$.

The results of this section provide a critical cornerstone for the representation of the interval $[\mathcal{S}, \mathcal{CR}]$ that will constitute the main result (the Polák theorem) of this chapter.

© The Author(s), under exclusive license to Springer Nature Switzerland AG 2024
M. Petrich and N. R. Reilly, *Completely Regular Semigroup Varieties*, Synthesis Lectures on Mathematics & Statistics, https://doi.org/10.1007/978-3-031-42891-3_4

Theorem 4.1.1 *Let $\mathcal{V} \in [\mathcal{S}, \mathcal{CR}]$, $\tau \in \Theta$ be such that $t(\tau) = T_r$ and $\mathcal{V}_\tau \notin \mathbb{N}_3$. Then $\mathcal{V}_\tau = (\mathcal{V}_{\tau K} \vee \mathcal{V}_{\tau T_\ell})^{K_\ell}$. In particular, if $\mathcal{LRB} \subseteq \mathcal{V}$ and $\mathcal{V} = \mathcal{V}_{T_r}$, then $\mathcal{V} = (\mathcal{V}_{T_\ell} \vee \mathcal{V}_K)^{K_\ell}$.*

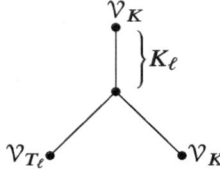

Proof We begin by proving the claim for the particular case. So let $\mathcal{V} \in [\mathcal{S}, \mathcal{CR}]$ be such that $\mathcal{LRB} \subseteq \mathcal{V}$ and $\mathcal{V} = \mathcal{V}_{T_r}$. Let $\mathcal{W} = (\mathcal{V}_K \vee \mathcal{V}_{T_\ell})^{K_\ell}$.

$$\mathcal{S} \subseteq \mathcal{V} \implies \mathcal{S} \subseteq \mathcal{V}_{T_\ell} \implies \mathcal{LRB} = \mathcal{S}^{K_\ell} \subseteq \mathcal{W}.$$

By Theorem 2.6.2(xii), we also have $\mathcal{W}_{T_r} = \mathcal{W}$. In addition, from the fact that $\mathcal{V}_K, \mathcal{V}_{T_\ell} \subseteq \mathcal{W} \subseteq \mathcal{V}^{K_\ell} = \mathcal{V}^K \cap \mathcal{V}^{T_\ell}$ we see that $\mathcal{V} (K \cap T_\ell) \mathcal{W}$ and, in particular, $\mathcal{V}_K = \mathcal{W}_K, \mathcal{V}_{T_\ell} = \mathcal{W}_{T_\ell}$. Hence

$$\zeta_\mathcal{V}^K = \zeta_{\mathcal{V}_K} = \zeta_{\mathcal{W}_K} = \zeta_\mathcal{W}^K. \tag{4.1.1}$$

Since $\mathcal{LRB} \subseteq \mathcal{V} \cap \mathcal{W} = \mathcal{V}_{T_r} \cap \mathcal{W}_{T_r}$, we have by Theorem 1.5.6 that $\zeta_\mathcal{V}^s = \zeta_{\mathcal{V}_{T_r}} = \zeta_\mathcal{V}$ and, likewise, $\zeta_\mathcal{W}^s = \zeta_{\mathcal{W}_{T_r}} = \zeta_\mathcal{W}$. Hence

$$\zeta_\mathcal{V}^{s^k} = \zeta_\mathcal{V} \quad \text{and} \quad \zeta_\mathcal{W}^{s^k} = \zeta_\mathcal{W} \quad \text{for all } k \geq 0. \tag{4.1.2}$$

On the other hand, since $\mathcal{V}_{T_\ell} = \mathcal{W}_{T_\ell}$, we have $\mathcal{V} T_\ell \mathcal{W}$ and $\mathcal{V}_{T_\ell} \in \mathbb{N}_3 \iff \mathcal{W}_{T_\ell} \in \mathbb{N}_3$. From the duals of Lemma 1.5.2(vii) and Theorems 1.5.5(i), 1.5.6, we have

$$\zeta_\mathcal{V}^e = \zeta_\mathcal{W}^e. \tag{4.1.3}$$

though this may not be a fully invariant congruence.

Let $\pi \in \Pi$ and $\pi^* = \pi \varphi_{\Pi\Theta}$, in the notation of Sects. 3.1 and 3.2. First assume that $\mathcal{V}_{\pi^*} \notin \mathbb{N}_3$ (equivalently, $\zeta_\mathcal{V}^\pi \in \Delta$). Then we can write $\pi = s^k \pi_1$ where $k \geq 0$ and either $\pi_1 = \emptyset$ or $\pi_1 = e\pi_2$ for some $\pi_2 \in \Pi^1$. By (4.1.2), this means that $\zeta_\mathcal{V}^\pi = \zeta_\mathcal{V}^{\pi_1}$ and $\zeta_\mathcal{W}^\pi = \zeta_\mathcal{W}^{\pi_1}$. If $\pi_1 = \emptyset$, then this leads to

$$\zeta_\mathcal{V}^{\pi K} = \zeta_\mathcal{V}^K = \zeta_{\mathcal{V}_K} = \zeta_{\mathcal{W}_K} = \zeta_\mathcal{W}^K = \zeta_\mathcal{W}^{\pi K} \tag{4.1.4}$$

while, if $\pi_1 \neq \emptyset$ then we obtain by (4.1.3) that

$$\zeta_\mathcal{V}^{\pi K} = (\zeta_\mathcal{V}^e)^{\pi_2 K} = (\zeta_\mathcal{W}^e)^{\pi_2 K} = \zeta_\mathcal{W}^{\pi K}. \tag{4.1.5}$$

From (4.1.1), (4.1.4) and (4.1.5) it follows that

$$\zeta_{\mathcal{V}}^{\pi K} = \zeta_{\mathcal{W}}^{\pi K} \quad \text{for all } \pi \in \Pi^1 \tag{4.1.6}$$

(and, in particular, for $\pi = \emptyset$ or π such that $\mathcal{V}_{\pi*}$ or $\mathcal{W}_{\pi*} \notin \mathbb{N}_3$).

Now consider the case where $\mathcal{V}_{\pi*} \in \mathbb{N}_3$, $\pi \in \Pi$. We know that $\mathcal{V}_{\pi*} \in \mathbb{N}_3$ if and only if $\mathcal{W}_{\pi*} \in \mathbb{N}_3$. Also, for this to happen, π must be of the form $\pi = s^k e \pi_2$ for some $k \geq 0$, $\pi_2 \in \Pi^1$. We then have

$$\zeta_{\mathcal{V}}^\pi = \left(\zeta_{\mathcal{V}}^{s^k}\right)^{e\pi_2} = \left(\zeta_{\mathcal{V}}^{e}\right)^{\pi_2} = \left(\zeta_{\mathcal{W}}^{e}\right)^{\pi_2} = \left(\zeta_{\mathcal{W}}^{s^k}\right)^{e\pi_2} = \zeta_{\mathcal{W}}^\pi. \tag{4.1.7}$$

Now let $u, v \in U$. By (4.1.6), (4.1.7) and Theorem 3.1.7(ii), we have

$$u \zeta_\mathcal{V} v \iff \begin{cases} u\pi \left(\zeta_{\mathcal{V}}^{\pi K} \cap \eta\right) v\pi & \text{for all } \pi \in \Pi^1 \text{ such that} \\ & \text{either } \pi = \emptyset \text{ or } \mathcal{V}_{\pi*} \notin \mathbb{N}_3 \\ u\pi \; \zeta_{\mathcal{V}}^\pi \; v\pi & \text{for all } \pi \in \Pi \text{ such that} \\ & \mathcal{V}_{\pi*} \in \mathbb{N}_3 \end{cases}$$

$$\iff \begin{cases} u\pi \left(\zeta_{\mathcal{W}}^{\pi K} \cap \eta\right) v\pi & \text{for all } \pi \in \Pi^1 \text{ such that} \\ & \text{either } \pi = \emptyset \text{ or } \mathcal{W}_{\pi*} \notin \mathbb{N}_3 \\ u\pi \; \zeta_{\mathcal{W}}^\pi \; v\pi & \text{for all } \pi \in \Pi \text{ such that} \\ & \mathcal{W}_{\pi*} \in \mathbb{N}_3 \end{cases}$$

$$\iff u \zeta_\mathcal{W} v.$$

Consequently, $\mathcal{V} = \mathcal{W}$ as claimed.

We now consider the general case. Since $t(\tau) = T_r$ and $\mathcal{V}_\tau \notin \mathbb{N}_3$, it follows that $\mathcal{LRB} \subseteq \mathcal{V}_\tau$. Also $(\mathcal{V}_\tau)_{T_r} = \mathcal{V}_\tau T_r = \mathcal{V}_\tau$ and so we can apply the particular case to $\mathcal{V}_\tau$ and obtain the desired result. □

Innocent as Theorem 4.1.1 may appear, it is actually the linchpin in the powerful Polák Theorem to come in Sect. 4.4. Using the formula in Theorem 4.1.1, the diagram there may be iterated, as illustrated in the accompanying diagrams, on the varieties $\mathcal{V}_{T_\ell}, \mathcal{V}_{T_r}, \mathcal{V}_{T_\ell T_r}, \ldots$ until ultimately arriving at Diagram 4.1. In the diagrams below, the varieties at the unlabelled vertices (and also the variety $\mathcal{V}$) are completely determined by the varieties at the vertices on the next level down in the diagram (where we start with $\mathcal{V}$ at level zero and count down). In particular, if $\mathcal{P}$ and $\mathcal{Q}$ are the varieties represented by the lower and upper vertices, respectively, of a vertical edge then $\mathcal{Q} = \mathcal{P}^{K_\ell}$ or $\mathcal{Q} = \mathcal{P}^{K_r}$. If $\mathcal{R}$ is the variety represented by a vertex covering two vertices representing varieties $\mathcal{P}$ and $\mathcal{Q}$, then $\mathcal{R} = \mathcal{P} \vee \mathcal{Q}$.

step 1: $\mathcal{V} = \mathcal{V}_{T_\ell} \vee \mathcal{V}_K \vee \mathcal{V}_{T_r}$.

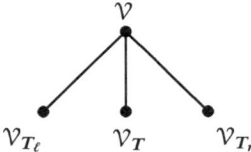

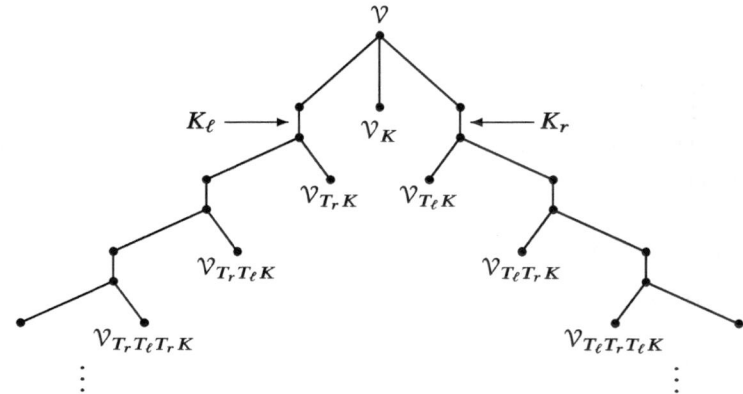

Diagram 4.1 *Case*: $\mathcal{V}_\tau \notin \mathbb{N}_3$ for all $\tau \in \Theta$

step 2: $\mathcal{V}_{T_\ell} = \left(\mathcal{V}_{T_\ell K} \vee \mathcal{V}_{T_\ell T_r}\right)^{K_r}$; $\quad \mathcal{V}_{T_r} = \left(\mathcal{V}_{T_r K} \vee \mathcal{V}_{T_r T_\ell}\right)^{K_\ell}$.

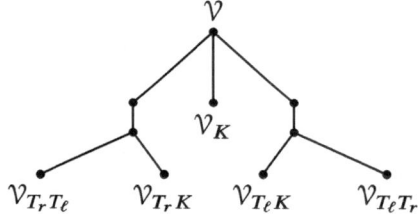

step 3: $\mathcal{V}_{T_r T_\ell} = \left(\mathcal{V}_{T_r T_\ell K} \vee \mathcal{V}_{T_r T_\ell T_r}\right)^{K_r}$; $\quad \mathcal{V}_{T_\ell T_r} = \left(\mathcal{V}_{T_\ell T_r K} \vee \mathcal{V}_{T_\ell T_r T_\ell}\right)^{K_\ell}$.

One loose end in the above discussion is the case where $\mathcal{V}_\tau \in \mathbb{N}_3$ for some $\tau \in \Theta$, that is $\mathcal{V}_\tau = \mathcal{LNB}$, $\mathcal{S}$ or $\mathcal{RNB}$. In these cases, we have

$$(\mathcal{LNB}_K \vee \mathcal{LNB}_{T_\ell})^{K_\ell} = (\mathcal{J} \vee \mathcal{S})^{K_\ell} = \mathcal{S}^{K_\ell} = \mathcal{LRB}$$

and similarly $(\mathcal{S}_K \vee \mathcal{S}_{T_\ell})^{K_\ell} = \mathcal{LRB}$, while

$$(\mathcal{RNB}_K \vee \mathcal{RNB}_{T_\ell})^{K_r} = (\mathcal{J} \vee \mathcal{S})^{K_r} = \mathcal{S}^{K_r} = \mathcal{RRB}.$$

In other words, the varieties $\mathcal{LNB}$, $\mathcal{S}$ and $\mathcal{RNB}$ cannot be recovered from (possibly) smaller varieties in the style of Theorem 4.1.1. On the other hand, there is not much that is not known about these varieties and it is reasonable to consider them as "primitives" in our efforts to describe varieties in terms of simpler varieties.

However, this does mean that the above diagrams need some adjustment in order to accomodate the possibility that $\mathcal{V}_\tau \in \mathbb{N}_3$. We adapt the construction of the diagram as follows. Whenever $\mathcal{V}_\tau \in \mathbb{N}_3$, instead of progressing to $\mathcal{V}_{\tau K}$, we simply retain $\mathcal{V}_\tau$. Note that the vertices in Diagram 4.2 down the extreme left-hand side represent the varieties $\mathcal{V}$, $\mathcal{V}_{T_r}$, $\mathcal{V}_{T_r T_\ell}$, ... obtained by applying the T_r-chain and dually for the vertices on the extreme right-hand side

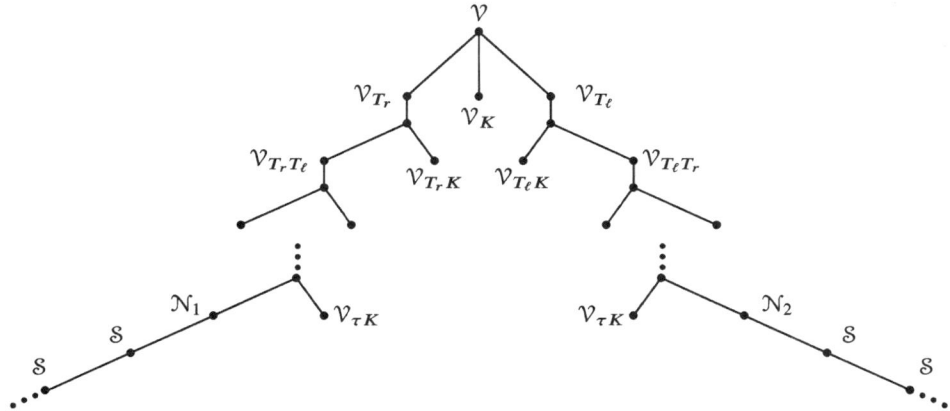

Diagram 4.2 $\mathcal{N}_i \in \mathbb{N}_3$, $i = 1, 2$

by applying the T_ℓ-chain. So suppose that for some τ we have $\mathcal{V}_\tau \in \mathbb{N}_3$. If $t(\tau) = T_\ell$, then we have $\mathcal{V}_{\tau T_r} = \mathcal{S} \in \mathbb{N}_3$. Once $\mathcal{V}_\tau \in \mathbb{N}_3$, the sequence $\{\mathcal{V}_\tau\}$ quickly stabilizes since

$$\mathcal{V}_\tau \in \mathbb{N}_3, \ |\sigma| \geq |\tau| + 1 \implies \mathcal{V}_\sigma = \mathcal{S}.$$

Thus, in order to accomodate this possible transition in Diagram 4.1, we must indicate the possible occurrence of $\mathcal{LNB}$, $\mathcal{S}$ and $\mathcal{RNB}$. Note that the values $\mathcal{LNB}$ and $\mathcal{RNB}$ can appear on either side of the diagram. However they cannot appear on the same side.

This focusses the spotlight clearly on the following two sequences of subvarieties of any $\mathcal{V} \in [\mathcal{S}, \mathcal{CR}]$:

$$\mathcal{V}_K, \ \mathcal{V}_{T_r K}, \ \mathcal{V}_{T_r T_\ell K}, \ \mathcal{V}_{T_r T_\ell T_r K}, \ \ldots$$
$$\mathcal{V}_K, \ \mathcal{V}_{T_\ell K}, \ \mathcal{V}_{T_\ell T_r K}, \ \mathcal{V}_{T_\ell T_r T_\ell K}, \ \ldots.$$

Definition 4.1.2 We call the partially labelled diagram obtained from the variety $\mathcal{V} \in [\mathcal{S}, \mathcal{CR}]$ and with the labels $\mathcal{V}_K, \mathcal{V}_{T_r K}, \mathcal{V}_{T_r T_\ell K}, \ldots$ and $\mathcal{V}_{T_\ell K}, \mathcal{V}_{T_\ell T_r K}, \ldots$ at the labelled vertices the *frame* of $\mathcal{V}$. We count the levels starting with level 0 at the top for $\mathcal{V}$.

Since the varieties to be placed at the unlabelled vertices in the frame of $\mathcal{V}$, at level n say, and the variety $\mathcal{V}$ itself, are determined by the varieties at the vertices of the frame at level $n + 1$, it is intuitively clear that each $\mathcal{V} \in [\mathcal{S}, \mathcal{CR}]$ is uniquely determined by the varieties strictly below $\mathcal{V}$ in its frame.

Lemma 4.1.3 *Suppose that* $\mathcal{U}, \mathcal{V} \in [\mathcal{S}, \mathcal{CR}]$ *satisfy the conditions:*

(i) $\mathcal{U}_\tau \in \mathbb{N}_3 \Leftrightarrow \mathcal{V}_\tau \in \mathbb{N}_3$, *for* $\tau \in \Theta$,
(ii) $\mathcal{U}_\tau, \mathcal{V}_\tau \in \mathbb{N}_3 \implies \mathcal{U}_\tau = \mathcal{V}_\tau$, *for* $\tau \in \Theta$,
(iii) $\mathcal{U}_{\tau K} = \mathcal{V}_{\tau K}$ *if either* $\mathcal{U}_\tau, \mathcal{V}_\tau \notin \mathbb{N}_3$ *or* $\tau = \emptyset$.

Then $\mathcal{U} = \mathcal{V}$

Proof Let $\pi \in \Pi^1$ and $\tau = \pi^*$.

Case: $\pi = \emptyset$. By condition (iii), $\mathcal{U}_K = \mathcal{V}_K$ whence $\zeta_{\mathcal{U}_K} = \zeta_{\mathcal{V}_K}$.

Case: $\pi \in \Pi$, $\mathcal{U}_\tau, \mathcal{V}_\tau \notin \mathbb{N}_3$. By condition (iii), $\mathcal{U}_{\tau K} = \mathcal{V}_{\tau K}$. Hence $\zeta_{\mathcal{U}}^{\pi K} = \zeta_{\mathcal{V}}^{\pi K}$.

Case: $\pi \in \Pi$, $\mathcal{U}_\tau, \mathcal{V}_\tau \in \mathbb{N}_3$. By part (ii), we then have $\mathcal{U}_\tau = \mathcal{V}_\tau$. Hence $\zeta_{\mathcal{U}}^\pi = \zeta_{\mathcal{V}}^\pi$, by Lemma 3.1.9 and its dual.

From the above cases and Theorem 3.1.7(ii), it follows that, for any $u, v \in U$, $u \zeta_{\mathcal{U}} v \Leftrightarrow u \zeta_{\mathcal{V}} v$. Consequently, $\zeta_{\mathcal{U}} = \zeta_{\mathcal{V}}$. □

Theorem 4.1.4 *Every variety in* $[\mathcal{S}, \mathcal{CR}]$*, is uniquely determined by its frame.*

Proof This follows immediately from Lemma 4.1.3. □

From Theorem 4.1.4 we obtain the following description of any variety in $[\mathcal{S}, \mathcal{CR}]$ as an intersection of related varieties.

Corollary 4.1.5 *For any* $\mathcal{V} \in [\mathcal{S}, \mathcal{CR}]$,

$$\mathcal{V} = \mathcal{V}^K \cap \left(\bigcap_{\tau \in \Theta, \, \mathcal{V}_\tau \notin \mathbb{N}_3} (\mathcal{V}_{\tau K})^{K\bar{\tau}} \right) \cap \left(\bigcap_{\tau \in \Theta, \, \mathcal{V}_\tau \mathcal{LNB}} \mathcal{LNB}^{\bar{\tau}} \right)$$

$$\cap \left(\bigcap_{\tau \in \Theta, \, \mathcal{V}_\tau = \mathcal{RNB}} \mathcal{RNB}^{\bar{\tau}} \right) \cap \left(\bigcap_{\tau \in \Theta, \, \mathcal{V}_\tau = \mathcal{S}} \mathcal{S}^{\bar{\tau}} \right) \tag{4.1.8}$$

$$= \mathcal{V}^K \cap \left(\bigcap_{\tau \in \Theta, \, \mathcal{V}_\tau \notin \mathbb{N}_3} (\mathcal{V}_{\tau K})^{K\bar{\tau}} \right) \cap \mathcal{V}^B. \tag{4.1.9}$$

Proof Let $\mathcal{W}$ denote the variety on the right side of equation (4.1.8). We apply Lemma 4.1.3 to $\mathcal{V}$ and $\mathcal{W}$ and begin by considering the conditions (i) and (ii) in that lemma.

Let $\sigma \in \Theta$ be such that $\mathcal{V}_\sigma \in \mathbb{N}_3$, say $\mathcal{V}_\sigma = \mathcal{LNB}$. Since $\mathcal{V}_\sigma$ is the least element in its $t(\sigma)$-class, it follows that $t(\sigma) = T_r$. We examine $\mathcal{V}^K$ and the other (bracketed) terms in $\mathcal{W}$ in (4.1.8) in turn.

Case: $\mathcal{V}^K$. By Theorem 2.5.2(xii), we have $(\mathcal{V}^K)_\sigma = \mathcal{V}^K \supseteq \mathcal{LNB}$.

Case: $\mathcal{V}_\tau \notin \mathbb{N}_3$. By Theorem 2.5.2(xii),

$$\left((\mathcal{V}_{\tau K})^{K\bar{\tau}} \right)_\sigma \supseteq \left((\mathcal{V}_\tau)^K \right)_\sigma = (\mathcal{V}_\tau)^K \supseteq \mathcal{B} \supseteq \mathcal{LNB}.$$

Case: $\tau = \sigma$ and $\mathcal{V}_\tau = \mathcal{LNB}$. Then, by Theorem 2.5.2(xiii) and its dual,

$$\left((\mathcal{LNB})^{\bar{\tau}} \right)_\sigma = (\mathcal{LNB}^{\bar{\sigma}})_\sigma = \mathcal{LNB}_{t(\sigma)} = \mathcal{LNB}_{T_r} = \mathcal{LNB}.$$

Case: $\mathcal{V}_\tau \in \mathbb{N}_3$, $\tau \neq \sigma$. By Lemma 3.1.4, $\mathcal{V} \subseteq (\mathcal{V}_\tau)^{\overline{\tau}}$ so that $\mathcal{V}_\sigma \subseteq \left((\mathcal{V}_\tau)^{\overline{\tau}}\right)_\sigma$.

Consequently, for each term $\mathcal{P}$ in $\mathcal{W}$, we have $\mathcal{P}_\sigma \supseteq \mathcal{V}_\sigma$ while for the term $\mathcal{P} = (\mathcal{V}_\sigma)^{\overline{\sigma}}$, we have exactly $\mathcal{P}_\sigma = \mathcal{V}_\sigma$. Therefore $\mathcal{W}_\sigma = \mathcal{V}_\sigma$ and we have shown that $\mathcal{V}_\sigma = \mathcal{LNB} \implies \mathcal{W}_\sigma = \mathcal{LNB}$.

Similar arguments apply to the case where $\mathcal{V}_\sigma = \mathcal{RNB}$. Now assume that $\mathcal{V}_\sigma = \mathcal{S}$. Then there exists a term $\mathcal{S}^{\overline{\sigma}}$ in (4.1.8). Since $(\mathcal{S}^{\overline{\sigma}})_\sigma = \mathcal{S}$, the smallest possible value for all the terms, it follows that $\mathcal{W}_\sigma = \mathcal{S} = \mathcal{V}_\sigma$. Thus $\mathcal{V}_\sigma \in \mathbb{N}_3 \implies \mathcal{W}_\sigma = \mathcal{V}_\sigma$.

Now suppose that $\mathcal{W}_\sigma \in \mathbb{N}_3$. For every term $\mathcal{P}$ in (4.1.8) we have $\mathcal{V} \subseteq \mathcal{P}$ so that $\mathcal{V} \subseteq \mathcal{W}$. Hence $\mathcal{V}_\sigma \subseteq \mathcal{W}_\sigma$ which forces $\mathcal{V}_\sigma \in \mathbb{N}_3$. From the preceding argument this then implies that $\mathcal{V}_\sigma = \mathcal{W}_\sigma$. Accordingly the conditions (i) and (ii) of Lemma 4.1.3 are satisfied.

In regard to the condition (iii) in Lemma 4.1.3, the first term in the definition of $\mathcal{W}$ is $\mathcal{V}^K$. On the other hand, $\mathcal{V}$ is contained in every term that appears in the definition of $\mathcal{W}$. Hence $\mathcal{V} \subseteq \mathcal{W} \subseteq \mathcal{V}^K$ so that $\mathcal{V} \, K \, \mathcal{W}$. Therefore $\mathcal{V}_{\varnothing K} = \mathcal{W}_{\varnothing K}$.

Now consider $\sigma \in \Theta$ such that $\mathcal{V}_\sigma \notin \mathbb{N}_3$ (and therefore $\mathcal{W}_\sigma \notin \mathbb{N}_3$). Again we consider each of the terms $\mathcal{P}$ in the definition of $\mathcal{W}$.

Case: $\mathcal{V}^K$. Here $(\mathcal{V}^K)_{\sigma K} = (\mathcal{V}^K)_K = \mathcal{V}_K \supseteq \mathcal{V}_{\sigma K}$.

Case: $\mathcal{V}_\tau \notin \mathbb{N}_3$. Then $\mathcal{V} \subseteq (\mathcal{V}_\tau)^{\overline{\tau}} \subseteq (\mathcal{V}_{\tau K})^{K\overline{\tau}}$ so that $\mathcal{V}_{\sigma K} \subseteq \left((\mathcal{V}_{\tau K})^{K\overline{\tau}}\right)_{\sigma K}$ while

$$\left((\mathcal{V}_{\sigma K})^{K\overline{\sigma}}\right)_{\sigma K} = \left((\mathcal{V}_\sigma)^{K\overline{\sigma}}\right)_{\sigma K} = \left((\mathcal{V}_\sigma)^K\right)_{t(\sigma)K} = \left((\mathcal{V}_\sigma)^K\right)_K = \mathcal{V}_{\sigma K}.$$

Case: $\mathcal{V}_\tau \in \mathbb{N}_3$. Then $\mathcal{V} \subseteq (\mathcal{V}_\tau)^{\overline{\tau}}$ so that $\mathcal{V}_{\sigma K} \subseteq ((\mathcal{V}_\tau)^{\overline{\tau}})_{\sigma K}$.

Consequently, for all terms $\mathcal{P}$ in $\mathcal{W}$, we have $\mathcal{V}_{\sigma K} \subseteq \mathcal{P}_{\sigma K}$ while for one term we have $\mathcal{V}_{\sigma K} = \mathcal{P}_{\sigma K}$. Hence, if we let $\{\mathcal{U}_i \mid i \in I\}$ denote the set of all varieties appearing in the definition of $\mathcal{W}$, then we have $\bigcap_{i \in I}(\mathcal{U}_i)_{\sigma K} = \mathcal{V}_{\sigma K}$. Therefore

$$\mathcal{W}_{\sigma K} = \left[\bigcap_{i \in I}(\mathcal{U}_i)_\sigma\right]_K K \bigcap_{i \in I}(\mathcal{U}_i)_{\sigma K} = \mathcal{V}_{\sigma K}.$$

However, $\mathcal{V}_{\sigma K}$ and $\mathcal{W}_{\sigma K}$ are both the least element in their K-class and so $\mathcal{V}_{\sigma K} = \mathcal{W}_{\sigma K}$. Thus $\mathcal{V}$ and $\mathcal{W}$ also satisfy condition (iii) of Lemma 4.1.3. Accordingly $\mathcal{W} = \mathcal{V}$.

The second equality now follows from Theorem 3.6.4(ix).   □

The results of this section are based on Polák [1, 2].

4.2 A Faithful Representation of $[\mathcal{S}, \mathcal{CR}]$

In this section we will derive a faithful representation of the lattice $[\mathcal{S}, \mathcal{CR}]$ as a sublattice of the direct product $(\mathcal{L}(\mathcal{CR})/K)^{\Theta^1} \times [\mathcal{S}, \mathcal{B}]$, thereby highlighting the prominent role played by the lattice of varieties of bands in the representations of $\mathcal{L}(\mathcal{CR})$ to come. Instead of working with the lattice $\mathcal{L}(\mathcal{CR})/K$, however, it will be more convenient, both notationally and intuitively, to deal with a specific representative from each K-class. We choose to work with the class $\mathcal{K}_0$ of all least elements from the K-classes of $\mathcal{L}(\mathcal{CR})$. Since $\mathcal{K}_0$ and $[\mathcal{S}, \mathcal{B}]$ are both complete lattices so also is $\mathcal{K}_0^{\Theta^1} \times [\mathcal{S}, \mathcal{B}]$ with respect to the product order. Our representation will focus on a complete sublattice of this lattice.

Notation 4.2.1 Let $\mathcal{K}_0 = \{\mathcal{V}_K \mid \mathcal{V} \in \mathcal{L}(\mathcal{CR})\}$, $\mathcal{K}_1 = \{\mathcal{V} \vee \mathcal{S} \mid \mathcal{V} \in \mathcal{K}_0\}$.

It is important to note that, for $\mathcal{W} \in \mathcal{K}_1$, we have $\mathcal{W} = \mathcal{W}_K \vee \mathcal{S}$.

By Theorem 1.3.4, [PR1, Lemma I.2.2] and the dual of [PR1, Lemma I.2.1], we know that $\mathcal{K}_0$ is a complete $\vee$-sublattice of $\mathcal{L}(\mathcal{CR})$. It is also a complete lattice with respect to the order inherited from $\mathcal{L}(\mathcal{CR})$. By Theorem 2.1.3(vi), we know that $\mathcal{K}_0$ is not an $\cap$-subsemilattice of $\mathcal{L}(\mathcal{CR})$. We thereby obtain the following.

Lemma 4.2.2 (i) *Let $\mathcal{K}_0$ be endowed with the join operation $\vee$ of $\mathcal{L}(\mathcal{CR})$ and the meet operation defined by $\mathcal{U} \wedge \mathcal{V} = (\mathcal{U} \cap \mathcal{V})_K$, for $\mathcal{U}, \mathcal{V} \in \mathcal{L}(\mathcal{CR})$. Then $(\mathcal{K}_0, \vee, \wedge)$ is a complete lattice isomorphic to $\mathcal{L}(\mathcal{CR})/K$.*

(ii) *Let $\mathcal{K}_1$ be endowed with the join operation $\vee$ of $\mathcal{L}(\mathcal{CR})$ and the meet operation defined by $\mathcal{U} \wedge \mathcal{V} = (\mathcal{U} \cap \mathcal{V})_K \vee \mathcal{S}$. Then $(\mathcal{K}_1, \vee, \wedge)$ is a complete lattice isomorphic to $\mathcal{L}(\mathcal{CR})/K$.*

Proof (i) It is clear that $(\mathcal{K}_0, \vee, \wedge)$, henceforth simply $\mathcal{K}_0$, is isomorphic to $\mathcal{L}(\mathcal{CR})/K$ since the join and meet operations have been designed to make it so.

(ii) For all $\mathcal{V} \in \mathcal{L}(\mathcal{CR})$, we have $\mathcal{V} K \mathcal{V} \vee \mathcal{S}$. The claim now follows as in part (i).   □

Notation 4.2.3 Let Ψ denote the class of all pairs $(\varphi, \mathcal{U}) \in (\mathcal{K}_1)^{\Theta^1} \times [\mathcal{S}, \mathcal{B}]$ (the direct product of the lattices $(\mathcal{K}_1)^{\Theta^1}$ and $[\mathcal{S}, \mathcal{B}]$) satisfying the following conditions.

(C1) φ is order preserving,
(C2) $\tau\varphi \subseteq (\mathcal{U}^B)_\tau$ for all $\tau \in \Theta^1$,
(C3) $(\emptyset\varphi)_{\tau K} \subseteq \tau\varphi$ for all $\tau \in \Theta^1$,
(C4) $\sigma, \tau \in \Theta$, $t(\sigma) \neq h(\tau)$ implies that $(\sigma\varphi)_{\tau K} \subseteq (\sigma\tau)\varphi$.

Our focus will be on the interval $[\mathcal{S}, \mathcal{CR}]$. However, not all elements of $\mathcal{K}_0$ lie in this interval (for example, $\mathcal{G}$, $\mathcal{CS}$) and so it will be more convenient here to work with $\mathcal{K}_1$. Note that, by Theorem 2.1.3(vi), for any $\mathcal{U} \in \mathcal{K}_0$, we have $(\mathcal{U} \vee \mathcal{S})_K = \mathcal{U}_K \vee \mathcal{S}_K = \mathcal{U}_K \vee \mathcal{T} = \mathcal{U}_K$.

In what follows, we will often use the fact that the mapping $\mathcal{V} \mapsto \mathcal{V} \cap \mathcal{B}$ is a complete homomorphism of $\mathcal{L}(\mathcal{CR})$ (Theorem 3.2.8(v)).

Lemma 4.2.4 Ψ *is a complete sublattice of $(\mathcal{K}_1)^{\Theta^1} \times [\mathcal{S}, \mathcal{B}]$, where $(\mathcal{K}_1)^{\Theta^1} \times [\mathcal{S}, \mathcal{B}]$ is endowed with the componentwise order.*

Proof Let $(\varphi_\alpha, \mathcal{V}_\alpha) \in \Psi$, $\alpha \in A$, and

$$\xi = \bigvee_{\alpha \in A} (\varphi_\alpha, \mathcal{V}_\alpha) = \left(\bigvee_{\alpha \in A} \varphi_\alpha, \bigvee_{\alpha \in A} \mathcal{V}_\alpha \right),$$

$$\eta = \bigwedge_{\alpha \in A} (\varphi_\alpha, \mathcal{V}_\alpha) = \left(\bigwedge_{\alpha \in A} \varphi_\alpha, \bigcap_{\alpha \in A} \mathcal{V}_\alpha \right).$$

Our goal is to show that $\xi, \eta \in \Psi$.

(C1) Clearly, the first components of ξ and η are order preserving, that is, ξ, η satisfy (C1).

(C2) Let $\tau \in \Theta^1$. Then

$$\tau \bigvee_{\alpha \in A} \varphi_\alpha = \bigvee_{\alpha \in A} \tau\varphi_\alpha \subseteq \bigvee_{\alpha \in A} (\mathcal{V}_\alpha^B)_\tau = \left(\bigvee_{\alpha \in A} \mathcal{V}_\alpha^B \right)_\tau \subseteq \left(\left(\bigvee_{\alpha \in A} \mathcal{V}_\alpha \right)^B \right)_\tau,$$

where the first inclusion follows from the assumption that all pairs $(\varphi_\alpha, \mathcal{V}_\alpha)$ satisfy condition (C2) and from the fact that the join operation in the lattice $(\mathcal{K}_1, \vee, \wedge)$ coincides with this operation in the whole lattice $\mathcal{L}(\mathcal{CR})$. Furthermore,

$$\tau \bigwedge_{\alpha \in A} \varphi_\alpha = \bigwedge_{\alpha \in A} \tau\varphi_\alpha \subseteq \bigcap_{\alpha \in A} \tau\varphi_\alpha \subseteq \bigcap_{\alpha \in A} ((\mathcal{V}_\alpha)^B)_\tau$$

$$= \left(\bigcap_{\alpha \in A} (\mathcal{V}_\alpha^B) \right)_\tau = \left(\left(\bigcap_{\alpha \in A} \mathcal{V}_\alpha \right)^B \right)_\tau,$$

where the first inclusion follows from the definition of the meet operation in the lattice $(\mathcal{K}, \vee, \wedge)$, the second inclusion follows from the assumption that all the pairs $(\varphi_\alpha, \mathcal{V}_\alpha)$ satisfy condition (C2), the next equality follows from Theorem 2.5.2(viii) and its dual and the last equality follows from Proposition 3.6.3(ii). Hence ξ, η satisfy condition (C2).

(C3) Let $\tau \in \Theta^1$. Then, recalling that the join operation in $(\mathcal{K}_1, \vee, \wedge)$ coincides with that in $\mathcal{L}(\mathcal{CR})$ and with the help of Theorem 2.5.2(viii), we obtain

$$\emptyset \bigvee_{\alpha \in A} \varphi_\alpha = \left(\bigvee_{\alpha \in A} \emptyset\varphi_\alpha \right)_{\tau K} = \bigvee_{\alpha \in A} (\emptyset\varphi_\alpha)_{\tau K} \subseteq \bigvee_{\alpha \in A} \tau\varphi_\alpha = \tau \bigvee_{\alpha \in A} \varphi_\alpha.$$

Next we have

$$\left(\emptyset \bigwedge_{\alpha \in A} \varphi_\alpha \right)_{\tau K} = \left(\bigwedge_{\alpha \in A} \emptyset\varphi_\alpha \right)_{\tau K} \subseteq \left(\bigcap_{\alpha \in A} \emptyset\varphi_\alpha \right)_{\tau K} = \left(\bigcap_{\alpha \in A} (\emptyset\varphi_\alpha)_\tau \right)_K$$

$$= \left(\bigcap_{\alpha \in A} (\emptyset\varphi_\alpha)_\tau \right)_{KK} = \left(\bigcap_{\alpha \in A} (\emptyset\varphi_\alpha)_{\tau K} \right)_K \subseteq \left(\bigcap_{\alpha \in A} (\emptyset\varphi_\alpha)_{\tau K} \vee \mathcal{S} \right)_K \vee \mathcal{S}$$

$$= \bigwedge_{\alpha \in A} (\emptyset\varphi_\alpha)_{\tau K} \vee \mathcal{S} \subseteq \bigwedge_{\alpha \in A} \tau\varphi_\alpha = \tau \bigwedge_{\alpha \in A} \varphi_\alpha.$$

Here the first inclusion follows by the definition of the meet operation in the lattice $(\mathcal{K}_1, \vee, \wedge)$, the next equality follows by Theorem 2.5.2(viii), the third equality follows by Theorem 2.1.3(i), the last equality but one follows again by the definition of the meet operation in the lattice $(\mathcal{K}_1, \vee, \wedge)$ and the final inclusion follows from the assumption that all pairs $(\varphi_\alpha, \mathcal{V}_\alpha)$ satisfies condition (C3). Thus ξ and η both satisfy condition (C3).

(C4) Let $\sigma, \tau \in \Theta$, $\tau(\sigma) \neq h(\tau)$. Then

$$\left(\sigma \bigvee_{\alpha \in A} \varphi_\alpha\right)_{\tau K} = \bigvee_{\alpha \in A} (\sigma\varphi_\alpha)_{\tau K} \subseteq \bigvee_{\alpha \in A} (\sigma\tau)\varphi_\alpha = (\sigma\tau) \bigvee_{\alpha \in A} \varphi_\alpha$$

where the first equality follows from Theorems 2.5.2(viii) and 2.1.3(i), the containment follows from condition (C4) on each φ_α and the final equality follows from the definition of the join operation in the lattice $(\mathcal{K}_1, \vee, \wedge)$. Furthermore,

$$\left(\sigma \bigwedge_{\alpha \in A} \varphi_\alpha\right)_{\tau K} = \left(\bigwedge_{\alpha \in A} \sigma\varphi_\alpha\right)_{\tau K} \subseteq \left(\bigcap_{\alpha \in A} \sigma\varphi_\alpha\right)_{\tau K}$$
$$= \left(\bigcap_{\alpha \in A} (\sigma\varphi_\alpha)_\tau\right)_K = \left(\bigcap_{\alpha \in A} (\sigma\varphi_\alpha)_\tau\right)_{KK}$$
$$\subseteq \left(\bigcap_{\alpha \in A} (\sigma\varphi_\alpha)_{\tau K}\right)_K \subseteq \left(\bigcap_{\alpha \in A} ((\sigma\varphi_\alpha)_{\tau K}) \vee \mathcal{S}\right)_K \vee \mathcal{S}$$
$$= \bigwedge_{\alpha \in A} ((\sigma\varphi_\alpha)_{\tau K} \vee \mathcal{S}) \subseteq \bigwedge_{\alpha \in A} (\sigma\tau)\varphi_\alpha = \sigma\tau \bigwedge_{\alpha \in A} \varphi_\alpha.$$

Thus ξ, η satisfy condition (C4) and therefore $\xi, \eta \in \Psi$. □

Theorem 4.2.5 *For each $\mathcal{V} \in [\mathcal{S}, \mathcal{CR}]$, define a function $\gamma_\mathcal{V}$ by*

$$\gamma_\mathcal{V} : \tau \mapsto \mathcal{V}_{\tau K} \vee \mathcal{S} \qquad (\tau \in \Theta^1).$$

Then the mappings $\gamma : \mathcal{V} \mapsto (\gamma_\mathcal{V}, \mathcal{V} \cap \mathcal{B})$ and

$$\xi_{\mathcal{K}_1} : (\varphi, \mathcal{U}) \mapsto (\emptyset\varphi)^K \cap \bigcap_{\tau \in \Theta} (\tau\varphi)^{K\bar{\tau}} \cap \mathcal{U}^B$$

are mutually inverse isomorphisms between the lattices $[\mathcal{S}, \mathcal{CR}]$ and Ψ.

Proof (1) γ *maps* $[\mathcal{S}, \mathcal{CR}]$ *into* Ψ. Let $\mathcal{V} \in [\mathcal{S}, \mathcal{CR}]$. We must show that $(\gamma_\mathcal{V}, \mathcal{V} \cap \mathcal{B})$ satisfies the conditions (C1)–(C4).

(C1) This is clear.

(C2) Let $\tau \in \Theta^1$. Then

$$\tau\gamma v = \mathcal{V}_{\tau K} \vee \mathcal{S} \subseteq \big((\mathcal{V} \cap \mathcal{B})^B\big)_{\tau K} \vee \mathcal{S}$$
$$\subseteq \big((\mathcal{V} \cap \mathcal{B})^B\big)_\tau \vee \mathcal{S} = \big((\mathcal{V} \cap \mathcal{B})^B \vee \mathcal{S}\big)_\tau = \big((\mathcal{V} \cap \mathcal{B})^B\big)_\tau$$

as required.

(C3) Let $\tau \in \Theta^1$. Then

$$(\emptyset\gamma v)_{\tau K} \vee \mathcal{S} = (\mathcal{V}_K \vee \mathcal{S})_{\tau K} \vee \mathcal{S} = \mathcal{V}_{K\tau K} \vee \mathcal{S}_{\tau K} \vee \mathcal{S} \subseteq \mathcal{V}_{\tau K} \vee \mathcal{S} = \tau\gamma v$$

as required.

(C4) Let $\sigma, \tau \in \Theta$, $t(\sigma) \neq h(\tau)$. Then

$$(\sigma\gamma v)_{\tau K} \vee \mathcal{S} = (\mathcal{V}_{\sigma K} \vee \mathcal{S})_{\tau K} \vee \mathcal{S} \subseteq (\mathcal{V}_\sigma \vee \mathcal{S})_{\tau K} \vee \mathcal{S}$$
$$= \mathcal{V}_{\sigma\tau K} \vee \mathcal{S}_{\tau K} \vee \mathcal{S}$$
$$= \mathcal{V}_{\sigma\tau K} \vee \mathcal{S} = (\sigma\tau)\gamma v.$$

Thus $(\gamma v, \mathcal{V} \cap \mathcal{B})$ satisfies conditions (C1)–(C4) and therefore γ maps $[\mathcal{S}, \mathcal{CR}]$ into Ψ.

(2) $\xi_{\mathcal{K}_1}$ *maps* Ψ *to* $[\mathcal{S}, \mathcal{CR}]$. This is clear.

(3) γ *and* $\xi_{\mathcal{K}_1}$ *are order preserving.* This is clear.

(4) $\gamma\xi_{\mathcal{K}_1} = \iota_{[\mathcal{S},\mathcal{CR}]}$. Let $\mathcal{V} \in [\mathcal{S}, \mathcal{CR}]$. Then

$$\mathcal{V}\gamma\xi_{\mathcal{K}_1} = (\gamma v, \mathcal{V} \cap \mathcal{B})\xi_{\mathcal{K}_1}$$
$$= (\emptyset\gamma v)^K \cap \bigcap_{\tau\in\Theta} (\tau\gamma v)^{K\overline{\tau}} \cap (\mathcal{V} \cap \mathcal{B})^B$$
$$= (\mathcal{V}_K \vee \mathcal{S})^K \cap \bigcap_{\tau\in\Theta} (\mathcal{V}_{\tau K} \vee \mathcal{S})^{K\overline{\tau}} \cap \mathcal{V}^B$$
$$= \mathcal{V}^K \cap \bigcap_{\tau\in\Theta} (\mathcal{V}_{\tau K})^{K\overline{\tau}} \cap \mathcal{V}^B$$
$$\subseteq \mathcal{V} \qquad\qquad \text{by Corollary 4.1.5.}$$

But clearly $\mathcal{V} \subseteq \mathcal{V}^K$ and, by Lemma 3.1.4(iii), we have $\mathcal{V} \subseteq (\mathcal{V}_\tau)^{K\overline{\tau}} = (\mathcal{V}_{\tau K})^{K\overline{\tau}}$ for all $\tau \in \Theta$ and $\mathcal{V} \subseteq \mathcal{V}^B$. Therefore $\mathcal{V} \subseteq \mathcal{V}\gamma\xi_{\mathcal{K}_1}$ and equality follows. Thus $\gamma\xi_{\mathcal{K}_1} = \iota_{[\mathcal{S},\mathcal{CR}]}$.

(5) $\xi_{\mathcal{K}_1}$ *is one-to-one.* Let $(\theta, \mathcal{U})$, $(\varphi, \mathcal{V})$ be distinct elements of Ψ. Let $(\theta, \mathcal{U})\xi_{\mathcal{K}_1} = \mathcal{P}$, $(\varphi, \mathcal{V})\xi_{\mathcal{K}_1} = \mathcal{Q}$ so that

$$\mathcal{P} = (\emptyset\theta)^K \cap \left(\bigcap_{\tau\in\Theta} (\tau\theta)^{K\overline{\tau}}\right) \cap \mathcal{U}^B,$$
$$\mathcal{Q} = (\emptyset\varphi)^K \cap \left(\bigcap_{\tau\in\Theta} (\tau\varphi)^{K\overline{\tau}}\right) \cap \mathcal{V}^B.$$

For any $\mathcal{W} \in \mathcal{L}(\mathcal{CR})$, we have $\mathcal{B} \subseteq \mathcal{W}^K$. Hence

$$\mathcal{P} \cap \mathcal{B} = \mathcal{U}^B \cap \mathcal{B} = \mathcal{U}, \quad \mathcal{Q} \cap \mathcal{B} = \mathcal{V}^B \cap \mathcal{B} = \mathcal{V}.$$

Hence, if $\mathcal{U} \neq \mathcal{V}$, then $\mathcal{P} \neq \mathcal{Q}$. So let us assume that $\mathcal{U} = \mathcal{V}$, but $\theta \neq \varphi$. Then there exists $\sigma \in \Theta^1$ such that $\sigma\theta \neq \sigma\varphi$.

First, we consider the possibility that $\sigma = \emptyset$, so that $\emptyset\theta \neq \emptyset\varphi$. Now $\emptyset\theta \subseteq (\emptyset\theta)^K$ while

$$\begin{aligned} \emptyset\theta &\subseteq \mathcal{U}^B & &\text{by condition (C2),} \\ \emptyset\theta &\subseteq \left((\emptyset\theta)_{\tau K}\right)^{K\overline{\tau}} \subseteq (\tau\theta)^{K\overline{\tau}} & &\text{by condition (C3).} \end{aligned}$$

Consequently, $\emptyset\theta \subseteq \mathcal{P}$. But clearly $\mathcal{P} \subseteq (\emptyset\theta)^K$. Hence $\emptyset\theta\ K\ \mathcal{P}$ so that $\mathcal{P}_K = (\emptyset\theta)_K$. Since $\emptyset\theta \in \mathcal{K}_1$, it follows that $\emptyset\theta = \mathcal{W}_\theta \vee \mathcal{S}$ for some $\mathcal{W}_\theta \in \mathcal{K}_0$. Then we have $\mathcal{P}_K = (\mathcal{W}_\theta \vee \mathcal{S})_K = \mathcal{W}_\theta$. Similarly $\emptyset\varphi = \mathcal{W}_\varphi \vee \mathcal{S}$, for some $\mathcal{W}_\varphi \in \mathcal{K}_0$, and $\mathcal{Q}_K = \mathcal{W}_\varphi$. In the case under consideration, $\mathcal{W}_\theta \vee \mathcal{S} = \emptyset\theta \neq \emptyset\varphi = \mathcal{W}_\varphi \vee \mathcal{S}$. Hence $\mathcal{P}_K = \mathcal{W}_\theta \neq \mathcal{W}_\varphi = \mathcal{Q}_K$ which implies that $\mathcal{P} \neq \mathcal{Q}$.   □

We can henceforth assume that $\sigma \in \Theta$. Let $\sigma\theta = \mathcal{W}_\theta \vee \mathcal{S}$ where $\mathcal{W}_\theta \in \mathcal{K}_0$. Our goal is to show that $\mathcal{P}_{\sigma K} \vee \mathcal{S} = \sigma\theta$. We consider, in the following sequence of lemmas, each of the terms in the above description of $\mathcal{P}$, one at a time. Note that we have already dealt with the case where $\sigma = \emptyset$ and can therefore assume that $\sigma \neq \emptyset$ throughout the lemmas below.

Lemma 4.2.6 *We have* $\sigma\theta \subseteq ((\emptyset\theta)^K)_{\sigma K} \vee \mathcal{S}$.

Proof We have

$$\begin{aligned} \left((\emptyset\theta)^K\right)_{\sigma K} \vee \mathcal{S} &\supseteq \left((\sigma\theta)^K\right)_{\sigma K} \vee \mathcal{S} & &\text{since } \theta \text{ is order preserving} \\ &= \left((\sigma\theta)^K\right)_K \vee \mathcal{S} = (\sigma\theta)_K \vee \mathcal{S} \\ &= (\mathcal{W}_\theta \vee \mathcal{S})_K \vee \mathcal{S} = \mathcal{W}_\theta \vee \mathcal{S} \\ &= \sigma\theta. \end{aligned}$$ □

Lemma 4.2.7 *We have* $\sigma\theta = (\sigma\theta^{K\overline{\sigma}})_{\sigma K} \vee \mathcal{S}$.

Proof We have

$$\begin{aligned} \left((\sigma\theta)^{K\overline{\sigma}}\right)_{\sigma K} \vee \mathcal{S} &= \left((\sigma\theta)^K\right)_{t(\sigma)K} \vee \mathcal{S} \\ &= \left((\sigma\theta)^K\right)_K \vee \mathcal{S} = (\sigma\theta)_K \vee \mathcal{S} \\ &= (\mathcal{W}_\theta \vee \mathcal{S})_K \vee \mathcal{S} = \mathcal{W}_\theta \vee \mathcal{S} = \sigma\theta. \end{aligned}$$ □

Lemma 4.2.8 *Let* $\tau \in \Theta$, $\tau \neq \sigma$. *Then* $\sigma\theta \subseteq ((\tau\theta)^{K\overline{\tau}})_{\sigma K} \vee \mathcal{S}$.

Proof As a first step, we will show that $\sigma\theta \subseteq ((\tau\theta)^{K\overline{\tau}})_\sigma$. We consider several cases.

Case: $h(\sigma) \neq h(\tau)$ *and* $|\sigma| \leq |\tau|$. Let $\sigma = h(\sigma)\sigma_1$ where $\sigma_1 \in \Theta^1$ and $|\sigma_1| < |\tau|$. Either $\sigma_1 = \emptyset$ or $h(\sigma_1) = h(\tau)$. Hence we can write $\tau = \sigma_1\tau_1$ for some $\tau_1 \neq \emptyset$. Since $t(\overline{\sigma}_1) = h(\sigma_1) = h(\tau) \neq h(\sigma)$, we have

$$\left((\tau\theta)^{K\overline{\tau}}\right)_\sigma = \left((\tau\theta)^{K\overline{\tau}_1\overline{\sigma}_1}\right)_{h(\sigma)\sigma_1} = (\tau\theta)^{K\overline{\tau}_1}$$

$$= \left((\sigma_1\tau_1)\theta\right)^{K\overline{\tau}_1}$$

$$\supseteq \left((\sigma\tau_1)\theta\right)^{K\overline{\tau}_1} \qquad \text{since } |\sigma\tau_1| = |h(\sigma)\sigma_1\tau_1| > |\sigma_1\tau_1|$$

$$\supseteq \left((\sigma\theta)_{\tau_1 K}\right)^{K\overline{\tau}_1} \qquad \text{by (C4)}$$

$$\supseteq \sigma\theta$$

as required.

Case: $h(\sigma) \neq h(\tau)$ *and* $|\sigma| > |\tau|$. Since $|\tau| < |\sigma|$, $\sigma\theta \subseteq \tau\theta$ and there exists $\sigma_1 \in \Theta^1$ such that $\sigma = h(\sigma)\tau\sigma_1$. Consequently,

$$\left((\tau\theta)^{K\overline{\tau}}\right)_\sigma = \left((\tau\theta)^{K\overline{\tau}}\right)_{h(\sigma)\tau\sigma_1} = \left((\tau\theta)^{K\overline{\tau}}\right)_{\tau\sigma_1}$$

$$= \left((\tau\theta)^K\right)_{t(\tau)\sigma_1} = (\tau\theta)^K \supseteq \tau\theta \supseteq \sigma\theta$$

as required.

Case: $h(\sigma) = h(\tau)$ *and* $|\sigma| < |\tau|$. Let $\tau = \sigma\tau_1$ where $t(\sigma) \neq h(\tau_1)$ and $\tau_1 \in \Theta$. Then

$$\left((\tau\theta)^{K\overline{\tau}}\right)_\sigma = \left((\tau\theta)^{K\overline{\tau}_1\overline{\sigma}}\right)_\sigma = (\tau\theta)^{K\overline{\tau}_1}$$

$$= \left((\sigma\tau_1)\theta\right)^{K\overline{\tau}_1}$$

$$\supseteq \left((\sigma\theta)_{\tau_1 K}\right)^{K\overline{\tau}_1} \qquad \text{by (C4)}$$

$$\supseteq \sigma\theta.$$

Case: $h(\sigma) = h(\tau)$ *and* $|\sigma| \geq |\tau|$. Since $h(\sigma) = h(\tau)$, it follows that either $\sigma = \tau$ or $|\tau| < |\sigma|$ so that, either way, $\sigma \leq \tau$ and $\sigma\theta \subseteq \tau\theta$. Let $\sigma = \tau\sigma_1$ where $\sigma_1 \in \Theta^1$ and $t(\tau) \neq h(\sigma_1)$ (if $\sigma_1 \neq \emptyset$). Then

$$\left((\tau\theta)^{K\overline{\tau}}\right)_\sigma = \left((\tau\theta)^{K\overline{\tau}}\right)_{\tau\sigma_1} = \left((\tau\theta)^K\right)_{t(\tau)\sigma_1} = (\tau\theta)^K \supseteq \tau\theta \supseteq \sigma\theta.$$

Thus, in all cases, we have $\sigma\theta \subseteq \left((\tau\theta)^{K\overline{\tau}}\right)_\sigma$. Therefore

$$\left((\tau\theta)^{K\overline{\tau}}\right)_{\sigma K} \vee \mathcal{S} \supseteq (\sigma\theta)_K \vee \mathcal{S} = \sigma\theta. \qquad \square$$

One last preliminary lemma.

Lemma 4.2.9 *We have* $\sigma\theta \subseteq (\mathcal{U}^B)_{\sigma K} \vee \mathcal{S}$.

Proof It follows from (C2) that $\sigma\theta = (\sigma\theta)_K \vee \mathcal{S} \subseteq (\mathcal{U}^B)_{\sigma K} \vee \mathcal{S}$.   □

We are now almost ready for the final step to establish that $\xi_{\mathcal{K}_1}$ is one-to-one. First note that, for any $\mathcal{V} \in \mathcal{L}(\mathcal{CR})$, we have $\mathcal{V} K \mathcal{V} \vee \mathcal{S}$ and $\mathcal{V}_K = (\mathcal{V} \vee \mathcal{S})_K$. From Lemmas 4.2.6–4.2.9, we have

$$\sigma\theta \subseteq \left((\emptyset\theta)^K\right)_{\sigma K} \vee \mathcal{S},$$
$$\sigma\theta = \left((\sigma\theta)^{K\bar\sigma}\right)_{\sigma K} \vee \mathcal{S},$$
$$\sigma\theta \subseteq \left((\tau\theta)^{K\bar\tau}\right)_{\sigma K} \vee \mathcal{S} \quad (\tau \in \Theta, \ \tau \neq \sigma),$$
$$\sigma\theta = (\sigma\theta)_K \vee \mathcal{S} \subseteq (\mathcal{U}^B)_{\sigma K} \vee \mathcal{S},$$

so that $\sigma\theta$ is the intersection of the expressions on the right-hand sides of the above equalities and containments. Hence, combining the second and third expression into one intersection ranging over the whole of Θ, we obtain

$$\sigma\theta = \left(\left((\emptyset\theta)^K\right)_{\sigma K} \vee \mathcal{S}\right) \cap \left(\bigcap_{\tau \in \Theta} \left((\tau\theta)^{K\bar\tau}\right)_{\sigma K} \vee \mathcal{S}\right) \cap \left((\mathcal{U}^B)_{\sigma K} \vee \mathcal{S}\right)$$

$$K \left((\emptyset\theta)^K\right)_\sigma \cap \bigcap_{\tau \in \Theta} \left((\tau\theta)^{K\bar\tau}\right)_\sigma \cap (\mathcal{U}^B)_\sigma$$

$$= \left((\emptyset\theta)^K \cap \bigcap_{\tau \in \Theta} (\tau\theta)^{K\bar\tau} \cap \mathcal{U}^B\right)_\sigma = \mathcal{P}_\sigma.$$

Therefore, for $\mathcal{W}_1 \in \mathcal{K}_0$ with $\sigma\theta = \mathcal{W}_1 \vee \mathcal{S}$, we have

$$\mathcal{P}_{\sigma K} \vee \mathcal{S} = (\sigma\theta)_K \vee \mathcal{S} = \mathcal{W}_1 \vee \mathcal{S} = \sigma\theta.$$

Likewise, $\mathcal{Q}_{\sigma K} \vee \mathcal{S} = \sigma\varphi$. But $\sigma\theta \neq \sigma\varphi$. Hence $\mathcal{P} \neq \mathcal{Q}$. Thus we have shown that if, for any value of σ we have $\sigma\theta \neq \sigma\varphi$, then we have $\mathcal{P} \neq \mathcal{Q}$, that is, $(\theta, \mathcal{U})\xi_{\mathcal{K}_1} \neq (\varphi, \mathcal{V})\xi_{\mathcal{K}_1}$. Therefore $\xi_{\mathcal{K}_1}$ is one-to-one. The careful reader might have noticed that along the way we have also established that $\xi_{\mathcal{K}_1}\gamma$ is the identity mapping on Ψ. However, we have now established explicitly that

γ maps $[\mathcal{S}, \mathcal{CR}]$ into Ψ,
$\xi_{\mathcal{K}_1}$ maps Ψ into $[\mathcal{S}, \mathcal{CR}]$,
γ and $\xi_{\mathcal{K}_1}$ are order preserving,
$\gamma\xi_{\mathcal{K}_1} = \iota_{[\mathcal{S},\mathcal{CR}]}$,
$\xi_{\mathcal{K}_1}$ is one-to-one.

By Lemma 3.2.7, this is sufficient to establish that γ and $\xi_{\mathcal{K}_1}$ are inverse isomorphisms.

Corollary 4.2.10 *Let* $\mathcal{Q} \in \mathcal{L}(\mathcal{CR})$, $\mathcal{Q} = \mathcal{Q}_K$ *and* $\mathcal{P} = \mathcal{Q}^K$. *If the mapping* $\pi_{\mathcal{Q}} : \mathcal{V} \mapsto \mathcal{V} \cap \mathcal{Q}$ *($\mathcal{V} \in \mathcal{L}(\mathcal{CR})$) is a complete retraction of* $\mathcal{L}(\mathcal{CR})$ *onto* $\mathcal{L}(\mathcal{Q})$, *then the mapping* $\pi_{\mathcal{P}} : \mathcal{V} \mapsto \mathcal{V} \cap \mathcal{P}$ *($\mathcal{V} \in \mathcal{L}(\mathcal{CR})$) is a complete retraction of* $\mathcal{L}(\mathcal{CR})$ *onto* $\mathcal{L}(\mathcal{P})$. *In particular, the mappings* $\pi_{\mathcal{O}}$ *and* $\pi_{L\mathcal{O}}$ *are complete retractions of* $\mathcal{L}(\mathcal{CR})$ *onto* $\mathcal{L}(\mathcal{O})$ *and* $\mathcal{L}(L\mathcal{O})$, *respectively.*

Proof Clearly it is only necessary to show that $\pi_{\mathcal{P}}$ respects arbitrary joins. Let $\mathcal{V}_\alpha$, $\alpha \in A$, be a family of varieties in $\mathcal{L}(\mathcal{CR})$.

Case: $\mathcal{V}_\alpha \in [\mathcal{S}, \mathcal{CR}]$ for all $\alpha \in A$. Let γ be as in Theorem 4.2.5, let $\mathcal{V}_\alpha \gamma = (\gamma \mathcal{V}_\alpha, \mathcal{V}_\alpha \cap B)$. Likewise, let $\mathcal{P}\gamma = (\gamma_{\mathcal{P}}, \mathcal{P} \cap B)$ and $\mathcal{Q}\gamma = (\gamma_{\mathcal{Q}}, \mathcal{Q} \cap B)$. Now consider $\left(\bigvee_{\alpha \in A} \mathcal{V}_\alpha\right) \cap \mathcal{P}$. It suffices to consider

$$\left(\bigvee_{\alpha \in A} (\gamma \mathcal{V}_\alpha, \mathcal{V}_\alpha \cap B)\right) \wedge (\gamma_{\mathcal{P}}, \mathcal{P} \cap B)$$

$$= \left(\bigvee_{\alpha \in A} \gamma \mathcal{V}_\alpha, \bigvee_{\alpha \in A} (\mathcal{V}_\alpha \cap B)\right) \wedge (\gamma_{\mathcal{P}}, \mathcal{P} \cap B)$$

$$= \left(\left(\bigvee_{\alpha \in A} \gamma \mathcal{V}_\alpha\right) \wedge \gamma_{\mathcal{P}}, \left(\bigvee_{\alpha \in A} (\mathcal{V}_\alpha \cap B)\right) \cap (\mathcal{P} \cap B)\right). \tag{4.2.1}$$

We begin with the first component: for $\tau \in \Theta^1$, we have

$$\begin{aligned}
\tau\left(\left(\bigvee_{\alpha \in A} \gamma \mathcal{V}_\alpha\right) \wedge \gamma_{\mathcal{P}}\right) &= \left(\bigvee_{\alpha \in A} \tau \gamma \mathcal{V}_\alpha\right) \wedge \tau \gamma_{\mathcal{P}} \\
&= \left(\bigvee_{\alpha \in A} (\mathcal{V}_\alpha)_{\tau K} \vee \mathcal{S}\right) \wedge (\mathcal{P}_{\tau K} \vee \mathcal{S}) \\
&= \left(\left(\bigvee_{\alpha \in A} (\mathcal{V}_\alpha)_{\tau K}\right) \cap \mathcal{P}_{\tau K}\right)_K \vee \mathcal{S} && \text{definition of } \wedge \text{ in } \mathcal{K}_1 \\
&= \left(\left(\bigvee_{\alpha \in A} (\mathcal{V}_\alpha)_{\tau K}\right) \cap (\mathcal{Q}^K)_{\tau K}\right)_K \vee \mathcal{S} \\
&= \left(\left(\bigvee_{\alpha \in A} (\mathcal{V}_\alpha)_{\tau K}\right) \cap \mathcal{Q}_K\right)_K \vee \mathcal{S} \\
&= \left(\left(\bigvee_{\alpha \in A} (\mathcal{V}_\alpha)_\tau\right) \cap \mathcal{Q}\right)_K \vee \mathcal{S} && K \text{ is a complete congruence} \\
&= \left(\bigvee_{\alpha \in A} ((\mathcal{V})_\tau \cap \mathcal{Q})\right)_K \vee \mathcal{S} && \text{by hypothesis on } \mathcal{Q} \\
&= \left(\bigvee_{\alpha \in A} ((\mathcal{V}_\alpha)_\tau \cap \mathcal{P}_{\tau K})\right)_K \vee \mathcal{S} \\
&= \left(\bigvee_{\alpha \in A} ((\mathcal{V}_\alpha)_\tau \cap \mathcal{P}_\tau)\right)_K \vee \mathcal{S} && K \text{ a complete congruence} \\
&= \left(\bigvee_{\alpha \in A} (\mathcal{V}_\alpha \cap \mathcal{P})_\tau\right)_K \vee \mathcal{S} \\
&= \bigvee_{\alpha \in A} (\mathcal{V}_\alpha \cap \mathcal{P})_{\tau K} \vee \mathcal{S} && \text{Theorem 2.1.3(vi)} \\
&= \bigvee_{\alpha \in A} \tau \gamma \mathcal{V}_\alpha \cap \mathcal{P} = \tau\left(\bigvee_{\alpha \in A} \gamma \mathcal{V}_\alpha \cap \mathcal{P}\right).
\end{aligned}$$

Hence

$$\left(\bigvee_{\alpha \in A} \gamma \mathcal{V}_\alpha\right) \wedge \gamma_{\mathcal{P}} = \bigvee_{\alpha \in A} \gamma \mathcal{V}_\alpha \cap \mathcal{P}. \tag{4.2.2}$$

Now consider the second component in (4.2.1). We have

$$\left(\bigvee_{\alpha \in A} (\mathcal{V}_\alpha \cap \mathcal{B}) \right) \cap (\mathcal{P} \cap \mathcal{B}) = \bigvee_{\alpha \in A} ((\mathcal{V}_\alpha \cap \mathcal{B}) \cap (\mathcal{P} \cap \mathcal{B}))$$

by Theorem 3.4.2(v)

$$= \bigvee_{\alpha \in A} (\mathcal{V}_\alpha \cap \mathcal{P}) \cap \mathcal{B}$$

$$= \left(\bigvee_{\alpha \in A} (\mathcal{V}_\alpha \cap \mathcal{P}) \right) \cap \mathcal{B}. \tag{4.2.3}$$

Hence, by (4.2.1), (4.2.2) and (4.2.3),

$$\left(\bigvee_{\alpha \in A} (\gamma \mathcal{V}_\alpha, \mathcal{V}_\alpha \cap \mathcal{B}) \right) \wedge (\gamma \mathcal{P}, \mathcal{P} \cap \mathcal{B})$$

$$= \left(\bigvee_{\alpha \in A} \gamma \mathcal{V}_\alpha \cap \mathcal{P}, \left(\bigvee_{\alpha \in A} (\mathcal{V}_\alpha \cap \mathcal{P}) \right) \cap \mathcal{B} \right)$$

$$= \bigvee_{\alpha \in A} (\gamma \mathcal{V}_\alpha \cap \mathcal{P}, (\mathcal{V}_\alpha \cap \mathcal{P}) \cap \mathcal{B}).$$

Consequently, under the isomorphism γ^{-1} we obtain

$$\left(\bigvee_{\alpha \in A} \mathcal{V}_\alpha \right) \cap \mathcal{P} = \bigvee_{\alpha \in A} (\mathcal{V}_\alpha \cap \mathcal{P}). \tag{4.2.4}$$

This completes the argument for the case where all the varieties $\mathcal{V}_\alpha$ contain $\mathcal{S}$. Two cases remain.

Case: There exists $\beta \in A$ with $\mathcal{S} \subseteq \mathcal{V}_\beta$ (but possibly some $\mathcal{V}_\alpha$ that do not contain $\mathcal{S}$). Then

$$\left(\bigvee_{\alpha \in A} \mathcal{V}_\alpha \right) \cap \mathcal{P} = \left(\bigvee_{\alpha \in A} (\mathcal{V}_\alpha \vee \mathcal{V}_\beta \vee \mathcal{S}) \right) \cap \mathcal{P}$$

$$= \left(\bigvee_{\alpha \in A} (\mathcal{V}_\alpha \vee \mathcal{S}) \right) \cap \mathcal{P}$$

$$= \bigvee_{\alpha \in A} ((\mathcal{V}_\alpha \vee \mathcal{S}) \cap \mathcal{P}) \qquad \text{by (4.2.4)}$$

$$= \bigvee_{\alpha \in A} ((\mathcal{V}_\alpha \cap \mathcal{P}) \vee \mathcal{S})$$

$$= \bigvee_{\alpha \in A} ((\mathcal{V}_\alpha \vee \mathcal{S}) \cap (\mathcal{P} \vee \mathcal{S}))$$

$$= \left(\bigvee_{\alpha \in A} (\mathcal{V}_\alpha \cap \mathcal{P}) \right) \vee \mathcal{S} \quad \text{by [PR1, Proposition IX.9.2]}$$

$$\bigvee_{\alpha \in A} (\mathcal{V}_\alpha \cap \mathcal{P}). \tag{4.2.5}$$

Case: $\mathcal{V}_\alpha \in \mathcal{L}(\mathcal{CS})$ for all $\alpha \in A$. Then

$$\left(\bigvee_{\alpha \in A} \mathcal{V}_\alpha \right) \cap \mathcal{P} = \left(\left(\left(\bigvee_{\alpha \in A} \mathcal{V}_\alpha \right) \cap \mathcal{P} \right) \vee \mathcal{S} \right) \cap \mathcal{CS}$$

$$= \left(\left(\bigvee_{\alpha \in A} (\mathcal{V}_\alpha \vee \mathcal{S}) \right) \cap \mathcal{P} \right) \cap \mathcal{CS}$$

$$\text{by [PR1, Proposition IX.9.2]}$$

$$= \left(\bigvee_{\alpha \in A} (\mathcal{V}_\alpha \vee \mathcal{S}) \cap \mathcal{P} \right) \cap \mathcal{CS} \quad \text{by (4.2.4)}$$

$$= \left(\bigvee_{\alpha \in A} (\mathcal{V}_\alpha \vee \mathcal{S}) \cap (\mathcal{P} \vee \mathcal{S}) \right) \cap \mathcal{CS}$$

$$= \left(\left(\bigvee_{\alpha \in A} (\mathcal{V}_\alpha \cap \mathcal{P}) \right) \vee \mathcal{S} \right) \cap \mathcal{CS}$$

$$\text{by [PR1, Proposition IX.9.2]}$$

$$= \bigvee_{\alpha \in A} (\mathcal{V}_\alpha \cap \mathcal{P}). \tag{4.2.6}$$

Thus, in all cases (4.2.4), (4.2.5) and (4.2.6) we see that $\left(\bigvee_{\alpha \in A} \mathcal{V}_\alpha \right) \cap \mathcal{P} = \bigvee_{\alpha \in A} (\mathcal{V}_\alpha \cap \mathcal{P})$ and the claim concerning $\pi_\mathcal{P}$ is established.

The final claim holds from the fact that, from Theorem 2.1.3(v), $\mathcal{O} = \mathcal{G}^K$ and $L\mathcal{O} = \mathcal{CS}^K$ combined with [PR1, Lemma VIII.1.1]. □

The results of this section are based on Polák [1, 2]. A result similar to Theorem 4.2.5 can be found in Petrich [7].

4.3   The Lattice Φ

In Theorem 3.2.8, we obtained a representation of the interval $[\mathcal{S}, \mathcal{CR}]$ as a lattice of mappings of the four element lattice D. In Theorem 4.2.5 we obtained another representation of this interval involving mappings into $\mathcal{K}_1$. In this and the next section we will combine these representations by amalgamating the lattices $\mathcal{K}_0$ and D. In addition, we characterize the lattice of mappings of this amalgamated lattice that will be of most interest to us.

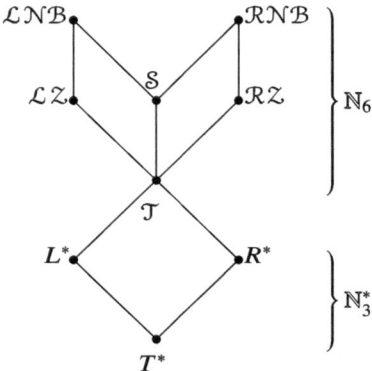

Diagram 4.3 $\mathbb{N}_3^* \cup \mathbb{N}_6$

Notation 4.3.1 Recall that $\mathbb{N}_3 = \{\mathcal{LNB}, \mathcal{S}, \mathcal{RNB}\}$. Let

$$\mathbb{N}_3^* = \{L^*, T^*, R^*\},$$
$$\mathbb{N}_6 = \{\mathcal{LZ}, \mathcal{T}, \mathcal{RZ}\} \cup \mathbb{N}_3 = \{\mathcal{LZ}, \mathcal{T}, \mathcal{RZ}, \mathcal{LNB}, \mathcal{S}, \mathcal{RNB}\},$$

and $\mathcal{K} = \mathcal{K}_0 \cup \mathbb{N}_3^*$. Let $\mathcal{K}$ be endowed with the order inherited from $\mathcal{L}(\mathcal{CR})$ in $\mathcal{K}_0$ together with $T^* < L^* < \mathcal{V}$, $T^* < R^* < \mathcal{V}$ for all $\mathcal{V} \in \mathcal{K}_0$.

The sets $\mathbb{N}_3^*$ and $\mathbb{N}_6$ are illustrated in Diagram 4.3.

When referring to the order relation between two generic elements $\mathcal{U}, \mathcal{V} \in \mathcal{K}$, we may still employ the set theoretic notation $\mathcal{U} \subseteq \mathcal{V}$ or $\mathcal{U} \subset \mathcal{V}$ as customary within $\mathcal{K}_0$, it being understood that should $\mathcal{U} \in \mathbb{N}_3^*$ then we mean $\mathcal{U} \leq \mathcal{V}$ or $\mathcal{U} < \mathcal{V}$.

Since $\mathcal{K}_0$ has $\mathcal{T}$ as its least element, the ordered set $\mathcal{K}$ can be pictured as in Diagram 4.3.

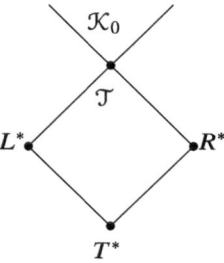

Since $\mathcal{K}_0$ is a complete lattice, it is easily seen that we have.

Lemma 4.3.2 $\mathcal{K}$ *is a complete lattice.*

It is important to recognize a complication that arises in dealing with $\mathcal{K}$ as a lattice. By Theorem 2.1.3(vi), the least upper bound of a subset of $\mathcal{K}$ is the same whether it is calculated in $\mathcal{K}$ or in $\mathcal{L}(\mathcal{CR}) \cup \mathbb{N}_3^*$ (with $\mathbb{N}_3^*$ as an ideal). However, the greatest lower bound or meet of a subset $\{\mathcal{U}_\alpha \mid \alpha \in A\}$ of $\mathcal{K}$, which we will denote by $\bigwedge_{\alpha \in A} \mathcal{U}_\alpha$ or, more explicitly, $\bigwedge_{\alpha \in A}^{\mathcal{K}} \mathcal{U}_\alpha$, is sometimes different in $\mathcal{K}$ from $\mathcal{L}(\mathcal{CR}) \cup \mathbb{N}_3^*$. We have

$$\bigwedge_{\alpha \in A}^{\mathcal{K}} \mathcal{U}_\alpha = \begin{cases} \left(\bigcap_{\alpha \in A} \mathcal{U}_\alpha\right)_K & \text{if } \mathcal{U}_\alpha \in \mathcal{L}(\mathcal{CR}) \text{ for all } \alpha \in A \\ \bigwedge_{\beta \in B} \mathcal{U}_\beta & \text{if } B = \{\beta \in A \mid \mathcal{U}_\beta \in \mathbb{N}_3^*\} \neq \emptyset. \end{cases}$$

We have available from our earlier discussions the operators $\mathcal{V} \mapsto \mathcal{V}_K$, $\mathcal{V} \mapsto \mathcal{V}^K$ ($\mathcal{V} \in \mathcal{L}(\mathcal{CR})$). We now introduce two associated operators.

Notation 4.3.3 For each $\mathcal{V} \in \mathcal{L}(\mathcal{CR}) \cup \mathbb{N}_3^*$ we define

$$\mathcal{V}_{K^*} = \begin{cases} \mathcal{V}_K & \text{if } \mathcal{V} \in \mathcal{L}(\mathcal{CR}) \setminus \mathbb{N}_6 \\ L^* & \text{if } \mathcal{V} \in \{L^*, \mathcal{LZ}, \mathcal{LNB}\} \\ T^* & \text{if } \mathcal{V} \in \{T^*, \mathcal{T}, \mathcal{S}\} \\ R^* & \text{if } \mathcal{V} \in \{R^*, \mathcal{RZ}, \mathcal{RNB}\} \end{cases} \qquad \mathcal{V}^{K^*} = \begin{cases} \mathcal{V}^K & \text{if } \mathcal{V} \in \mathcal{L}(\mathcal{CR}) \\ \mathcal{LNB} & \text{if } \mathcal{V} = L^* \\ \mathcal{S} & \text{if } \mathcal{V} = T^* \\ \mathcal{RNB} & \text{if } \mathcal{V} = R^*. \end{cases}$$

We also extend the domains of the mappings $\mathcal{V} \mapsto \mathcal{V}_{T_\ell}$, $\mathcal{V} \mapsto \mathcal{V}_{T_r}$ and consequently for the mapping $\mathcal{V} \mapsto \mathcal{V}_\tau$ for $\tau \in \Theta^1$, to include L^*, T^*, R^* by defining

$$(L^*)_{T_\ell} = (T^*)_{T_\ell} = T^* = (T^*)_{T_r} = (R^*)_{T_r}; \quad (L^*)_{T_r} = L^*; \quad (R^*)_{T_\ell} = R^*.$$

For any $\mathcal{V} \in \mathcal{L}(\mathcal{CR})$, respectively $\mathcal{V} \in \mathcal{L}(\mathcal{CR}) \cup \mathbb{N}_3^*$, and $\tau \in \Theta^1$, we write $\mathcal{V}_{\tau K^*} = (\mathcal{V}_\tau)_{K^*}$, respectively $\mathcal{V}^{\tau K^*} = (\mathcal{V}^\tau)^{K^*}$.

Clearly the mappings $\mathcal{V} \mapsto \mathcal{V}_{T_\ell}$, $\mathcal{V} \mapsto \mathcal{V}_{T_r}$ and $\mathcal{V} \mapsto \mathcal{V}_\tau$ continue to be complete endomorphisms on their enlarged domains. It is very important to note that the operator $\mathcal{V} \mapsto \mathcal{V}_{K^*}$ is *not* one-to-one on $\mathbb{N}_6$ and that the mapping $\mathcal{V} \mapsto \mathcal{V}^{K^*}$ sends each element $z \in \mathbb{N}_3^*$ onto the *maximal* pre-image of z under this mapping.

Notation 4.3.3 singles out the exceptional procedure to be followed as regards the varieties $\mathcal{LNB}$, $\mathcal{S}$ and $\mathcal{RNB}$. For any of these $\mathcal{V}_{K^*} \neq \mathcal{V}_K = \mathcal{T}$ but instead L^*, T^*, R^*, respectively. In the opposite direction, from L^*, T^*, R^* we get, under the mapping $\mathcal{V} \mapsto \mathcal{V}^{K^*}$, $\mathcal{LNB}$, $\mathcal{S}$, $\mathcal{RNB}$, respectively. For varieties in $\mathcal{K}_0$ we have $\mathcal{V}^{K^*} = \mathcal{V}^K$.

Convention 4.3.4 In Sect. 3.2, we introduced a lattice $D = \{1, L, T, R\}$. Here we have defined the lattice $\mathcal{K}$. Now $\mathcal{K}$ contains a sublattice/ideal $\{\mathcal{T}, L^*, T^*, R^*\} = \{\mathcal{T}\} \cup \mathbb{N}_3^*$ which is clearly isomorphic to D. We could have used the notation $\mathcal{T}, L^*, T^*, R^*$ in Sect. 13.2 but, for the sake of notational simplicity, we opted for $1, L, T, R$. Henceforth we will identify these two lattices under the correspondences:

$$1 \leftrightarrow \mathcal{T}, \quad L \leftrightarrow L^*, \quad T \leftrightarrow T^*, \quad R \leftrightarrow R^*.$$

Lemma 4.3.5 *The following statements hold.*

(i) *For $\mathcal{V} \in \mathcal{K}$, we have $(\mathcal{V}^{K^*})_{K^*} = \mathcal{V}$; $\mathcal{V} \in \mathbb{N}_3$, we have $(\mathcal{V}_{K^*})^{K^*} = \mathcal{V}$; $\mathcal{V} \in \mathcal{L}(\mathcal{CR})$, we have $(\mathcal{V}_{K^*})^{K^*} \supseteq \mathcal{V}$.*

(ii) *The mapping $\mathcal{V} \mapsto \mathcal{V}^{K^*}$, is order preserving on $\mathcal{L}(\mathcal{CR}) \cup \mathbb{N}_3^*$.*

(iii) *The restriction of the mapping $\mathcal{V} \mapsto \mathcal{V}_{K^*}$ to $\mathcal{L}(\mathcal{CR})$ is order preserving and, on $\mathbb{N}_3^*$, it is the identity mapping.*

(iv) *For any subset $\{\mathcal{V}_\alpha \mid \alpha \in A\} \subseteq \mathcal{L}(\mathcal{CR})$, we have*

$$\left(\bigcap_{\alpha \in A} \mathcal{V}_\alpha\right)_{K^*} \leq \bigwedge_{\alpha \in A} (\mathcal{V}_\alpha)_{K^*}.$$

(v) *For any subset $\{\mathcal{V}_\alpha \mid \alpha \in A\} \subseteq \mathcal{L}(\mathcal{CR})$, we have*

$$\left(\bigvee_{\alpha \in A} \mathcal{V}_\alpha\right)_{K^*} = \bigvee_{\alpha \in A} (\mathcal{V}_\alpha)_{K^*}.$$

Proof (i) The first two equalities follow immediately from the definitions and the third claim is obvious in each of the cases $\mathcal{V} \in \mathbb{N}_6$, $\mathcal{V} \notin \mathbb{N}_6$.

(ii) By Theorem 2.1.3(vii), we know that the mapping $\mathcal{V} \mapsto \mathcal{V}^K$ ($\mathcal{V} \in \mathcal{L}(\mathcal{CR})$) is a complete endomorphism and therefore order preserving. Since $\mathcal{T}^K = \mathcal{B} \supseteq \mathcal{LNB} \vee \mathcal{S} \vee \mathcal{RNB}$ it follows that the mapping $\mathcal{V} \mapsto \mathcal{V}^{K^*}$ is order preserving on $\mathcal{L}(\mathcal{CR}) \cup \mathbb{N}_3^*$.

(iii) By Theorem 2.1.3(vi), the mapping $\mathcal{V} \mapsto \mathcal{V}_K$ ($\mathcal{V} \in \mathcal{L}(\mathcal{CR})$) is order preserving. The claims then follow easily from the definition of the mapping $\mathcal{V} \mapsto \mathcal{V}_{K^*}$.

(iv) Since $\bigcap_{\alpha \in A} \mathcal{V}_\alpha \subseteq \mathcal{V}_\alpha$, for all $\alpha \in A$, it follows from (iii) that $(\bigcap_{\alpha \in A} \mathcal{V}_\alpha)_{K^*} \leq (\mathcal{V}_\alpha)_{K^*}$, for all $\alpha \in A$. Therefore $(\bigcap_{\alpha \in A} \mathcal{V}_\alpha)_{K^*} \leq \bigwedge_{\alpha \in A} (\mathcal{V}_\alpha)_{K^*}$.

(v) We divide the proof into several cases determined by the value of $\bigvee_{\alpha \in A} \mathcal{V}_\alpha$.

Case: $\bigvee_{\alpha \in A} \mathcal{V}_\alpha \in \mathbb{N}_6$. We consider only $\bigvee_{\alpha \in A} \mathcal{V}_\alpha = \mathcal{LNB}$. The arguments for the other possibilities are similar. If $\bigvee_{\alpha \in A} \mathcal{V}_\alpha = \mathcal{LNB}$, then we must have $\{\mathcal{V}_\alpha \mid \alpha \in A\} \subseteq \{\mathcal{T}, \mathcal{LZ}, \mathcal{S}, \mathcal{LNB}\}$ and must include one of $\mathcal{LZ}, \mathcal{LNB}$. This implies that $\{(\mathcal{V}_\alpha)_{K^*} \mid \alpha \in A\} \subseteq \{T^*, L^*\}$ and includes L^*. Hence $(\bigvee_{\alpha \in A} \mathcal{V}_\alpha)_{K^*} = L^* = \bigvee_{\alpha \in A} (\mathcal{V}_\alpha)_{K^*}$.

Case: $\bigvee_{\alpha \in A} \mathcal{V}_\alpha \notin \mathbb{N}_6$ and $\{\mathcal{V}_\alpha \mid \alpha \in A\} \subseteq \mathbb{N}_6$. We must have $\bigvee_{\alpha \in A} \mathcal{V}_\alpha \in \{\mathcal{RB}, \mathcal{NB}\}$. Suppose that $\bigvee_{\alpha \in A} \mathcal{V}_\alpha = \mathcal{NB}$. Then there must exist $\alpha, \beta, \gamma \in A$ (not necessarily distinct) such that $\mathcal{S} \subseteq \mathcal{V}_\alpha, \mathcal{LZ} \subseteq \mathcal{V}_\beta, \mathcal{RZ} \subseteq \mathcal{V}_\gamma$. Hence,

$$\left(\bigvee_{\alpha \in A} \mathcal{V}_\alpha\right)_{K^*} = \mathcal{NB}_{K^*} = \mathcal{T} = T^* \vee L^* \vee R^*$$

$$= \mathcal{S}_{K^*} \vee \mathcal{LZ}_{K^*} \vee \mathcal{RZ}_{K^*} = \bigvee_{\alpha \in A} (\mathcal{V}_\alpha)_{K^*}$$

A similar argument applies when $\bigvee_{\alpha \in A} \mathcal{V}_\alpha = \mathcal{RB}$.

Case: $\bigvee_{\alpha \in A} \mathcal{V}_\alpha \notin \mathbb{N}_6$ *and there exists* $\alpha \in A$ *such that* $\mathcal{V}_\alpha \notin \mathbb{N}_6$. If there happens to be $\beta \in A$ with $\mathcal{V}_\beta \in \mathbb{N}_6$, then

$$(\mathcal{V}_\alpha)_{K^*} \vee (\mathcal{V}_\beta)_{K^*} = (\mathcal{V}_\alpha)_K \vee (\mathcal{V}_\beta)_{K^*} = (\mathcal{V}_\alpha)_K$$

since $(\mathcal{V}_\beta)_{K^*} < \mathcal{T}$. Therefore, there is no loss in assuming that $\mathcal{V}_\alpha \in \mathcal{L}(\mathcal{CR}) \setminus \mathbb{N}_6$ for all $\alpha \in A$ so that

$$\begin{aligned}
\bigvee_{\alpha \in A} (\mathcal{V}_\alpha)_{K^*} &= \bigvee_{\alpha \in A} (\mathcal{V}_\alpha)_K \\
&= \left(\bigvee_{\alpha \in A} \mathcal{V}_\alpha \right)_K \qquad \text{by Theorem 2.1.3(vi)} \\
&= \left(\bigvee_{\alpha \in A} \mathcal{V}_\alpha \right)_{K^*}.
\end{aligned}$$

We have now covered all the possibilities.   □

In what follows we consider $\mathcal{K}^{\Theta^1}$ (the set of all mappings of Θ^1 into $\mathcal{K}$) as a partially ordered set with respect to the componentwise (or product) order. Since $\mathcal{K}$ is a complete lattice, so also is $\mathcal{K}^{\Theta^1}$.

Notation 4.3.6 Let Φ denote the set of all mappings $\varphi : \Theta^1 \mapsto \mathcal{K}$ satisfying the following conditions:

(P1) $\emptyset \varphi \in \mathcal{K}_0$,
(P2) φ is order preserving,
(P3) if $\tau \in \Theta$ and $\tau \varphi = L^*$, then $t(\tau) = T_r$,
(P4) if $\tau \in \Theta$ and $\tau \varphi = R^*$, then $t(\tau) = T_\ell$,
(P5) if $\tau \in \Theta$, then $(\emptyset \varphi)_{\tau K^*} \leq \tau \varphi$,
(P6) if $\sigma, \tau \in \Theta$ are such that $\sigma \varphi \in \mathcal{K}_0$ and $t(\sigma) \neq h(\tau)$, $(\sigma \varphi)_{\tau K^*} \leq (\sigma \tau) \varphi$.

By our convention we are identifying D with $\{\mathcal{T}\} \cup \mathbb{N}_3^*$. We can easily verify that $\Phi_D \subseteq \Phi$; that is, that elements of Φ_D satisfy all the conditions (P1)–(P6). In addition, Φ_D is then a sublattice, indeed an ideal, of Φ. See Corollary 4.4.7. The conditions (P1)–(P4) in the definition of Φ are usually easy to check. It is the last two conditions that can be tricky. However, as we will see later, there are contexts in which they can be greatly simplified.

Theorem 4.3.7 Φ *is a complete sublattice of* $\mathcal{K}^{\Theta^1}$.

Proof Let $\{\varphi_\alpha \mid \alpha \in A\}$ be a nonempty subset of Φ, $\xi = \bigwedge_{\alpha \in A} \varphi_\alpha$ and $\psi = \bigvee_{\alpha \in A} \varphi_\alpha$. We wish to show that ξ, $\psi \in \Phi$. Since each φ_α satisfies (P1)–(P4), it is easily verified ξ and ψ satisfy these conditions also. Let $\tau \in \Theta$. Then

$$(\emptyset\xi)_{\tau K^*} = \left(\emptyset \bigwedge_{\alpha \in A} \varphi_\alpha\right)_{\tau K^*} = \left(\bigwedge_{\alpha \in A} \emptyset\varphi_\alpha\right)_{\tau K^*}$$

$$= \left(\left(\bigcap_{\alpha \in A} \emptyset\varphi_\alpha\right)_K\right)_{\tau K^*} \qquad \text{since } \emptyset\varphi_\alpha \in \mathcal{K}_0 \text{ for all } \alpha \in A$$

$$\leq \left(\bigcap_{\alpha \in A} \emptyset\varphi_\alpha\right)_{\tau K^*} = \left(\bigcap_{\alpha \in A} (\emptyset\varphi_\alpha)_\tau\right)_{K^*}$$

$$\leq \bigwedge_{\alpha \in A} (\emptyset\varphi_\alpha)_{\tau K^*} \qquad \text{by Lemma 4.3.5(iv)}$$

$$\leq \bigwedge_{\alpha \in A} \tau\varphi_\alpha = \tau\xi.$$

Thus ξ satisfies the condition (P5).

Now consider (P6). Let σ, $\tau \in \Theta$ be such that $\sigma\xi \in \mathcal{K}_0$ and $t(\sigma) \neq h(\tau)$. Then $\bigwedge_{\alpha \in A} \sigma \varphi_\alpha = \sigma \bigwedge_{\alpha \in A} \varphi_\alpha = \sigma\xi \in \mathcal{K}_0$ and necessarily $\sigma\varphi_\alpha \in \mathcal{K}_0$ for all $\alpha \in A$.

Case: $(\sigma\varphi_\alpha)_{\tau K^*} \in \mathcal{K}_0$, *for all* $\alpha \in A$. It follows that $(\sigma\varphi_\alpha)_{\tau K^*} = (\sigma\varphi_\alpha)_{\tau K} \in \mathcal{L}(\mathcal{CR})$ and $\sigma\varphi_\alpha \in \mathcal{K}_0$. Hence

$$(\sigma\tau)\xi = (\sigma\tau) \bigwedge_{\alpha \in A} \varphi_\alpha = \bigwedge_{\alpha \in A} (\sigma\tau)\varphi_\alpha$$

$$\geq \bigwedge_{\alpha \in A} (\sigma\varphi_\alpha)_{\tau K^*} \qquad \text{by (P6) for each } \varphi_\alpha$$

$$\geq \left(\bigcap_{\alpha \in A} (\sigma\varphi_\alpha)_\tau\right)_{K^*} \qquad \text{by Lemma 4.3.5(iv)}$$

$$= \left(\bigcap_{\alpha \in A} \sigma\varphi_\alpha\right)_{\tau K^*} \qquad \text{by Theorem 2.5.2(viii) and its dual}$$

$$\geq \left(\left(\bigcap_{\alpha \in A} \sigma\varphi_\alpha\right)_K\right)_{\tau K^*} = \left(\bigwedge_{\alpha \in A} \sigma\varphi_\alpha\right)_{\tau K^*}$$

$$= \left(\sigma \bigwedge_{\alpha \in A} \varphi_\alpha\right)_{\tau K^*} = (\sigma\xi)_{\tau K^*}$$

as required.

Case: *There exists* $\beta \in A$ *such that* $(\sigma\varphi_\beta)_{\tau K^*} = L^*$ *and* $(\sigma\varphi_\alpha)_{\tau K^*} \neq T^*$, *for all* $\alpha \in A$.

Necessarily, $(\sigma\varphi_\beta)_\tau = \mathcal{LZ}$ or $\mathcal{LNB}$ so that $t(\tau) = T_r$. Hence it is not possible to have $(\sigma\varphi_\alpha)_\tau = \mathcal{RZ}$ or $\mathcal{RNB}$, for any $\alpha \in A$, or $(\sigma\varphi_\alpha)_{\tau K^*} = R^*$. Therefore

$$(\sigma\tau)\xi = (\sigma\tau)\bigwedge_{\alpha\in A}\varphi_\alpha = \bigwedge_{\alpha\in A}(\sigma\tau)\varphi_\alpha \geq \bigwedge_{\alpha\in A}(\sigma\varphi_\alpha)_{\tau K^*} = L^*.$$

On the other hand,

$$(\sigma\xi)_{\tau K^*} = \left(\sigma\bigwedge_{\alpha\in A}\varphi_\alpha\right)_{\tau K^*} = \left(\bigwedge_{\alpha\in A}\sigma\varphi_\alpha\right)_{\tau K^*}$$

$$= \left(\left(\bigcap_{\alpha\in A}\sigma\varphi_\alpha\right)_K\right)_{\tau K^*} \qquad \text{by the definition } \bigwedge \text{ in } \mathcal{K}$$

$$\leq \left(\bigcap_{\alpha\in A}\sigma\varphi_\alpha\right)_{\tau K^*} = \left(\bigcap_{\alpha\in A}(\sigma\varphi_\alpha)_\tau\right)_{K^*} \leq (\mathcal{LNB})_{K^*} = L^*.$$

Thus $(\sigma\xi)_{\tau K^*} \leq (\sigma\tau)\xi$, as required.

Case: There exists $\beta \in A$ such that $(\sigma\varphi_\beta)_{\tau K^*} = R^*$ and $(\alpha\varphi_\alpha)_{\tau K^*} \neq T^*$, for all $\alpha \in A$.
This is the dual of the preceding case.

Case: There exists $\beta \in A$ such that $(\sigma\varphi_\beta)_{\tau K^*} = T^*$. Then $(\sigma\varphi_\beta)_\tau = \mathcal{T}$ or $\mathcal{S}$ and

$$(\sigma\xi)_{\tau K^*} = \left(\sigma\bigwedge_{\alpha\in A}\varphi_\alpha\right)_{\tau K^*} = \left(\bigwedge_{\alpha\in A}\sigma\varphi_\alpha\right)_{\tau K^*}$$

$$= \left(\left(\bigcap_{\alpha\in A}\sigma\varphi_\alpha\right)_K\right)_{\tau K^*} \qquad \text{by the definition of } \bigwedge \text{ in } \mathcal{K}$$

$$\leq \left(\bigcap_{\alpha\in A}\sigma\varphi_\alpha\right)_{\tau K^*} = \left(\bigcap_{\alpha\in A}(\sigma\varphi_\alpha)_\tau\right)_{K^*} \leq \mathcal{S}_{K^*} = T^* \leq (\sigma\tau)\xi$$

and again (P6) is satisfied. Consequently ξ satisfies (P6) and $\xi \in \Phi$.

Consider ψ. We have $\emptyset\varphi_\alpha \in \mathcal{K}_0$, for all $\alpha \in A$, and so

$$(\emptyset\psi)_{\tau K^*} = \left(\emptyset\bigvee_{\alpha\in A}\varphi_\alpha\right)_{\tau K^*} = \left(\bigvee_{\alpha\in A}\emptyset\varphi_\alpha\right)_{\tau K^*}$$

$$= \bigvee_{\alpha\in A}(\emptyset\varphi_\alpha)_{\tau K^*} \qquad \text{by Lemmas 3.1.4(i) and 4.3.5(v)}$$

$$\leq \bigvee_{\alpha\in A}\tau\varphi_\alpha = \tau\psi \qquad \text{by (P5) applied to each } \varphi_\alpha$$

so that ψ satisfies (P5).

Finally, consider (P6). Let $\sigma, \tau \in \Theta$ be such that $\sigma\psi \in \mathcal{K}_0$ and $t(\sigma) \neq h(\tau)$. The assumption that $\sigma\psi \in \mathcal{K}_0$ means that $\bigvee_{\alpha\in A}\sigma\varphi_\alpha = \sigma\bigvee_{\alpha\in A}\varphi_\alpha \in \mathcal{K}_0$. Hence either

(i) there exists $\beta \in A$ such that $\sigma\varphi_\beta \in \mathcal{K}_0$ or
(ii) $\sigma\varphi_\alpha \in \mathbb{N}_3$ for all $\alpha \in A$ and there exist $\beta, \gamma \in A$ such that $\sigma\varphi_\beta = L^*$, $\sigma\varphi_\gamma = R^*$ so that $\sigma(\varphi_\beta \vee \varphi_\gamma) = \mathcal{T}$.

Case (i). Let $B = \{\alpha \in A \mid \sigma \varphi_\alpha \in \mathcal{K}_0\}$. If $\alpha \notin B$, then $\sigma \varphi_\alpha \in \mathbb{N}_3^*$ so that $(\sigma h(\tau))\varphi_\alpha \in \mathbb{N}_3^*$ since φ_α is order preserving by (P2). But by the choice of σ and τ, we have $t(\sigma) \neq h(\tau)$ whence $\sigma h(\tau)$ is in canonical form and $h(\tau) = t(\sigma h(\tau))$. Hence, by (P3) and (P4), we cannot have $\sigma \varphi_\alpha = (\sigma h(\tau))\varphi_\alpha = L^*$ or $\sigma \varphi_\alpha = (\sigma h(\tau))\varphi_\alpha = R^*$. Since $(\sigma h(\tau))\varphi_\alpha \subseteq \sigma \varphi_\alpha$, we must have $(\sigma h(\tau))\varphi_\alpha = T^*$ and also $(\sigma \tau)\varphi_\alpha = T^*$. Therefore

$$
\begin{aligned}
(\sigma \tau)\psi = (\sigma \tau) \bigvee_{\alpha \in A} \varphi_\alpha &= \bigvee_{\alpha \in A} (\sigma \tau)\varphi_\alpha && \text{by definition} \\
&= \bigvee_{\alpha \in B} (\sigma \tau)\varphi_\alpha && \\
&\geq \bigvee_{\alpha \in B} (\sigma \varphi_\alpha)_{\tau K^*} && \text{by (P6) applied to } \varphi_\alpha \\
&= \left(\bigvee_{\alpha \in B} (\sigma \varphi_\alpha)_\tau \right)_{K^*} && \text{by Lemma 4.3.5(v)} \\
&= \left(\bigvee_{\alpha \in B} \sigma \varphi_\alpha \right)_{\tau K^*} && \text{by Lemma 3.1.4(i)} \\
&= \left(\bigvee_{\alpha \in A} \sigma \varphi_\alpha \right)_{\tau K^*} && \\
&= \left(\sigma \left(\bigvee_{\alpha \in A} \varphi_\alpha \right) \right)_{\tau K^*} && \text{by definition} \\
&= (\sigma \psi)_{\tau K^*}.
\end{aligned}
$$

Thus the claim holds in Case (i).

Case (ii). Since $\sigma \psi \in \mathcal{K}_0$, it follows that $\sigma \bigvee_{\alpha \in A} \varphi_\alpha \in \mathcal{K}_0$, that is $\bigvee_{\alpha \in A} \sigma \varphi_\alpha \in \mathcal{K}_0$ whereas $\sigma \varphi_\alpha \in \mathbb{N}_3$ for all $\alpha \in A$, it therefore ensues that $\sigma \bigvee_{\alpha \in A} \varphi_\alpha = \mathcal{T}$. Using this fact, it is possible to conclude that

$$
(\sigma \psi)_{\tau K^*} = \left(\sigma \bigvee_{\alpha \in A} \varphi_\alpha \right)_{\tau K^*} = \left(\bigvee_{\alpha \in A} \sigma \varphi_\alpha \right)_{\tau K^*} = \mathcal{T}_{\tau K^*} = \mathcal{T}_{K^*} = \mathcal{T}^* \leq (\sigma \tau)\psi
$$

and the claim holds in case (ii).

Thus ψ satisfies (P6) and $\psi \in \Phi$. Hence $\xi, \psi \in \Phi$ and Φ is a complete sublattice of $\mathcal{K}^{\Theta^1}$.   □

The material in this section is based on Polák [2], but in the language of varieties, as in Reilly [3,4], rather than the language of fully invariant congruences.

4.4   Main Theorem

The remarkable result presented here has become the centerpiece in the study of the interval $[\mathcal{S}, \mathcal{CR}]$. The Polák theorem establishes an isomorphism between the lattices $[\mathcal{S}, \mathcal{CR}]$ and Φ. The domain Θ^1 of the elements of Φ is quite simple and the conditions (P1)–(P4) are straightforward. It is true that the conditions (P5) and (P6) and the lattice $\mathcal{K}$ are complicated. However, in certain special circumstances these simplify and make it possible to obtain considerable information concerning sublattices of $\mathcal{L}(\mathcal{CR})$.

Theorem 4.4.1 (Polák) *For every $\mathcal{V} \in [\mathcal{S}, \mathcal{CR}]$, define a function $\chi_\mathcal{V} : \Theta^1 \mapsto \mathcal{K}$ by*

$$\chi_\mathcal{V} : \tau \to \begin{cases} \mathcal{V}_{\tau K^*} & \text{if } \tau \in \Theta \\ \mathcal{V}_K & \text{if } \tau = \emptyset. \end{cases}$$

The mappings $\chi : \mathcal{V} \mapsto \chi_\mathcal{V}$, and $\xi : \varphi \mapsto \bigcap_{\tau \in \Theta^1} (\tau \varphi)^{K^* \bar{\tau}}$ are mutually inverse ismorphisms between the lattices $[\mathcal{S}, \mathcal{CR}]$ and Φ.

The definition of $\varphi\xi$ is, perhaps, more meaningful when written in "long form" with similar elements grouped together:

$$\varphi\xi = (\emptyset\varphi)^{K^*} \cap \bigcap_{\tau\varphi \in \mathcal{K}_0} (\tau\varphi)^{K^*\bar{\tau}} \cap \bigcap_{\tau\varphi = L^*} (L^*)^{K^*\bar{\tau}} \cap \bigcap_{\tau\varphi = R^*} (R^*)^{K^*\bar{\tau}} \cap \bigcap_{\tau\varphi = T^*} (T^*)^{K^*\bar{\tau}}$$

$$= (\emptyset\varphi)^{K} \cap \bigcap_{\tau\varphi \in \mathcal{K}_0} (\tau\varphi)^{K\bar{\tau}} \cap \bigcap_{\tau\varphi = L^*} \mathcal{LNB}^{\bar{\tau}} \cap \bigcap_{\tau\varphi = R^*} \mathcal{RNB}^{\bar{\tau}} \cap \bigcap_{\tau\varphi = T^*} \mathcal{S}^{\bar{\tau}}.$$

In this form, $\varphi\xi$ has five obvious components and some of the arguments to follow deal with each component in turn. We have encountered the last three of these components before, as they appeared in Theorem 3.6.4(ix), thereby revealing that the varieties of the form $\mathcal{V}^B$ have an important role to play in these considerations.

Definition 4.4.2 We will refer to the above theorem as *Polák's theorem* and to the isomorphism χ of Theorem 4.4.1 as the *Polák isomorphism* or *representation* and to the mapping $\chi_\mathcal{V}$ ($\mathcal{V} \in [\mathcal{S}, \mathcal{CR}]$) or the partially ordered set Θ^1 with the labels $\tau \chi_\mathcal{V}$ ($\tau \in \Theta^1$) at the corresponding vertices as the *(Polák) ladder* of $\mathcal{V}$.

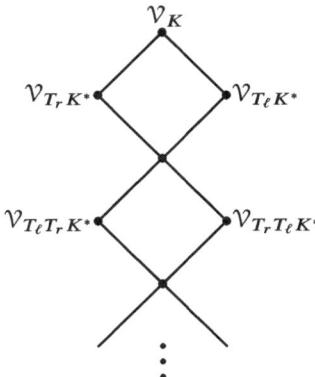

Diagram 4.4 Polák ladder of $\mathcal{V}$

For $\mathcal{V} \in [\mathcal{S}, \mathcal{CR}]$, the ladder $\chi_\mathcal{V}$ of $\mathcal{V}$ can be visualized as in Diagram 4.4 where this infinite sequence is interrupted when it becomes clear (or obscure) what the remaining labels of the vertices should be.

Proof (*of Theorem 4.4.1*) **1.** χ *maps* $[\mathcal{S}, \mathcal{CR}]$ *into* Φ. Let $\mathcal{V} \in [\mathcal{S}, \mathcal{CR}]$. By definition, $\emptyset \chi_\mathcal{V} = \mathcal{V}_K \in \mathcal{K}_0$ so that $\chi_\mathcal{V}$ satisfies (P1).

To see that $\chi_\mathcal{V}$ satisfies (P2), let $\sigma, \tau \in \Theta^1$ and $\sigma < \tau$. Since the mappings $\mathcal{U} \mapsto \mathcal{U}_{T_r}$, $\mathcal{U} \mapsto \mathcal{U}_{T_\ell}$ ($\mathcal{U} \in \mathcal{L}(\mathcal{CR})$) are order preserving and since $\mathcal{U}_{T_\ell}, \mathcal{U}_{T_r} \subseteq \mathcal{U}$ for all $\mathcal{U} \in \mathcal{L}(\mathcal{CR})$, it is clear that $\mathcal{V}_\sigma \subseteq \mathcal{V}_\tau$. By the definition of the mapping $\mathcal{U} \mapsto \mathcal{U}_{K^*}$ ($\mathcal{U} \in [\mathcal{S}, \mathcal{CR}]$) and Lemma 4.3.5(iii), we have that if $\tau = \emptyset$, then

$$\sigma \chi_\mathcal{V} = \mathcal{V}_{\sigma K^*} \leq \mathcal{V}_{K^*} \leq \mathcal{V}_K = \tau \chi_\mathcal{V} \tag{4.4.1}$$

while, if $\tau \neq \emptyset$, then

$$\sigma \chi_\mathcal{V} = \mathcal{V}_{\sigma K^*} \leq \mathcal{V}_{\tau K^*} = \tau \chi_{p\mathcal{V}}. \tag{4.4.2}$$

From (4.4.1) and (4.4.2) it follows that $\chi_\mathcal{V}$ satisfies (P2). Now suppose that $\tau \chi_\mathcal{V} = L^*$, that is, $\mathcal{V}_{\tau K^*} = L^*$. By the definition of the mapping $\chi_\mathcal{V}$, this implies that $\tau \neq \emptyset$. From the assumption that $\mathcal{V} \in [\mathcal{S}, \mathcal{CR}]$, it follows that $\mathcal{V}_\tau$ also lies in $[\mathcal{S}, \mathcal{CR}]$. Consequently, by the definition of the mapping $\mathcal{U} \mapsto \mathcal{U}_{K^*}$, we have that $\mathcal{V}_\tau = \mathcal{LNB}$. If $t(\tau) = T_\ell$, then since $T_\ell^2 = T_\ell$, we would have

$$\mathcal{LNB} = \mathcal{V}_\tau = \mathcal{V}_{\tau T_\ell} = \mathcal{LNB}_{T_\ell} = \mathcal{S},$$

a contradiction. Therefore τ does not end in T_ℓ and $\chi_\mathcal{V}$ satisfies (P3). By duality, $\chi_\mathcal{V}$ also satisfies (P4).

Now let $\tau \in \Theta$. Since the mappings $\mathcal{U} \mapsto \mathcal{U}_\tau, \mathcal{U} \mapsto \mathcal{U}_{K^*}$ are order preserving on $\mathcal{L}(\mathcal{CR})$, we obtain $\tau \chi_\mathcal{V} = \mathcal{V}_{\tau K^*} \geq (\mathcal{V}_K)_{\tau K^*} = (\emptyset \chi_\mathcal{V})_{\tau K^*}$. Thus $\chi_\mathcal{V}$ satisfies (P5).

For (P6), let $\sigma, \tau \in \Theta, \sigma \chi_\mathcal{V} \in \mathcal{K}_0$ and $t(\sigma) \neq h(\tau)$. Since $\mathcal{V}_{\sigma K^*} = \sigma \chi_\mathcal{V} \in \mathcal{K}_0$, we must have $\mathcal{V}_{\sigma K^*} = \mathcal{V}_{\sigma K}$ so that $\sigma \chi_\mathcal{V} = \mathcal{V}_{\sigma K}$. Hence, by Lemma 4.3.5(iii), $(\sigma \tau) \chi_\mathcal{V} = \mathcal{V}_{\sigma \tau K^*} = (\mathcal{V}_\sigma)_{\tau K^*} \geq (\mathcal{V}_{\sigma K})_{\tau K^*} = (\sigma \chi_\mathcal{V})_{\tau K^*}$. Thus $\chi_\mathcal{V}$ satisfies (P6) and consequently $\chi_\mathcal{V} \in \Phi$.

2. χ *is one-to-one.* This follows from Theorem 4.1.4.

3. χ *is order preserving.* By Lemma 3.1.4(i), the mapping $\mathcal{V} \mapsto \mathcal{V}_\tau$ ($\mathcal{V} \in \mathcal{L}(\mathcal{CR})$) is order preserving. Consequently, if $\mathcal{U} \subseteq \mathcal{V}$ and $\tau \in \Theta$, then $\mathcal{U}_\tau \subseteq \mathcal{V}_\tau$ so that by Lemma 4.3.5(iii) $\tau \chi_\mathcal{U} = \mathcal{U}_{\tau K^*} \leq \mathcal{V}_{\tau K^*} = \tau \chi_\mathcal{V}$ while, $\emptyset \chi_\mathcal{U} = \mathcal{U}_K \subseteq \mathcal{V}_K = \emptyset \chi_\mathcal{V}$ by Theorem 2.1.3(vi). Thus $\chi_\mathcal{U} \leq \chi_\mathcal{V}$ and χ is order preserving.

4. ξ *is order preserving.* By Lemma 3.1.4(ii), we know that the mapping $\mathcal{V} \mapsto \mathcal{V}^\tau$ is order preserving on $\mathcal{L}(\mathcal{CR})$ and by Lemma 4.3.5(ii), we know that the mapping $\mathcal{V} \mapsto \mathcal{V}^{K^*}$ ($\mathcal{V} \in \mathcal{K}$) is also order preserving and therefore ξ is order preserving.

5. $\xi \chi = \iota_\Phi$. Let $\theta \in \Phi$ and let

$$\mathcal{P} = \theta\xi = (\emptyset\theta)^K \cap \left(\bigcap_{\tau\theta\in\mathcal{K}_0} (\tau\theta)^{K\overline{\tau}} \right) \cap \left(\bigcap_{\tau\theta=L^*} (\mathcal{LNB}^{\overline{\tau}}) \right)$$

$$\cap \left(\bigcap_{\tau\theta=R^*} \mathcal{RNB}^{\overline{\tau}} \right) \cap \left(\bigcap_{\tau\theta=T^*} \mathcal{S}^{\overline{\tau}} \right).$$

We break the proof into a series of three rather technical, but necessary, lemmas. We will invoke Lemma 3.1.4 and Theorem 2.5.2(xiii) frequently and without comment.

Lemma 4.4.3 *For $\sigma \in \Theta$, we have $\left((\sigma\theta)^{K^*\overline{\sigma}}\right)_{\sigma K^*} = \sigma\theta.$* □

Proof We consider the different cases, depending on the value of $\sigma\theta$.

Case: $\sigma\theta \in \mathcal{K}_0$. Then

$$\left((\sigma\theta)^{K^*\overline{\sigma}}\right)_{\sigma K^*} = \left((\sigma\theta)^{K\overline{\sigma}}\right)_{\sigma K^*} = \left((\sigma\theta)^K\right)_{t(\sigma)K^*}$$
$$= \left((\sigma\theta)^K\right)_{K^*} = \left((\sigma\theta)^K\right)_K = (\sigma\theta)_K = \sigma\theta.$$

Case: $\sigma\theta = L^*$. Then $t(\sigma) = T_r$, by (P3), and

$$\left((\sigma\theta)^{K^*\overline{\sigma}}\right)_{\sigma K^*} = \left((L^*)^{K^*\overline{\sigma}}\right)_{\sigma K^*} = \left((\mathcal{LNB})^{\overline{\sigma}}\right)_{\sigma K^*}$$
$$= (\mathcal{LNB})_{t(\sigma)K^*} = (\mathcal{LNB})_{T_r K^*} = \mathcal{LNB}_{K^*} = L^* = \sigma\theta.$$

Case: $\sigma\theta = R^*$. This is the dual of the preceding case.

Case: $\sigma\theta = T^*$. Then

$$\left((\sigma\theta)^{K^*\overline{\sigma}}\right)_{\sigma K^*} = \left((T^*)^{K^*\overline{\sigma}}\right)_{\sigma K^*} = (\mathcal{S}^{\overline{\sigma}})_{\sigma K^*}$$
$$= \left(\mathcal{S}_{t(\sigma)}\right)_{K^*} = \mathcal{S}_{K^*} = T^* = \sigma\theta.$$

□

Lemma 4.4.4 *If $\sigma \in \Theta$, $\tau \in \Theta^1$ and $\sigma\theta \in \mathcal{K}_0$, then $\sigma\theta \subseteq \left((\tau\theta)^{K^*\overline{\tau}}\right)_\sigma.$*

Proof First suppose that $\tau = \emptyset$. Then $\sigma\theta \leq \emptyset\theta$ since θ is order preserving. Hence

$$\sigma\theta \leq \emptyset\theta \subseteq (\emptyset\theta)^K = \left((\emptyset\theta)^K\right)_\sigma = \left((\emptyset\theta)^{K^*}\right)_\sigma$$

and the claim holds.

For the remainder of the proof we will assume that $\tau \neq \emptyset$. Note that, by the hypothesis, we also have $\sigma \neq \emptyset$. There are several cases and subcases.

Case: $\sigma, \tau \neq \emptyset$, $h(\sigma) \neq h(\tau)$ and $|\sigma| \leq |\tau|$. Let $\sigma = h(\sigma)\sigma_1$ where $\sigma_1 \in \Theta^1$ and $|\sigma_1| < |\tau|$. Either $\sigma_1 = \emptyset$ or $h(\sigma_1) = h(\tau)$. Hence we can write $\tau = \sigma_1\tau_1$ for some $\tau_1 \neq \emptyset$. Since $t(\overline{\sigma}_1) = h(\sigma_1) = h(\tau) \neq h(\sigma)$ we have

$$\left((\tau\theta)^{K^*\bar\tau}\right)_\sigma = \left((\tau\theta)^{K^*\bar\tau_1\bar\sigma_1}\right)_{h(\sigma)\sigma_1} = (\tau\theta)^{K^*\bar\tau_1} = \left((\sigma_1\tau_1)\theta\right)^{K^*\bar\tau_1}$$

$$\supseteq \left((\sigma\tau_1)\theta\right)^{K^*\bar\tau_1} \qquad \text{since } |\sigma\tau_1| = |h(\sigma)\sigma_1\tau_1| > |\sigma_1\tau_1|$$

$$\supseteq \left((\sigma\theta)_{\tau_1 K^*}\right)^{K^*\bar\tau_1} \qquad \text{by (P6) and Lemma 3.6(ii)}$$

$$\supseteq \sigma\theta$$

as required.

Case: $\sigma, \tau \neq \emptyset$, $h(\sigma) \neq h(\tau)$ *and* $|\sigma| > |\tau|$. Since $|\tau| < |\sigma|$, we have $\sigma\theta \subseteq \tau\theta$ and there exists $\sigma_1 \in \Theta^1$ such that $\sigma = h(\sigma)\tau\sigma_1$. Consequently,

$$\left((\tau\theta)^{K^*\bar\tau}\right)_\sigma = \left((\tau\theta)^{K^*\bar\tau}\right)_{h(\sigma)\tau\sigma_1} = \left((\tau\theta)^{K^*\bar\tau}\right)_{\tau\sigma_1} = \left((\tau\theta)^{K^*}\right)_{t(\tau)\sigma_1}. \qquad (4.4.3)$$

Since $\sigma\theta \subseteq \tau\theta$, it follows that $\tau\theta \in \mathcal{K}_0$. Hence $\tau\theta \notin \mathbb{N}_3^*$ and continuing, we have

$$\left((\tau\theta)^{K^*\bar\tau}\right)_\sigma = \left((\tau\theta)^K\right)_{t(\tau)\sigma_1} = (\tau\theta)^K \supseteq \tau\theta \supseteq \sigma\theta$$

as required.

Case: $\sigma, \tau \neq \emptyset$, $h(\sigma) = h(\tau)$ *and* $|\sigma| < |\tau|$. Let $\tau = \sigma\tau_1$ where $t(\sigma) \neq h(\tau_1)$ and $\tau_1 \in \Theta$. Since $\sigma\theta \in \mathcal{K}_0$, we have

$$\left((\tau\theta)^{K^*\bar\tau}\right)_\sigma = \left((\tau\theta)^{K^*\bar\tau_1\bar\sigma}\right)_\sigma = (\tau\theta)^{K^*\bar\tau_1}$$

$$= \left((\sigma\tau_1)\theta\right)^{K^*\bar\tau_1}$$

$$\supseteq \left((\sigma\theta)_{\tau_1 K^*}\right)^{K^*\bar\tau_1} \qquad \text{by (P6)}$$

$$\supseteq \sigma\theta \qquad \text{by Lemma 4.3.5(i)}$$

and the claim holds in this case.

Case: $\sigma, \tau \neq \emptyset$, $h(\sigma) = h(\tau)$ *and* $|\sigma| \geq |\tau|$. Since $h(\sigma) = h(\tau)$, it follows that either $\sigma = \tau$ or $|\tau| < |\sigma|$ so that, either way, $\sigma \leq \tau$ and thus $\sigma\theta \leq \tau\theta$. Let $\sigma = \tau\sigma_1$ where $\sigma_1 \in \Theta^1$ and $t(\tau) \neq h(\sigma_1)$ (if $\sigma_1 \neq \emptyset$). Then,

$$\left((\tau\theta)^{K^*\bar\tau}\right)_\sigma = \left((\tau\theta)^{K^*\bar\tau}\right)_{\tau\sigma_1} = \left((\tau\theta)^{K^*}\right)_{t(\tau)\sigma_1}. \qquad (4.4.4)$$

Since $\sigma\theta \in \mathcal{K}_0$ and θ is order preserving, it follows that $\tau\theta \in \mathcal{K}_0$. Hence, by (4.4.4) and Theorem 2.5.2(xii),

$$\left((\tau\theta)^{K^*\bar\tau}\right)_\sigma = \left((\tau\theta)^K\right)_{t(\tau)\sigma_1} = (\tau\theta)^K \supseteq \tau\theta \supseteq \sigma\theta. \qquad (4.4.5)$$

We have now covered all possibilities and the proof of Lemma 4.4.4 is complete. $\square$

Lemma 4.4.5 *Let* $\sigma \in \Theta, \tau \in \Theta^1$.

(i) $\mathcal{P}_K = \emptyset\theta$.
(ii) *If* $\sigma\theta \in \mathcal{K}_0$ *and* $t(\sigma) = T_r$, *then* $\mathcal{LRB} \subseteq \big((\tau\theta)^{K^*\overline{\tau}}\big)_\sigma$.
(iii) *If* $\sigma\theta = L^*$, *then* $\mathcal{LNB} \subseteq \big((\tau\theta)^{K^*\overline{\tau}}\big)_\sigma$.
(iv) *For* $\sigma \in \Theta$, *we have* $\mathcal{P}_{\sigma K^*} = \sigma\theta$.

Proof (i) Obviously $\mathcal{P} = \theta\xi \subseteq (\emptyset\theta)^K$. Every term in the definition of $\mathcal{P}$ other than the first is of the form $(\tau\theta)^{K^*\overline{\tau}}$, for some $\tau \in \Theta$. For these terms, using (P5) and Lemma 4.3.5(i), we have $(\tau\theta)^{K^*\overline{\tau}} \supseteq \big((\emptyset\theta)_{\tau K^*}\big)^{K^*\overline{\tau}} \supseteq \emptyset\theta$. Hence $\emptyset\theta \subseteq \mathcal{P} \subseteq (\emptyset\theta)^K$ so that, since $\emptyset\theta \in \mathcal{K}_0$, we have $\emptyset\theta = (\emptyset\theta)_K = \mathcal{P}_K$.

(ii) We consider several cases.

Case: $\tau\theta \supseteq \mathcal{J}$. Then $\big((\tau\theta)^{K^*\overline{\tau}}\big)_\sigma \supseteq \mathcal{B}_\sigma = \mathcal{B} \supset \mathcal{LRB}$. *Case:* $\tau\theta = L^*$. By (P3) $t(\tau) = T_r$ and, since θ is order preserving, we must have $|\tau| > |\sigma|$. Hence $\tau = \tau_1\sigma^d T_r$ for some $\tau_1 \in \Theta^1$. Therefore

$$\big((\tau\theta)^{K^*\overline{\tau}}\big)_\sigma \supseteq (\mathcal{LNB}^{T_r\overline{\sigma^d}})_\sigma \supseteq \mathcal{LNB}^{T_r T_\ell} \supseteq \mathcal{LRB}.$$

Case: $\tau\theta = R^*$. We must have $t(\tau) = T_\ell$ and either $\tau = \sigma^d$ or $|\tau| > |\sigma|$. Either way

$$\big((\tau\theta)^{K^*\overline{\tau}}\big)_\sigma = (\mathcal{RNB}^{\overline{\tau}})_\sigma \supseteq \mathcal{RNB}^{t(\tau)} = \mathcal{RNB}^{T_\ell} \supseteq \mathcal{LRB}.$$

Case: $\tau\theta = T^*$. We must have either $\tau = \sigma^d$ or $|\tau| > |\sigma|$. If $\tau = \sigma^d$, then

$$\big((\tau\theta)^{K^*\overline{\tau}}\big)_\sigma = (\mathcal{S}^{\overline{\tau}})_\sigma = \mathcal{S}^{t(\tau)} = \mathcal{S}^{T_\ell} \supseteq \mathcal{LRB},$$

while if $|\tau| > |\sigma|$, then we have $\tau = \tau_1\sigma^d\tau_2$, for some $\tau_1, \tau_2 \in \Theta^1$ so that

$$\big((\tau\theta)^{K^*\overline{\tau}}\big)_\sigma = (\mathcal{S}^{\overline{\tau_2\sigma^d\tau_1}})_\sigma \supseteq (\mathcal{S}^{\overline{\sigma^d}})_\sigma = \mathcal{S}^{t(\sigma^d)} = \mathcal{S}^{T_\ell} \supseteq \mathcal{LRB}.$$

This covers all possible cases. Consequently, if $\sigma\theta \in \mathcal{K}_0$ and $t(\sigma) = T_r$, then we have $\big((\tau\theta)^{K^*\overline{\tau}}\big)_\sigma \supseteq \mathcal{LRB}$ as claimed.

(iii) By (P3), we have $t(\sigma) = T_r$. If $\tau\theta \supseteq \mathcal{J}$, then $((\tau\theta)^{K^*\overline{\tau}})_\sigma \supseteq \mathcal{B} \supseteq \mathcal{LNB}$. So now we may assume that $\tau\theta$ lies in $\mathbb{N}_3^*$.

Case: $\sigma\theta = L^*, \tau\theta = L^*$. We must have $t(\tau) = T_r$. Since θ is order preserving, it follows that either $\tau = \sigma, \tau = T_x\sigma$ or $\sigma = T_y\tau$ where $x, y \in \{\ell, r\}$.
    If $\tau = \sigma$, then $((\tau\theta)^{K^*\overline{\tau}})_\sigma = (\mathcal{LNB}^{\overline{\sigma}})_\sigma = \mathcal{LNB}_{T_r} = \mathcal{LNB}$.
    If $\tau = T_x\sigma$, then $((\tau\theta)^{K^*\overline{\tau}})_\sigma = (\mathcal{LNB}^{\overline{\sigma}T_x})_\sigma \supseteq \mathcal{LNB}$.
    If $\sigma = T_y\tau$, then $((\tau\theta)^{K^*\overline{\tau}})_\sigma = (\mathcal{LNB}^{\overline{\tau}})_{T_y\tau} = \mathcal{LNB}_{T_r} = \mathcal{LNB}$.

Case: $\sigma\theta = L^*$, $\tau\theta = R^*$. In this case, we must have $\tau = \sigma^d$ and $t(\tau) = T_\ell$. Hence

$$((\tau\theta)^{K^*\overline{\tau}})_\sigma = ((\mathcal{R}\mathcal{N}\mathcal{B}^{\overline{\tau}})_\sigma = \mathcal{R}\mathcal{N}\mathcal{B}^{t(\tau)} \supseteq \mathcal{R}\mathcal{N}\mathcal{B}^{T_\ell} \supseteq \mathcal{L}\mathcal{N}\mathcal{B}.$$

Case: $\sigma\theta = L^*$, $\tau\theta = T^*$. It follows that either $\tau = \sigma^d$ or $|\tau| > |\sigma|$. If $\tau = \sigma^d$, then

$$((\tau\theta)^{K^*\overline{\tau}})_\sigma = (\mathcal{S}^{\overline{\tau}})_\sigma = \mathcal{S}^{t(\tau)} = \mathcal{S}^{T_\ell} \supseteq \mathcal{L}\mathcal{R}\mathcal{B} \supseteq \mathcal{L}\mathcal{N}\mathcal{B}.$$

If $|\tau| > |\sigma|$, then either $\tau \le T_x\sigma$ for some $x \in \{\ell, r\}$ or $\tau \le \sigma T_\ell$. In the former case, we deduce that

$$((\tau\theta)^{K^*\overline{\tau}})_\sigma = (\mathcal{S}^{\overline{\tau}})_\sigma \supseteq (\mathcal{S}^{\overline{\sigma}T_x})_\sigma = \mathcal{S}^{T_r T_\ell} \supseteq \mathcal{L}\mathcal{N}\mathcal{B}.$$

In the latter case, we obtain $((\tau\theta)^{K^*\overline{\tau}})_\sigma = (\mathcal{S}^{\overline{\tau}})_\sigma \supseteq (\mathcal{S}^{T_\ell\overline{\sigma}})_\sigma = \mathcal{S}^{T_\ell} \supseteq \mathcal{L}\mathcal{N}\mathcal{B}$. This completes the proof of part (iii).

(iv) We prove first that for $\sigma \in \Theta$, we have $\mathcal{P}_\sigma K^* \le \sigma\theta$. By Lemma 3.1.4,

$$\mathcal{P}_\sigma = ((\emptyset\theta)^K)_\sigma \cap \bigcap_{\tau\theta\in\mathcal{K}_0}((\tau\theta)^{K\overline{\tau}})_\sigma \cap \bigcap_{\tau\theta=L^*}(\mathcal{L}\mathcal{N}\mathcal{B}^{\overline{\tau}})_\sigma \cap \bigcap_{\tau\theta=R^*}(\mathcal{R}\mathcal{N}\mathcal{B}^{\overline{\tau}})_\sigma \cap \bigcap_{\tau\theta=T^*}(\mathcal{S}^{\overline{\tau}})_\sigma$$

that is, $\mathcal{P}_\sigma = \bigcap_{\tau\in\Theta^1}((\tau\theta)^{K^*\overline{\tau}})_\sigma$.

Consider $\sigma \in \Theta$. Since $(\sigma\theta)^{K^*\overline{\sigma}}$ is one of the terms in the definition of $\mathcal{P}$, it follows that $\mathcal{P} \subseteq (\sigma\theta)^{K^*\overline{\sigma}}$ so that, by Lemma 4.4.3,

$$\mathcal{P}_\sigma K^* \le ((\sigma\theta)^{K^*\overline{\sigma}})_{\sigma K^*} = \sigma\theta.$$

Now consider the reverse inequality. By parts (ii) and (iii), for any $\sigma \in \Theta$ such that $t(\sigma) = T_r$ and any $\tau \in \Theta^1$, we have that

$$\sigma\theta \in \mathcal{K}_0 \implies ((\tau\theta)^{K^*\overline{\tau}})_\sigma \supseteq \mathcal{L}\mathcal{R}\mathcal{B},$$
$$\sigma\theta = L^* \implies ((\tau\theta)^{K^*\overline{\tau}})_\sigma \supseteq \mathcal{L}\mathcal{N}\mathcal{B}.$$

Hence, with the help of Theorem 2.1.3(vi),

$$\sigma\theta \in \mathcal{K}_0 \implies \mathcal{P}_\sigma \supseteq \mathcal{L}\mathcal{R}\mathcal{B} \implies \mathcal{P}_\sigma K = \mathcal{P}_\sigma K^*$$
$$\implies \sigma\theta = (\sigma\theta)_K \subseteq \mathcal{P}_\sigma K = \mathcal{P}_\sigma K^* \quad \text{(by Lemma 4.4.4)}$$
$$\sigma\theta = L^* \implies \mathcal{P}_\sigma \supseteq \mathcal{L}\mathcal{N}\mathcal{B}$$
$$\implies \sigma\theta = L^* = \mathcal{L}\mathcal{N}\mathcal{B}_{K^*} \le \mathcal{P}_\sigma K^* \quad \text{(by Lemma 4.3.5(iii))}.$$

By duality, if $\sigma\theta = R^*$ (and $t(\sigma) = T_\ell$), then $\sigma\theta \le \mathcal{P}_\sigma K^*$. It is also obvious that if $\sigma\theta = T^*$, then $\sigma\theta \le \mathcal{P}_\sigma K^*$. We can now conclude that for all $\sigma \in \Theta$, such that $t(\sigma) = T_r$), we have $\sigma\theta \le \mathcal{P}_\sigma K^*$. However, by duality, we also have $\sigma\theta \le \mathcal{P}_\sigma K^*$ for all $\sigma \in \Theta$ such that $t(\sigma) = T_\ell$. Consequently, this inequality holds for all $\sigma \in \Theta$ and therefore equality follows: $\sigma\theta = \mathcal{P}_\sigma K^*$.   □

We can now complete the proof of claim 5. Let $\theta \in \Phi$, $\varphi = \theta(\xi \chi)$ and $\mathcal{P} = \theta \xi$ be as above. Then, by Lemma 4.4.5(i),

$$\emptyset \varphi = \emptyset \chi_{\theta \xi} = (\theta \xi)_K = \mathcal{P}_K = \emptyset \theta.$$

Next consider any $\sigma \in \Theta$. By Lemma 4.4.5(iv), we have

$$\sigma \varphi = \sigma \chi_{\theta \xi} = (\theta \xi)_{\sigma K^*} = \mathcal{P}_{\sigma K^*} = \sigma \theta.$$

Therefore $\varphi = \theta$ and $\xi \chi = \iota_\Phi$, as claimed.

We continue now with part six of the proof of Theorem 4.4.1.

6. $\chi \xi = \iota_{[\mathcal{S}, \mathcal{CR}]}$. Let $\mathcal{V} \in [\mathcal{S}, \mathcal{CR}]$. Then

$$\mathcal{V} \chi \xi = \chi_\mathcal{V} \xi$$

$$= (\emptyset \chi_\mathcal{V})^K \cap \bigcap_{\tau \chi_\mathcal{V} \in \mathcal{K}_0} (\tau \chi_\mathcal{V})^{K\overline{\tau}} \cap \bigcap_{\tau \chi_\mathcal{V} = L^*} \mathcal{LNB}^{\overline{\tau}} \cap \bigcap_{\tau \chi_\mathcal{V} = R^*} \mathcal{RNB}^{\overline{\tau}} \cap \bigcap_{\tau \chi_\mathcal{V} = T^*} \mathcal{S}^{\overline{\tau}}$$

$$= \mathcal{V}^K \cap \bigcap_{\mathcal{V}_\tau \notin \mathbb{N}_3} (\mathcal{V}_{\tau K})^{K\overline{\tau}} \cap \bigcap_{\mathcal{V}_\tau = \mathcal{LNB}} \mathcal{LNB}^{\overline{\tau}} \cap \bigcap_{\mathcal{V}_\tau = \mathcal{RNB}} \mathcal{RNB}^{\overline{\tau}} \cap \bigcap_{\mathcal{V}_\tau = \mathcal{S}} \mathcal{S}^{\overline{\tau}}$$

$$= \mathcal{V} \qquad \text{by Corollary 4.1.5.}$$

Therefore, $\chi \xi = \iota_{[\mathcal{S}, \mathcal{CR}]}$.

We are finally in a position to address the claim of Theorem 4.4.1. We have shown that:

(i) χ is an order preserving mapping of $[\mathcal{S}, \mathcal{CR}]$ into Φ.
(ii) ξ is an order preserving mapping of Φ into $[\mathcal{S}, \mathcal{CR}]$.
(iii) $\xi \chi = \iota_\Phi$, by part 5 of the proof, and $\chi \xi = \iota_{[\mathcal{S}, \mathcal{CR}]}$, by part 6 of the proof.

It follows that χ and ξ are mutually inverse isomorphisms.   □

The conditions (P5) and (P6) are the conditions that present the most problems in the Polák theorem and when applying the Polák theorem to more difficult situations. To see how these conditions do imply real constraints, consider the question: is there $\theta \in \Phi$ such that $\emptyset \theta = \mathcal{CS}$, and $T_r \theta = T^*$? Condition (P5) requires that

$$L^* = \mathcal{LZ}_{K^*} = (\mathcal{CS})_{T_r K^*} = (\emptyset \theta)_{T_r K^*} \leq T_r \theta.$$

Consequently, $T_r \theta$ must exceed L^* and cannot equal T^*. This illustration is simple enough that we can see why this must be so in another way. Since $\emptyset \theta = \mathcal{CS}$, we must have $\mathcal{RB} \subseteq \theta \xi$. Since we are only discussing ladders of varieties in the interval $[\mathcal{S}, \mathcal{CR}]$, we also have $\mathcal{S} \subseteq \theta \xi$. Hence $\mathcal{NB} \subseteq \theta \xi$ where $L^* = T_r \chi_{\mathcal{NB}} \leq T_r \theta$ so that again we see that we must have $L^* \leq T_r\theta$ and $T_r\theta \neq T^*$. Similar examples can be constructed to show the impact of (P6).

The lattice of varieties of bands plays a central role in the study of $\mathcal{L}(\mathcal{CR})$ and applications of the Polák theorem. It is not difficult to identify it in the Polák theorem, especially if we look back at Theorem 3.2.8.

Notation 4.4.6 Let $\Phi_{\mathcal{B}}$ denote the class of all mappings $\varphi : \Theta^1 \to \{L^*, T^*, R^*, \mathcal{T}\}$ that satisfy, in addition to (P2), (P3) and (P4), the condition:

(P1^*) $\emptyset\varphi = \mathcal{T}$.

Corollary 4.4.7 $\chi\big|_{[\mathcal{S},\mathcal{B}]}$ *is an isomorphism of* $[\mathcal{S}, \mathcal{B}]$ *onto* $\Phi_{\mathcal{B}}$.

Proof This follows easily from the observation that $\emptyset\chi_{\mathcal{B}} = \mathcal{T}$ and the fact that if $\varphi \in \Phi$ satisfies (P1*), then it automatically satisfies conditions (P5) and (P6). $\qquad\square$

There are some features of the Polák theorem that we should take note of for future reference. For instance, it is important to be able to recognize the **B**-relation in the context of the Polák theorem.

Lemma 4.4.8 *Let* $\mathcal{U}, \mathcal{V} \in [\mathcal{S}, \mathcal{CR}]$. *Then the following statements are equivalent.*

(i) $\mathcal{U}$ **B** $\mathcal{V}$.

(ii) *For all* $\mathcal{P} \in \{\mathcal{S}, \mathcal{LNB}, \mathcal{RNB}\}, \tau \in \Theta$, *we have* $\mathcal{U}_\tau = \mathcal{P} \Leftrightarrow \mathcal{V}_\tau = \mathcal{P}$.

(iii) *For all* $\mathcal{P} \in \{L^*, T^*, R^*\}, \tau \in \Theta$, *we have* $\tau\chi_{\mathcal{U}} = \mathcal{P} \Leftrightarrow \tau\chi_{\mathcal{V}} = \mathcal{P}$.

Proof (i) *implies* (ii) Let $\tau \in \Theta$ be such that $\mathcal{U}_\tau = \mathcal{P} \in \{\mathcal{S}, \mathcal{LNB}, \mathcal{RNB}\}$. Then

$$V_\tau \cap \mathcal{B} = (\mathcal{V} \cap \mathcal{B})_\tau = (\mathcal{U} \cap \mathcal{B})_\tau = \mathcal{U}_\tau \cap \mathcal{B} = \mathcal{U}_\tau = \mathcal{P} \in \{\mathcal{S}, \mathcal{LNB}, \mathcal{RNB}\}.$$

By Lemma 3.6.6, we have $\mathcal{V}_\tau = \mathcal{P} = \mathcal{U}_\tau$. By symmetry, the claim holds.

(ii) *implies* (i). Now suppose that $\mathcal{U}, \mathcal{V}$ are related as in part (ii). We then have, by Theorem 3.6.4(ix), that

$$\mathcal{U}^{\mathbf{B}} = \mathcal{CR} \cap \bigcap_{\mathcal{U}_\tau = \mathcal{S}} \mathcal{S}^{\overline{\tau}} \cap \bigcap_{\mathcal{U}_\tau = \mathcal{LNB}} \mathcal{LNB}^{\overline{\tau}} \cap \bigcap_{\mathcal{U}_\tau = \mathcal{RNB}} \mathcal{RNB}^{\overline{\tau}}$$

$$= \mathcal{CR} \cap \bigcap_{\mathcal{V}_\tau = \mathcal{S}} \mathcal{S}^{\overline{\tau}} \cap \bigcap_{\mathcal{V}_\tau = \mathcal{LNB}} \mathcal{LNB}^{\overline{\tau}} \cap \bigcap_{\mathcal{V}_\tau = \mathcal{RNB}} \mathcal{RNB}^{\overline{\tau}}$$

$$= \mathcal{V}^{\mathbf{B}}.$$

Therefore $\mathcal{U}$ **B** $\mathcal{V}$.

(ii) *if and only if* (iii). Recall that $\tau \chi_{\mathcal{U}} = \mathcal{U}_{\tau K^*}$ and, likewise, $\tau \chi_{\mathcal{V}} = \mathcal{V}_{\tau K^*}$. Then part (iii) is a simple reformulation of part (ii). $\qquad\square$

It is also important to be able to recognize K-, T-, T_ℓ- and T_r-related varieties via the Polák theorem.

Lemma 4.4.9 *Let* $\mathcal{U}, \mathcal{V} \in [\mathcal{S}, \mathcal{CR}]$.

(i) $\mathcal{U} K \mathcal{V} \Leftrightarrow \emptyset\chi_{\mathcal{U}} = \emptyset\chi_{\mathcal{V}}$.
(ii) $\mathcal{U} T \mathcal{V} \Leftrightarrow \tau\chi_{\mathcal{U}} = \tau\chi_{\mathcal{V}}$ *for all* $\tau \in \Theta$.
(iii) $\mathcal{U} T_\ell \mathcal{V} \Leftrightarrow \tau\chi_{\mathcal{U}} = \tau\chi_{\mathcal{V}}$ *for all* $\tau \in \Theta$ *with* $h(\tau) = T_\ell$.
(iv) $\mathcal{U} T_r \mathcal{V} \Leftrightarrow \tau\chi_{\mathcal{U}} = \tau\chi_{\mathcal{V}}$ *for all* $\tau \in \Theta$ *with* $h(\tau) = T_r$.

Now assume, in addition, that $\mathcal{V} \notin \{\mathcal{LNB}, \mathcal{S}, \mathcal{RNB}\}$.

(v) $\mathcal{V} = \mathcal{V}_{T_\ell} \Longleftrightarrow \tau\chi_{\mathcal{V}} = T_\ell\tau\chi_{\mathcal{V}}$ *for all* $\tau \in \Theta^1$.
(vi) $\mathcal{V} = \mathcal{V}_{T_r} \Longleftrightarrow \tau\chi_{\mathcal{V}} = T_r\tau\chi_{\mathcal{V}}$ *for all* $\tau \in \Theta^1$.

Proof (i) We have $\emptyset\chi_{\mathcal{U}} = \emptyset\chi_{\mathcal{V}} \Longleftrightarrow \mathcal{U}_K = \mathcal{V}_K \Longleftrightarrow \mathcal{U} K \mathcal{V}$.

(ii) This follows from parts (iii) and (iv).

(iii) This is the dual of part (iv).

(iv) If $\mathcal{U} T_r \mathcal{V}$, then $\mathcal{U}_{T_r} = \mathcal{V}_{T_r}$ so that $\mathcal{U}_\tau = \mathcal{V}_\tau$ for all $\tau \in \Theta$ with $h(\tau) = T_r$, and for such values of τ, $\tau\chi_{\mathcal{U}} = \mathcal{U}_{\tau K^*} = \mathcal{V}_{\tau K^*} = \tau\chi_{\mathcal{V}}$ so that the direct implication holds.

Conversely, let $\tau\chi_{\mathcal{U}} = \tau\chi_{\mathcal{V}}$ for all $\tau \in \Theta$ with $h(\tau) = T_r$. We will show that $\chi_{\mathcal{U}_{T_r}} = \chi_{\mathcal{V}_{T_r}}$. For all $\sigma \in \Theta$, we get

$$\sigma\chi_{\mathcal{U}_{T_r}} = \mathcal{U}_{T_r\sigma K^*} = (T_r\sigma)\chi_{\mathcal{U}} = (T_r\sigma)\chi_{\mathcal{V}} = \mathcal{V}_{T_r\sigma K^*} = \sigma\chi_{\mathcal{V}_{T_r}}.$$

It remains to show that $\emptyset\chi_{\mathcal{U}_{T_r}} = \emptyset\chi_{\mathcal{V}_{T_r}}$. We have $\mathcal{U}_{T_r K^*} = T_r\chi_{\mathcal{U}} = T_r\chi_{\mathcal{V}} = \mathcal{V}_{T_r K^*}$. Hence

$$\mathcal{U}_{T_r} \in \mathcal{L}(\mathcal{B}) \Longleftrightarrow \mathcal{U}_{T_r K^*} \in \{\mathcal{T}, L^*, T^*, R^*\}$$
$$\Longleftrightarrow \mathcal{V}_{T_r K^*} \in \{\mathcal{T}, L^*, T^*, R^*\}$$
$$\Longleftrightarrow \mathcal{V}_{T_r} \in \mathcal{L}(\mathcal{B}).$$

Consequently, for $\mathcal{U}_{T_r} \in \mathcal{L}(\mathcal{B})$, we have $\mathcal{V}_{T_r} \in \mathcal{L}(\mathcal{B})$ and $\mathcal{U}_{T_r K} = \mathcal{T} = \mathcal{V}_{T_r K}$. Thus $\mathcal{U}_{T_r} \in \mathcal{L}(\mathcal{B}) \Longrightarrow \emptyset\chi_{\mathcal{U}_{T_r}} = \emptyset\chi_{\mathcal{V}_{T_r}}$. On the other hand, $\mathcal{U}_{T_r} \notin \mathcal{L}(\mathcal{B}) \Longrightarrow \mathcal{V}_{T_r} \notin \mathcal{L}(\mathcal{B})$ so that

$$\emptyset\chi_{\mathcal{U}_{T_r}} = \mathcal{U}_{T_r K} = \mathcal{U}_{T_r K^*} = T_r\chi_{\mathcal{U}} = T_r\chi_{\mathcal{V}} = \mathcal{V}_{T_r K^*} = \mathcal{V}_{T_r K} = \emptyset\chi_{\mathcal{V}_{T_r}}.$$

Thus, in all cases, we have $\emptyset\chi_{\mathcal{U}_{T_r}} = \emptyset\chi_{\mathcal{V}_{T_r}}$ so that $\tau\chi_{\mathcal{U}_{T_r}} = \tau\chi_{\mathcal{V}_{T_r}}$ for all $\tau \in \Theta^1$. Therefore, by the Polák theorem, $\mathcal{U}_{T_r} = \mathcal{V}_{T_r}$ so that $\mathcal{U} T_r \mathcal{V}$.

(v), (vi). By duality, if suffices to prove (vi). First assume that $\mathcal{V} = \mathcal{V}_{T_r}$. For $\tau \in \Theta$ we have

$$(T_r\tau)\chi_{\mathcal{V}} = \mathcal{V}_{T_r\tau K^*} = \left(\mathcal{V}_{T_r}\right)_{T_r\tau K^*} = \left(\mathcal{V}_{T_r}\right)_{\tau K^*} = \mathcal{V}_{\tau K^*} = \tau\chi_{\mathcal{V}}$$

while, for $\tau = \emptyset$, since $\mathcal{V} \notin \{\mathcal{LNB}, \mathcal{S}, \mathcal{RNB}\}$,

$$(T_r\tau)\chi_{\mathcal{V}} = T_r\chi_{\mathcal{V}} = \mathcal{V}_{T_r K^*} = \left(\mathcal{V}_{T_r}\right)_{T_r K^*} = \left(\mathcal{V}_{T_r}\right)_{K^*} = \mathcal{V}_{K^*} = \mathcal{V}_K = \emptyset\chi_{\mathcal{V}}.$$

Thus, for all $\tau \in \Theta^1$ we have $(T_r\tau)\chi_{\mathcal{V}} = \tau\chi_{\mathcal{V}}$.

Conversely, assume that $(T_r\tau)\chi_{\mathcal{V}} = \tau\chi_{\mathcal{V}}$ for all $\tau \in \Theta^1$. From $\tau = \emptyset$ we obtain

$$\mathcal{V}_K = \emptyset\chi_{\mathcal{V}} = T_r\chi_{\mathcal{V}} = \mathcal{V}_{T_r K^*} \leq \mathcal{V}_{T_r K} \subseteq \mathcal{V}_K$$

so that $\mathcal{V}_{T_r K} = \mathcal{V}_K$. Thus $\emptyset\chi_{\mathcal{V}} = \emptyset\chi_{\mathcal{V}_{T_r}}$. On the other hand, if $\tau \in \Theta$, then

$$\tau\chi_{\mathcal{V}} = (T_r\tau)\chi_{\mathcal{V}} = \mathcal{V}_{T_r\tau K^*} = \left(\mathcal{V}_{T_r}\right)_{\tau K^*} = \tau\chi_{\mathcal{V}_{T_r}}.$$

Thus $\tau\chi_{\mathcal{V}} = \tau\chi_{\mathcal{V}_{T_r}}$ for all $\tau \in \Theta^1$. Therefore, by the Polák theorem, $\mathcal{V} = \mathcal{V}_{T_r}$. $\square$

Lemma 4.4.10 *Let $\mathcal{V} \in [\mathcal{S}, \mathcal{CR}]$, $\mathcal{U} = \mathcal{V}_{T_r}$ and $\tau \in \Theta^1$. Then*

$$\tau\chi_{\mathcal{U}} = \begin{cases} T_r\chi_{\mathcal{V}} & \text{if } \tau = \emptyset, \ T_r\chi_{\mathcal{V}} \in \mathcal{K}_0 \\ \mathcal{T} & \text{if } \tau = \emptyset, \ T_r\chi_{\mathcal{V}} \in \mathbb{N}_3^* \\ \tau\chi_{\mathcal{V}} & \text{if } h(\tau) = T_r \\ (T_r\tau)\chi_{\mathcal{V}} & \text{if } h(\tau) = T_\ell. \end{cases}$$

Proof On the one hand, we have

$$\emptyset\chi_{\mathcal{U}} = \mathcal{U}_K = \mathcal{V}_{T_r K} = \begin{cases} T_r\chi_{\mathcal{V}} & \text{if } T_r\chi_{\mathcal{V}} \in \mathcal{K}_0 \\ \mathcal{T} & \text{if } T_r\chi_{\mathcal{V}} \in \mathbb{N}_3^*. \end{cases}$$

On the other hand, if $h(\tau) = T_r$, then $T_r\tau = \tau$ so that

$$\tau\chi_{\mathcal{U}} = \mathcal{U}_{\tau K^*} = \mathcal{V}_{T_r\tau K^*} = \begin{cases} \mathcal{V}_{\tau K^*} & \text{if } h(\tau) = T_r \\ \mathcal{V}_{T_r\tau K^*} & \text{if } h(\tau) = T_\ell \end{cases} = \begin{cases} \tau\chi_{\mathcal{V}} & \text{if } h(\tau) = T_r \\ T_r\tau\chi_{\mathcal{V}} & \text{if } h(\tau) = T_\ell \end{cases} \quad \square$$

The next lemma treats varieties that are fixed points under our familiar operators. See also Proposition 2.5.5.

Lemma 4.4.11 *The following conditions on $\mathcal{V} \in [\mathcal{S}, \mathcal{CR}] \setminus \mathbb{N}_6$ are equivalent.*

(i) $\mathcal{V} = \mathcal{V}_{T_\ell} = \mathcal{V}_{T_r}$.   (ii) $\mathcal{V} = \mathcal{V}^K$.   (iii) $\chi_{\mathcal{V}}$ *is constant.*

Proof (i) *implies* (iii). Recall that $\chi_{\mathcal{V}} : \tau \mapsto \mathcal{V}_{\tau K^*}$ ($\tau \in \Theta$) and $\emptyset \mapsto \mathcal{V}_K$. Since $\mathcal{V} \notin \mathbb{N}_6$ and $\mathcal{V} = \mathcal{V}_{T_\ell} = \mathcal{V}_{T_r}$, we get for $\tau \in \Theta$, $\mathcal{V}_{\tau K^*} = \mathcal{V}_{K^*} = \mathcal{V}_K$. Hence the ladder of $\mathcal{V}$ has the form

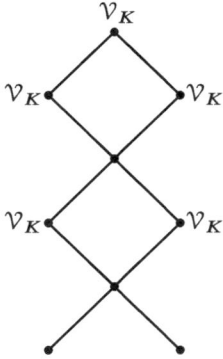

and $\chi_{\mathcal{V}}$ is constant.

(iii) *implies* (ii). Since $\emptyset\chi_{\mathcal{V}} = \mathcal{V}_K$, the hypothesis implies that $\tau\chi_{\mathcal{V}} = \mathcal{V}_K$, for all $\tau \in \Theta^1$. But we also have $\tau\chi_{\mathcal{V}^K} = \mathcal{V}_K$ for all $\tau \in \Theta^1$ since $(\mathcal{V}^K)_K = \mathcal{V}_K$ and $(\mathcal{V}^K)_{T_\ell} = (\mathcal{V}^K)_{T_r} = \mathcal{V}^K$ by Theorem 2.5.2(xii). Since χ is injective, we must have $\mathcal{V} = \mathcal{V}^K$.

(ii) *implies* (i). See Proposition 2.5.5.   □

The Polák theorem is particularly useful for computing joins of varieties whose ladders we are able to calculate, as we will see in the next section.

The results in this section are based on Polák [1, 2], though this particular treatment is drawn from Reilly [7, 8]. The authors are grateful to Kad'ourek for finding an error in the original proof of Lemma 4.4.5 and for suggesting the current version. See also Pastijn [2].

Jones and Szendrei [1] applied Theorem 4.4.1 to the study of local varieties. Pastijn [3] extended several major results and basic concepts in this and the preceding chapters to pseudovarieties of completely regular semigroups; in [5] he proved an analogue of the Polák theorem for pseudovarieties of completely regular monoids. Vachuska [1] proved the analogue of the Polák theorem for varieties of completely regular monoids. Ladders were employed by Jones [10] to establish the locality of certain varieties of completely regular monoids. See Petrich [6] for some commentary on Polák's theorem.

4.5   Applications to Joins

In this section we begin by showing that the mapping $\mathcal{V} \mapsto \mathcal{V} \vee \mathcal{B}$ is a complete homomorphism on $\mathcal{L}(\mathcal{CR})$. We continue with a small library of Polák ladders for special varieties. We then show how the Polák theorem can be used to identify the joins of certain varieties and also to describe certain ideals in $\mathcal{L}(\mathcal{CR})$ as subdirect products of better known lattices.

The first result is the complement of Theorem 3.2.8(v).

Proposition 4.5.1 (i) *The mapping* $\mathcal{V} \mapsto \mathcal{V} \vee \mathcal{T}$ *(*$\mathcal{V} \in \mathcal{K}$*) is a complete retraction of* $\mathcal{K}$ *onto* $\mathcal{K}_0$.

(ii) *The mapping* $\nu_{\mathcal{B}} : \mathcal{V} \mapsto \mathcal{V} \vee \mathcal{B}$ *(*$\mathcal{V} \in \mathcal{L}(\mathcal{CR})$*) is a complete retraction of* $\mathcal{L}(\mathcal{CR})$ *onto* $[\mathcal{B}, \mathcal{CR}]$.

(iii) *The mapping* $\theta_{\mathcal{B}} : \mathcal{V} \mapsto (\mathcal{V} \cap \mathcal{B}, \mathcal{V} \vee \mathcal{B})$ *is a complete monomorphism of* $\mathcal{L}(\mathcal{CR})$ *to a subdirect product of* $\mathcal{L}(\mathcal{B}) \times [\mathcal{B}, \mathcal{CR}]$.

Proof (i) The proof is straightforward.

(ii) Clearly $\nu_{\mathcal{B}}$ maps $\mathcal{L}(\mathcal{CR})$ onto $[\mathcal{B}, \mathcal{CR}]$ and respects arbitrary joins. For $\alpha \in A$, let $\mathcal{V}_\alpha \in \mathcal{CR}$.

Case: $\mathcal{S} \subseteq \mathcal{V}_\alpha$ for all $\alpha \in A$. Then with the isomorphism χ as in the Polák theorem we have, for all $\tau \in \Theta^1$,

$$\tau\left(\left(\bigwedge_{\alpha \in A} \chi \mathcal{V}_\alpha\right) \vee \chi_{\mathcal{B}}\right) = \left(\tau \bigwedge_{\alpha \in A} \chi \mathcal{V}_\alpha\right) \vee \tau \chi_{\mathcal{B}}$$

$$= \left(\bigwedge_{\alpha \in A} \tau \chi \mathcal{V}_\alpha\right) \vee \mathcal{T}$$

$$= \bigwedge_{\alpha \in A} (\tau \chi \mathcal{V}_\alpha \vee \mathcal{T}) \quad \text{by part (i)}$$

$$= \bigwedge_{\alpha \in A} \tau (\chi \mathcal{V}_\alpha \vee \chi_{\mathcal{B}})$$

$$= \tau \bigwedge_{\alpha \in A} (\chi \mathcal{V}_\alpha \vee \chi_{\mathcal{B}}).$$

Therefore $(\bigwedge_{\alpha \in A} \mathcal{V}_\alpha) \vee \mathcal{B} = \bigwedge_{\alpha \in A} (\mathcal{V}_\alpha \vee \mathcal{B})$, as required.

We can now consider the general case. We have

$$\left(\bigwedge_{\alpha \in A} \mathcal{V}_\alpha\right) \vee \mathcal{B} = \left(\bigwedge_{\alpha \in A} \mathcal{V}_\alpha\right) \vee \mathcal{S} \vee \mathcal{B}$$

$$= \left(\bigwedge_{\alpha \in A} (\mathcal{V}_\alpha \vee \mathcal{S})\right) \vee \mathcal{B} \quad \text{by [PR1, Proposition IX.9.2]}$$

$$= \bigwedge_{\alpha \in A} (\mathcal{V}_\alpha \vee \mathcal{S} \vee \mathcal{B}) \qquad \text{by above}$$

$$= \bigwedge_{\alpha \in A} (\mathcal{V}_\alpha \vee \mathcal{B}),$$

as required.

(iii) By Theorem 3.2.8(v) and part (ii), it is clear that the mapping $\theta_{\mathcal{B}}$ is a complete homomorphism of $\mathcal{L}(\mathcal{CR})$ to a subdirect product of $\mathcal{L}(\mathcal{B}) \times [\mathcal{B}, \mathcal{CR}]$. It only remains to show that $\theta_{\mathcal{B}}$ is injective. Towards that end, let $\mathcal{U}$, $\mathcal{V}$ be distinct varieties in $\mathcal{L}(\mathcal{CR})$. If one of $\mathcal{U}$, $\mathcal{V}$ contains $\mathcal{S}$ and the other does not, then $\mathcal{U}$, $\mathcal{V}$ are distinguished by $\mu_{\mathcal{B}}$ and therefore by $\theta_{\mathcal{B}}$. Indeed, if $\mathcal{U}\mu_{\mathcal{B}} \neq \mathcal{V}\mu_{\mathcal{B}}$ for any $\mathcal{U}$, $\mathcal{V}$, then $\mathcal{U}\theta_{\mathcal{B}} \neq \mathcal{U}\theta_{\mathcal{B}}$ and so we may assume that either $\mathcal{U}$, $\mathcal{V} \in \mathcal{L}(\mathcal{CS})$ or $\mathcal{S} \subseteq \mathcal{U}$, $\mathcal{V}$ and, in addition, that $\mathcal{U}\mu_{\mathcal{B}} = \mathcal{V}\mu_{\mathcal{B}}$.

Now assume in addition that $\mathcal{S} \subseteq \mathcal{U}$, $\mathcal{V}$. Then, from the Polák theorem, we know that $\chi_{\mathcal{U}} \neq \chi_{\mathcal{V}}$ and therefore there exists $\tau \in \Theta^1$ such that $\tau\chi_{\mathcal{U}} \neq \tau\chi_{\mathcal{V}}$. By Lemma 3.6.6(i), $\tau\chi_{\mathcal{U}}, \tau\chi_{\mathcal{V}} \notin \mathbb{N}_6$ or, equivalently, $\tau\chi_{\mathcal{U}}, \tau\chi_{\mathcal{V}} \in \mathcal{K}_0$. Hence,

$$\tau\chi_{\mathcal{U}\vee\mathcal{B}} = \tau\chi_{\mathcal{U}} \vee \tau\chi_{\mathcal{B}} = \tau\chi_{\mathcal{U}} \vee \mathcal{I} = \tau\chi_{\mathcal{U}} \neq \tau\chi_{\mathcal{V}} = \cdots = \tau\chi_{\mathcal{V}\vee\mathcal{B}}.$$

Hence $\mathcal{U}\nu_{\mathcal{B}} \neq \mathcal{V}\nu_{\mathcal{B}}$ and therefore $\mathcal{U}\theta_{\mathcal{B}} \neq \mathcal{V}\theta_{\mathcal{B}}$.

That leaves just the case where $\mathcal{U}$, $\mathcal{V} \in \mathcal{L}(\mathcal{CS})$. Since $\mathcal{U}$ **B** $\mathcal{V}$, we have that $\mathcal{RB} \subseteq \mathcal{U}$ if and only if $\mathcal{RB} \subseteq \mathcal{V}$. So let us assume that $\mathcal{RB} \not\subseteq \mathcal{U}$, $\mathcal{V}$. Then, by [PR1, Corollary VIII.1.4], we have

$$(\mathcal{U} \cap \mathcal{G}) \vee (\mathcal{U} \cap \mathcal{RB}) = \mathcal{U} \neq \mathcal{V} = (\mathcal{V} \cap \mathcal{G}) \vee (\mathcal{V} \cap \mathcal{RB}).$$

But $\mathcal{U} \cap \mathcal{RB} = \mathcal{V} \cap \mathcal{RB}$. Hence we must have $\mathcal{U} \cap \mathcal{G} \neq \mathcal{V} \cap \mathcal{G}$ so that

$$(\mathcal{U} \vee \mathcal{B}) \cap \mathcal{G} = \mathcal{U} \cap \mathcal{G} \neq \mathcal{V} \cap \mathcal{G} = (\mathcal{V} \vee \mathcal{B}) \cap \mathcal{G}.$$

Consequently, $\mathcal{U} \vee \mathcal{B} \neq \mathcal{V} \vee \mathcal{B}$ and $\mathcal{U}\theta_{\mathcal{B}} \neq \mathcal{V}\theta_{\mathcal{B}}$. Finally, assume that $\mathcal{RB} \subseteq \mathcal{U}$, $\mathcal{V}$. Then, for $\mathcal{P} \in \{\mathcal{U}, \mathcal{V}\}$, we have

$$(\mathcal{P} \vee \mathcal{B}) \cap \mathcal{CS} = (\mathcal{P} \cap \mathcal{CS}) \vee (\mathcal{B} \cap \mathcal{CS}) = \mathcal{P} \vee \mathcal{RB} = \mathcal{P}.$$

Hence $\mathcal{U}\nu_{\mathcal{B}} = \mathcal{U} \vee \mathcal{B} \neq \mathcal{V} \vee \mathcal{B} = \mathcal{V}\nu_{\mathcal{B}}$. Therefore $\theta_{\mathcal{B}}$ is injective and the proof is complete. $\square$

In order to take advantage of the Polák theorem, it will be useful to have a number of ladders of well-known varieties available for subsequent calculations.

Theorem 4.5.2 *Diagram 4.5 illustrates the ladders for the varieties indicated.*

Proof The information required to calculate all of these ladders (with the exception of $\mathcal{BG}$) can be found in Theorem 1.5.5 and its dual (for $\mathcal{V}_{T_\ell}$, $\mathcal{V}_{T_r}$) and in Theorem 2.1.3(v) (for $\mathcal{V}_K$). To illustrate the calculations we consider just one example: $\mathcal{RRO} = [xa = a^0 xa]$. By Theorem 2.1.3(v), we get $\mathcal{RRO}_K = \mathcal{G}$. By the dual of [PR1, Lemma V.3.1], $\mathcal{RRO}_{T_r} = \mathcal{S}$ so that $\mathcal{RRO}_{T_r K^*} = T^*$. By Theorem 1.5.5 and the dual of Theorem 2.5.2(xiii), $\mathcal{RRO}_{T_\ell} = \mathcal{RRO}$ so that $\mathcal{RRO}_{T_\ell K^*} = \mathcal{RRO}_K = \mathcal{G}$. All other points on the ladder satisfy $\tau \leq T_r$ so that $\mathcal{RRO}_{\tau K^*} \leq \mathcal{RRO}_{T_r K^*} = T^*$ whence $\mathcal{RRO}_{\tau K^*} = T$.

The case of $\mathcal{BG}$ requires the fact that $\mathcal{BG}_K = (\mathcal{B}^T)_K = \mathcal{B}^T = \mathcal{BG}$ (see Theorem 2.4.5(ii)). $\square$

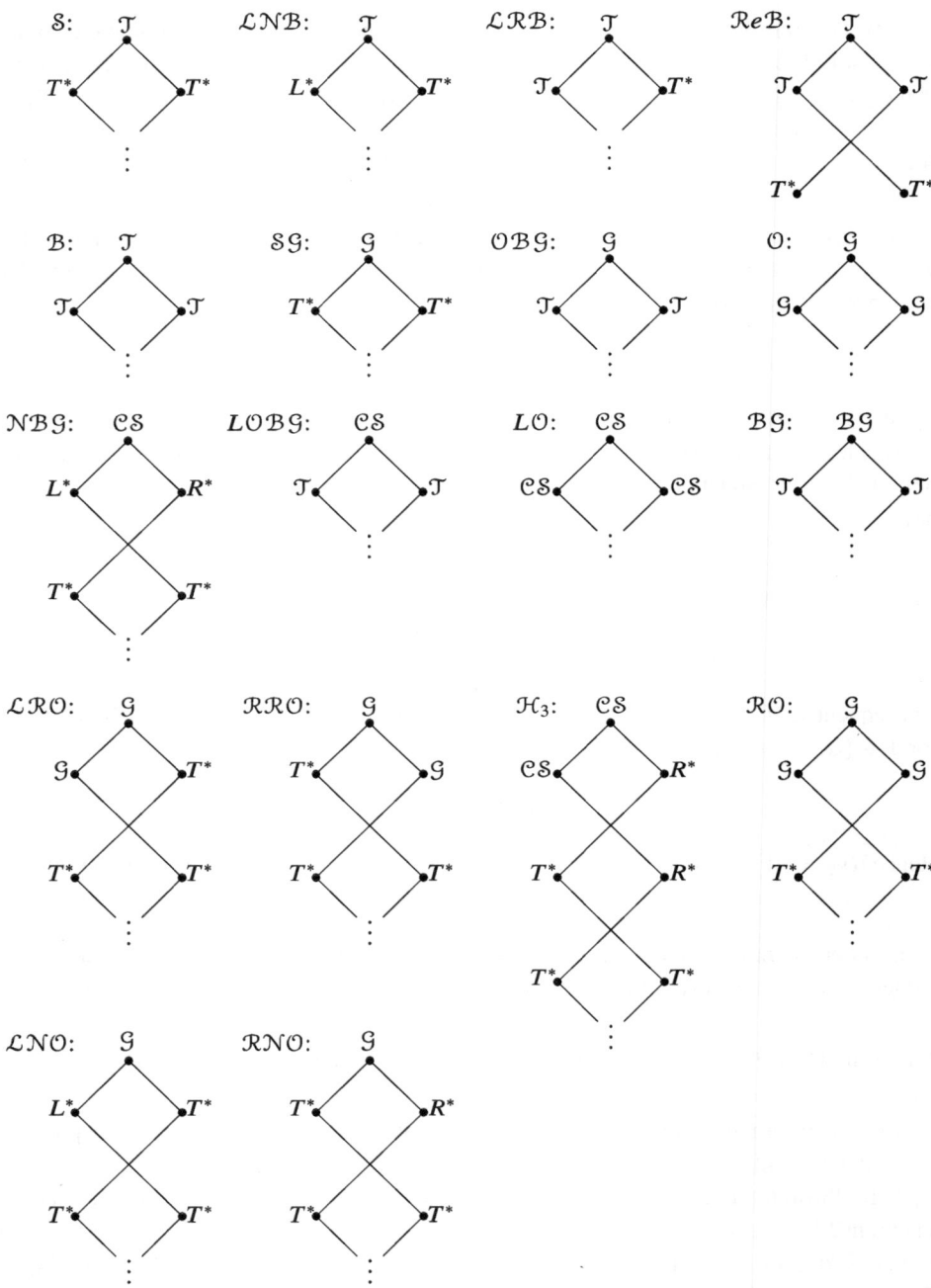

Diagram 4.5 Ladders for some common varieties

We can immediately combine the Polák theorem with the observations in Theorem 4.5.2 to gain insight into just how complex the kernel structure is in $\mathcal{L}(\mathcal{CR})$.

Corollary 4.5.3 (i) $\mathcal{SGK}_r$ *contains a subset of cardinality* $2^{\aleph_0}$ *consisting of pairwise incomparable varieties.*

(ii) $(\mathcal{GK})/K_\ell$ *contains a subset of cardinality* $2^{\aleph_0}$ *consisting of pairwise incomparable* K_ℓ-*classes.*

(iii) $|\mathcal{GK}| = 2^{\aleph_0}$.

Proof (i) Let χ, ξ be as in Theorem 4.4.1. By Olshansky [1], there exists a subset, $\mathcal{Q}$ say, of $\mathcal{L}(\mathcal{G})$ of cardinality $2^{\aleph_0}$ consisting of pairwise incomparable varieties of groups. Let $\varphi = \chi_{\mathcal{SG}}$, as displayed in Diagram 4.5. For each $\mathcal{U} \in \mathcal{Q}$, we define a ladder $\varphi_\mathcal{U}$ as follows:

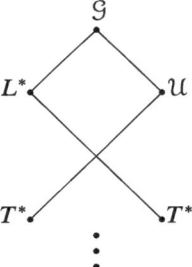

It is straightforward to verify that $\varphi_\mathcal{U} \in \Phi$. Then $\varphi_\mathcal{U}\xi \in [\mathcal{S}, \mathcal{CR}]$. Moreover, since $\emptyset\varphi_\mathcal{U} = \mathcal{G}$, it follows that $\varphi_\mathcal{U}\xi \in \mathcal{GK}$ and since $\tau\varphi_\mathcal{U} = \tau\chi_{\mathcal{SG}}$ for all $\tau \in \Theta$ with $h(\tau) = T_r$, it follows that $\varphi_\mathcal{U}\xi \in (\mathcal{SG})T_r$. Hence we actually have $\varphi_\mathcal{U}\xi \in (\mathcal{SG})K_r$. Since the elements of $\mathcal{Q}$ are pairwise incomparable, so are the ladders $\varphi_\mathcal{U}, \mathcal{U} \in \mathcal{Q}$ and therefore the varieties $\varphi_\mathcal{U}\xi, \mathcal{U} \in \mathcal{Q}$ are pairwise incomparable varieties in $(\mathcal{SG})K_r$.

(ii) For each $\mathcal{U} \in \mathcal{Q}$, let $\mathcal{U}^* \in (\varphi_\mathcal{U}\xi)K_\ell$. Then $(\mathcal{U}^*, \varphi_\mathcal{U}\xi) \in T_\ell$. Hence $T_\ell\chi_{\mathcal{U}^*} = T_\ell\chi_{\varphi_\mathcal{U}\xi} = \mathcal{U}$. Consequently, for distinct varieties $\mathcal{U}, \mathcal{V} \in \mathcal{Q}$, we have that $\chi_{\mathcal{U}^*}$ and $\chi_{\mathcal{V}^*}$ are incomparable and therefore $(\varphi_\mathcal{U}\xi)K_\ell$ and $(\varphi_\mathcal{V}\xi)K_\ell$ are incomparable K_ℓ-classes in the lattice of K_ℓ-classes $(\mathcal{GK})/K_\ell = (\mathcal{SGK})/K_\ell$.

(iii) Since $\mathcal{GK} = \mathcal{SGK}$, the claim follows immediately from part (i). □

The examples below show the procedure of applying the Polák theorem to determine the join of two varieties $\mathcal{U}$ and $\mathcal{V}$. The pattern is quite analogous to using logarithms for computing the product of two numbers: we find the log of the two factors, add them up, and find its antilog. Here we construct the ladders of $\mathcal{U}$ and $\mathcal{V}$, find their join and from it compute the "anti ladder" in the form of a meet of, hopefully, known varieties.

Proposition 4.5.4 *We have that* $\mathcal{B} \vee \mathcal{G} = \mathcal{OBG}$.

Proof Since $\mathcal{G}$ is a heterotypical variety, we "borrow" the variety $\mathcal{S}$ from $\mathcal{B}$ forming $\mathcal{S} \vee \mathcal{G} = \mathcal{SG}$. Hence instead of considering $\mathcal{B} \vee \mathcal{G}$ we may take $\mathcal{B} \vee \mathcal{SG}$. From Diagram 4.5 we have the ladders for $\mathcal{B}$, $\mathcal{SG}$ and the ladder for $\mathcal{B} \vee \mathcal{G}$ is then the join of the ladders for $\mathcal{B}$ and $\mathcal{SG}$.

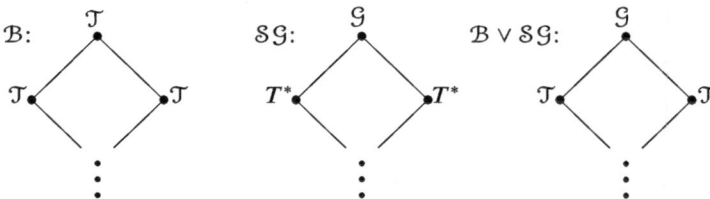

If we admit the verb "*to ladder*", we must now "*anti-ladder*" the third ladder in accordance with the definition of ξ in Theorem 4.4.1:

$$\mathcal{B} \vee \mathcal{G} = \mathcal{G}^K \cap \mathcal{J}^{KT_\ell} \cap \mathcal{J}^{KT_r} \cap \cdots = \mathcal{O} \cap \mathcal{B}^{T_\ell} \cap \mathcal{B}^{T_r} \cap \cdots = \mathcal{O} \cap \mathcal{B}^T = \mathcal{OBG}. \qquad \square$$

The only possible problems in any particular case are "*laddering*" and "*anti-laddering*" (or "*evaluating*"). We continue with

Proposition 4.5.5 *We have that* $\mathcal{B} \vee \mathcal{CS} = \mathcal{LOBG}$.

Proof Again we "borrow" $\mathcal{S}$ from $\mathcal{B}$ getting $\mathcal{B} \vee \mathcal{CS} = \mathcal{B} \vee (\mathcal{S} \vee \mathcal{CS}) = \mathcal{B} \vee \mathcal{NBG}$. The ladder of $\mathcal{B}$ is in Diagram 4.5, and

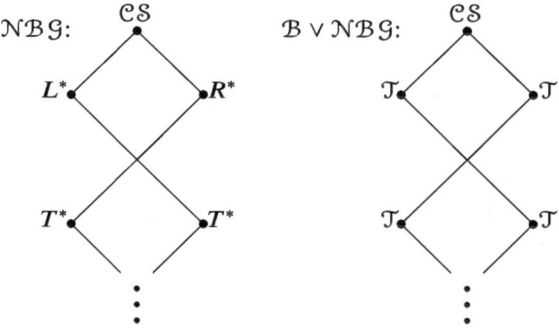

and anti-laddering gives

$$\mathcal{B} \vee \mathcal{CS} = \mathcal{CS}^K \cap \mathcal{J}^{KT_\ell} \cap \mathcal{J}^{KT_r} \cap \cdots = \mathcal{LO} \cap \mathcal{BG} = \mathcal{LOBG}. \qquad \square$$

Proposition 4.5.6 *We have that* $\mathcal{O} \vee \mathcal{CS} = L\mathcal{O} \cap \mathcal{O}^T$.

Proof Again we "borrow" $\mathcal{S}$ from $\mathcal{B}$ to get $\mathcal{O} \vee \mathcal{NBS}$. From Theorem 4.5.2, we obtain the ladders for $\mathcal{O}$, $\mathcal{NBS}$ and then $\mathcal{O} \vee \mathcal{NBS}$.

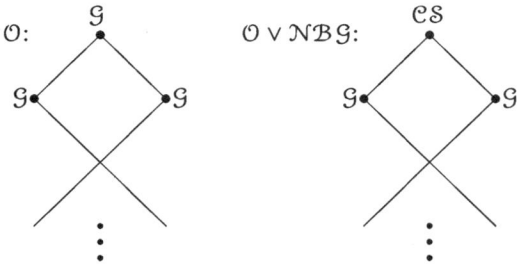

Anti-laddering, we obtain

$$\mathcal{O} \vee \mathcal{CS} = \mathcal{O} \vee \mathcal{NBS} = \mathcal{CS}^K \cap \mathcal{S}^{KT_\ell} \cap \mathcal{S}^{KT_r} \cap \cdots = L\mathcal{O} \cap \mathcal{O}^{T_\ell} \cap \mathcal{O}^{T_r} = L\mathcal{O} \cap \mathcal{O}^T. \quad \square$$

As we climb up the lattice, the calculation of the ladder becomes more involved.

Proposition 4.5.7 *We have that* $\mathcal{O} \vee \mathcal{BS} = \mathcal{BS}^K \cap \mathcal{O}^T$.

Proof We have the ladders of $\mathcal{O}$ and $\mathcal{BS}$ from Diagram 4.5. Hence the relevant ladders are

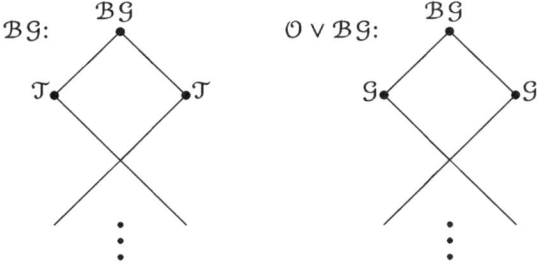

Hence

$$\mathcal{O} \vee \mathcal{BS} = (\mathcal{BS}))^K \cap \mathcal{S}^{KT_\ell} \cap \mathcal{S}^{KT_r} \cap \cdots = (\mathcal{BS})^K \cap \mathcal{O}^{T_\ell} \cap \mathcal{O}^{T_r} \cap \cdots$$
$$= (\mathcal{BS})^K \cap \mathcal{O}^T. \quad \square$$

Proposition 4.5.8 *We have that* $L\mathcal{O} \vee \mathcal{BS} = \mathcal{BS}^K \cap (L\mathcal{O})^T$.

Proof From Theorem 4.5.2, we know the ladders of $L\mathcal{O}$ and $\mathcal{BS}$. Hence that of $L\mathcal{O} \vee \mathcal{BS}$ is:

$LO \vee BG$:

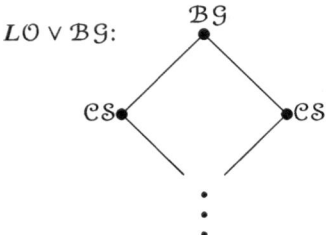

and the evaluation of the ladder, using $\mathcal{V}^T = \mathcal{V}^{T_\ell} \cap \mathcal{V}^{T_r}$, becomes

$$LO \vee BG = BG^K \cap \mathcal{CS}^{KT_r} \cap \mathcal{CS}^{KT_\ell} \cap \cdots = c BG^K \cap (LO)^T. \qquad \square$$

As we have seen above when calculating the join of a heterotypical variety $\mathcal{U}$ and a homotypical variety $\mathcal{V}$, we can still use the Polák theorem as follows: We "borrow" the variety $\mathcal{S}$ from $\mathcal{V}$ and join it with $\mathcal{U}$, thereby obtaining $\mathcal{U} \vee \mathcal{V} = (\mathcal{U} \vee \mathcal{S}) \vee \mathcal{V}$ where $\mathcal{U} \vee \mathcal{S}$ and $\mathcal{V}$. are both homotypical varieties and the theorem applies. In the calculation appears the ladder of $\mathcal{U} \vee \mathcal{S}$ where this variety is in $\mathcal{L}(\mathcal{NBG})$ and generally causes no problems.

The material in this section is implicit in Polák [1, 2, 3].

4.6   Applications to $\mathcal{L}(LO)$

It is possible to study specific sublattices of $[\mathcal{S}, \mathcal{CR}]$ by identifying the corresponding sublattices of Φ. The most difficult conditions to deal with in such endeavours are (P5) and (P6). However, in this section we will show how these conditions can be simplified or even disappear altogether in some instances. We will also include some remarks on the cardinalities of certain K-classes. Each variety of completely regular semigroups is defined by a set of identities of the form $u = v$ where $u, v \in U_X$ and $|U_X| = \aleph_0$. Thus $|\mathcal{L}(\mathcal{CR})| \leq 2^{\aleph_0}$ and so $|\mathcal{V}K| \leq 2^{\aleph_0}$ for all $\mathcal{V} \in \mathcal{L}(\mathcal{CR})$. Clearly $|\mathcal{B}K| = |\mathcal{L}(\mathcal{B})| = \aleph_0$. On the other hand, taking advantage of the fact that $|\mathcal{L}(\mathcal{G})| = 2^{\aleph_0}$, we will see that $|(LO)K| = |OK| = |\mathcal{LROK}| = 2^{\aleph_0}$.

Throughout this section, χ will denote the Polák isomorphism.

Notation 4.6.1 Let $\mathcal{K}_{\mathcal{CS}} = \mathcal{K}_0 \cap \mathcal{L}(\mathcal{CS})$ and let Φ_{LO} denote the class of all mappings φ of Θ^1 into $\mathcal{K}_{\mathcal{CS}} \cup \mathbb{N}_3^*$ satisfying the following conditions

(P1') $\emptyset\varphi \in \mathcal{K}_{\mathcal{CS}}$.
(P2) φ is order preserving.
(P3)   If $\tau \in \Theta$, $\tau\varphi = L^*$, then $t(\tau) = T_r$.
(P4)   If $\tau \in \Theta$, $\tau\varphi = R^*$, then $t(\tau) = T_\ell$.

(P5') If $\tau \in \{T_\ell, T_r\}$, then $(\emptyset\varphi)_{\tau K^*} \subseteq \tau\varphi$.
(P6') If $\sigma \in \Theta, \tau \in \{T_\ell, T_r\}$ are such that $\sigma\varphi \in \mathcal{K}_{\mathcal{CS}}$ and $t(\sigma) \neq \tau$, $(\sigma\varphi)_{\tau K^*} \subseteq (\sigma\tau)\varphi$.

The definition of $\Phi_{L\mathcal{O}}$ is considerably simpler than that of Φ in two important respects. The range of $\varphi \in \Phi_{L\mathcal{O}}$ is restricted to a subclass $\mathcal{K}_{\mathcal{CS}} \cup \mathbb{N}_3^*$ of $\mathcal{K}_0 \cup \mathbb{N}_3^*$ that is much better understood than the whole class. In addition, the values of τ that need to be considered in conditions (P5') and (P6') are restricted to simply T_ℓ and T_r.

Theorem 4.6.2 (i) $\Phi_{L\mathcal{O}}$ *is an ideal in* Φ *with greatest element* $\chi_{L\mathcal{O}}$.

(ii) $\chi\big|_{[\mathcal{S}, L\mathcal{O}]}$ *is an isomorphism of* $[\mathcal{S}, L\mathcal{O}]$ *onto* $\Phi_{L\mathcal{O}}$.

(iii) *There exists a subset of* $\mathcal{CS}K$ *consisting of pairwise incomparable elements of cardinality* $2^{\aleph_0}$.

Proof (i) Let $\varphi \in \Phi_{L\mathcal{O}}$. Since $\mathcal{K}_{\mathcal{CS}} \subseteq \mathcal{K}_0$, we have immediately that φ satisfies the conditions (P1)–(P4). Let $\tau \in \Theta$. If $|\tau| \geq 2$, then $(\emptyset\varphi)_\tau \subseteq \mathcal{CS}_\tau = \mathcal{J}$ so that $(\emptyset\varphi)_{\tau K^*} \subseteq \mathcal{J}_{K^*} = T^* \subseteq \tau\varphi$ and (P5) is satisfied. On the other hand, if $|\tau| = 1$ and $\tau \in \{T_\ell, T_r\}$ then (P5) is satisfied by (P5'). Thus φ satisfies (P5).

Let $\sigma, \tau \in \Theta, \sigma\varphi \in \mathcal{K}_0$ and $t(\sigma) \neq h(\tau)$. By hypothesis, $\sigma\varphi \subseteq \emptyset\varphi \subseteq \mathcal{CS}$. Hence if $|\tau| \geq 2$, then $(\sigma\varphi)_\tau = \mathcal{J}$ so that $(\sigma\varphi)_{\tau K^*} = T^* \subseteq (\sigma\tau)\varphi$ and (P6) is satisfied. If $|\tau| = 1$ and $\tau \in \{T_\ell, T_r\}$, then (P6) is satisfied on account of (P6'). Thus φ satisfies (P6). Therefore $\Phi_{L\mathcal{O}} \subseteq \Phi$.

The fact that $\Phi_{L\mathcal{O}}$ is an ideal in Φ follows easily from the fact that $\mathcal{K}_{\mathcal{CS}}$ is an ideal in $\mathcal{K}_0$ and that (P5') and (P6') are special cases of (P5) and (P6).

It is clear from the ladder $\chi_{L\mathcal{O}}$ of $L\mathcal{O}$ (see Diagram 4.5) that $\chi_{L\mathcal{O}}$ is the greatest element in $\Phi_{L\mathcal{O}}$.

(ii) Let $\mathcal{V} \in [\mathcal{S}, L\mathcal{O}]$. By Theorem 2.1.3(v), $\emptyset\chi_\mathcal{V} = \mathcal{V}_K \in \mathcal{K}_{\mathcal{CS}}$ and $\chi_\mathcal{V}$ satisfies (P1'). That $\chi_\mathcal{V}$ satisfies the conditions (P2)–(P4), (P5') and (P6') then follows immediately from the fact that $\chi_\mathcal{V} \in \Phi$. Hence $\chi_\mathcal{V} \in \Phi_{L\mathcal{O}}$.

Now let $\varphi \in \Phi_{L\mathcal{O}}$. By part (i) and the Polák theorem, we know that $\varphi = \chi_\mathcal{V}$ for some $\mathcal{V} \in [\mathcal{S}, \mathcal{CR}]$. Moreover $\mathcal{V}_K = \emptyset\chi_\mathcal{V} = \emptyset\varphi \in \mathcal{K}_{\mathcal{CS}}$ which, by Theorem 2.1.3(v) implies that $\mathcal{V} \in \mathcal{L}(L\mathcal{O})$. Thus χ maps $[\mathcal{S}, L\mathcal{O}]$ onto $\Phi_{L\mathcal{O}}$ and the claim holds.

(iii) To begin with, we might ask what the ladder is for the smallest element in $L\mathcal{O}K$ containing $\mathcal{S}$. Let $\mathcal{V} \in L\mathcal{O}K$. Then $\mathcal{V}_K = L\mathcal{O}_K = \mathcal{CS}$ so that the top element in $\chi_\mathcal{V}$ is $\mathcal{CS}$. What is the smallest possible value for $T_\ell\chi_\mathcal{V}$? This is where condition (P5') comes into play. That requires

$$R^* = \mathcal{RZ}_{K^*} = \mathcal{CS}_{T_\ell K^*} = (\emptyset\chi_\mathcal{V})_{T_\ell K^*} \leq T_\ell\chi_\mathcal{V}$$

so that the smallest possible value for $T_\ell\chi_\mathcal{V}$ is R^*. Likewise, the smallest possible value for $T_r\chi_\mathcal{V}$ is L^*. Since $\chi_\mathcal{V}$ is order preserving, with these values for $\chi_\mathcal{V}$, we must then have $\tau\chi_\mathcal{V} = T^*$ for any $\tau \in \Theta$ with $|\tau| \geq 2$. This leads to the ladder:

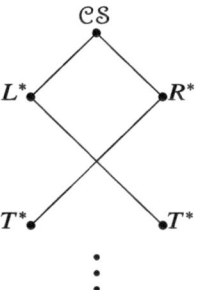

This, of course, is the Polák ladder for $\mathcal{NBG}$ which should come as no surprise since the smallest element in $L\mathcal{O}$ is $L\mathcal{O}_K = \mathcal{CS}$ and $\mathcal{CS} \vee \mathcal{S} = \mathcal{NBG}$.

Following the pattern of the argument in Corollary 4.5.3, let $\mathcal{U} \in \mathcal{L}(\mathcal{G})$ and suppose that we tweak the ladder for $\mathcal{NBG}$ ever so slightly to obtain the ladder

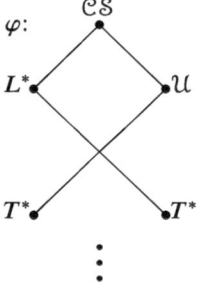

Since $\mathcal{U} \in \mathcal{K}_{\mathcal{CS}}$, it is clear that φ satisfies conditions (P1′), (P2)–(P4). It is also clear that the left arm $\mathcal{CS}, L^*, T^*, \ldots$ satisfies conditions (P5′) and (P6′). Consider the right-hand arm $\mathcal{CS}, \mathcal{U}, T^*, \ldots$. First we have $\mathcal{CS}_{T_\ell K^*} = \mathcal{RZ}_{K^*} = R^* \subseteq \mathcal{U} = T_\ell \varphi$ so that (P5′) is satisfied. With $\sigma = T_\ell$, $\tau = T_r$ we get $(\sigma\varphi)_{\tau K^*} = \mathcal{U}_{T_r K^*} = \mathcal{T}_{K^*} = T^* \subseteq \sigma\tau\varphi$. Thus φ is a Polák ladder. Different choices of $\mathcal{U}$ give different Polák ladders and therefore define different varieties in $(L\mathcal{O})K$. Since $\mathcal{L}(\mathcal{G})$ contains a subset of cardinality $2^{\aleph_0}$ consisting of incomparable elements, so also does $(\mathcal{CS})K$. $\qquad\square$

As we focus on smaller and more special sublattices of $\mathcal{L}(\mathcal{CR})$, the corresponding sub-lattice of Φ also becomes more special and, sometimes, simpler. Recall that $\mathcal{A}_n$ denotes the variety of abelian groups of exponent dividing n.

Notation 4.6.3 Let $\Phi_\mathcal{O}$ denote the class of all mappings φ of Θ^1 into $\mathcal{L}(\mathcal{G}) \cup \mathbb{N}_3^*$ such that

(P1″) $\emptyset\varphi \in \mathcal{L}(\mathcal{G})$,   (P3) $\tau \in \Theta$ and $\tau\varphi = L^* \Longrightarrow t(\tau) = T_r$,

(P2) φ is order preserving,   (P4) $\tau \in \Theta$ and $\tau\varphi = R^* \Longrightarrow t(\tau) = T_\ell$, and, for any $\mathcal{U} \in \mathcal{L}(\mathcal{G})$, let $\Phi_{\mathcal{O}H\mathcal{U}} = \{\varphi \in \Phi_{\mathcal{O}} \mid \emptyset\varphi = \mathcal{U}\}$, in particular, let $\Phi_{\mathcal{O}H A_p} = \{\varphi \in \Phi_{\mathcal{O}} \mid \emptyset\varphi = A_p\}$.

Theorem 4.6.4 (i) $\Phi_{\mathcal{O}}$ *is an ideal in* Φ.

(ii) $\chi|_{[S,\mathcal{O}]}$ *is an isomorphism of* $[S, \mathcal{O}]$ *onto* $\Phi_{\mathcal{O}}$.

(iii) $\chi|_{A_p K \cap [S,\mathcal{CR}]}$ *is an isomorphism of* $[S, \mathcal{O}H A_p] = A_p K \cap [S, \mathcal{CR}]$ *onto* $\Phi_{\mathcal{O}H A_p}$.

Proof (i) Let $\varphi \in \Phi_{\mathcal{O}}$. Since $\mathcal{L}(\mathcal{G}) \subseteq \mathcal{K}_0$, it follows immediately that φ satisfies (P1)–(P4). For any $\sigma \in \Theta^1$ and $\tau \in \Theta$, we have

$$(\sigma\varphi)_{\tau K^*} \leq (\emptyset\varphi)_{\tau K^*} \leq \mathcal{G}_{\tau K^*} = \mathcal{J}_{K^*} = T^*$$

and, consequently, both (P5) and (P6) are trivially satisfied. Thus $\varphi \in \Phi$ and $\Phi_{\mathcal{O}} \subseteq \Phi$. The fact that $\Phi_{\mathcal{O}}$ is an ideal in Φ now follows easily from the fact that $\mathcal{L}(\mathcal{G})$ is an ideal in $\mathcal{K}_0$.

(ii) Let $\mathcal{V} = [S, \mathcal{O}]$. By Theorem 2.1.3(v), $\mathcal{V}_K \in \mathcal{L}(\mathcal{G})$ so that $\emptyset\chi_{\mathcal{V}} \in \mathcal{L}(\mathcal{G})$ and, since $\chi_{\mathcal{V}}$ is order preserving, $\tau\chi_{\mathcal{V}} \in \mathcal{L}(\mathcal{G}) \cup \mathbb{N}_3^*$ for all $\tau \in \Theta$. It now follows from the Polák theorem and the definition of Φ that $\chi_{\mathcal{V}} \in \Phi_{\mathcal{O}}$. Let $\varphi \in \Phi_{\mathcal{O}}$. By part (i), $\varphi \in \Phi$ and therefore by the Polák theorem, there exists $\mathcal{V} \in [S, \mathcal{CR}]$ with $\varphi = \chi_{\mathcal{V}}$. Then $\mathcal{V}_K = \emptyset\chi_{\mathcal{V}} = \emptyset\varphi \in \mathcal{L}(\mathcal{G})$, by Theorem 2.1.3(v), so that $\mathcal{V} \in \mathcal{L}(\mathcal{O})$. Consequently, $\Phi_{\mathcal{V}} \subseteq \mathcal{L}(\mathcal{O})\chi$ and equality follows. Since $[S, \mathcal{O}]$ is an ideal in $[S, \mathcal{CR}]$ and χ is an isomorphism, we can conclude that $\chi|_{[S,\mathcal{O}]}$ is an isomorphism of $[S, \mathcal{O}]$ onto $\Phi_{\mathcal{O}}$.

(iii) This follows from part (ii) and the fact that, for $\varphi \in \Phi$ and $\mathcal{V} = \varphi\xi$, we have

$$\mathcal{V} \in A_p K \iff \mathcal{V}_K = A_p \iff \emptyset\varphi = \emptyset\chi_{\mathcal{V}} = \mathcal{V} = A_p. \qquad \square$$

As we consider ever smaller sublattices of $\mathcal{L}(\mathcal{CR})$ the corresponding sublattices of Φ become ever simpler.

Notation 4.6.5 Let $\Phi_{\mathcal{LRO}}$ denote the class of all mappings from Θ^1 into $\mathcal{L}(\mathcal{G}) \cup \mathbb{N}_3^*$ satisfying the following conditions

(P1''') $\emptyset\varphi \in \mathcal{L}(\mathcal{G})$, $T_{\ell}\varphi = T^*$ and (P2)–(P4).

Similar arguments to those applied above in the case of $\mathcal{L}(L\mathcal{O})$ and $\mathcal{L}(\mathcal{O})$ will now establish the following.

Theorem 4.6.6 (i) $\Phi_{\mathcal{LRO}}$ *is an ideal in* Φ.

(ii) $\chi|_{[S,\mathcal{LRO}]}$ *is an isomorphism of* $[S, \mathcal{LRO}]$ *onto* $\Phi_{\mathcal{LRO}}$. (iii) $|\mathcal{G}K \cap \mathcal{LRO}| = 2^{\aleph_0}$.

Proof (i) The proofs of parts (i) and (ii) follow a now familiar pattern.

For part (iii) it is useful to recall from the dual of Theorem 1.5.5 that $\mathcal{LRO} = S^{T_{\ell}}$ so that $(\mathcal{LRO})_{T_{\ell}} = S$ while $(\mathcal{LRO})_{T_r} = \mathcal{LRO}$. The details of the proof are left as an exercise.

From this it follows that the ladder for any variety in $\mathcal{LRO}K$ is of the form

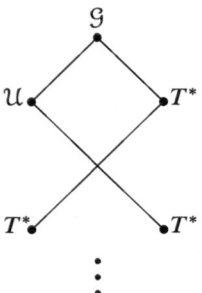

where $\mathcal{U} \in \mathcal{L}(\mathcal{G})$ and that any such ladder defines a variety in $\mathcal{G}K \cap \mathcal{LRO}$. The claim now follows.   □

Our final application is a simple illustration of the use of ladders to describe joins of varieties.

Corollary 4.6.7 *We have that* $\mathcal{RO} = \mathcal{LRO} \vee \mathcal{RRO}$.

Proof Diagram 4.5 provides the ladders for $\mathcal{LRO}$ and $\mathcal{RRO}$:

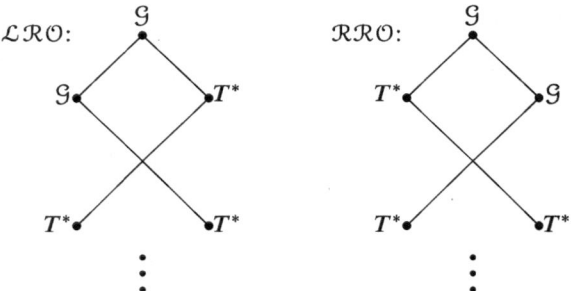

The join of these ladders is precisely the ladder for $\mathcal{RO}$.   □

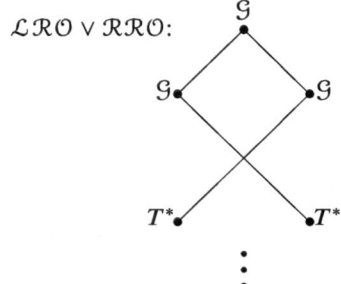

Theorem 4.6.4(iii) appears in Polák [3]. The fact that $\mathcal{L}(\mathcal{G})$ contains a subset of cardinality $2^{\aleph_0}$ consisting of pairwise incomparable elements can be found in Olshansky [1]. The remaining observations in this section are new.

Further applications of Polák's theorem will be found in [PR3].

4.7 Review

The Polák theorem is truly remarkable. It provides an isomorphic copy of the interval $[\mathcal{S}, \mathcal{CR}]$ in terms of functions on the lattice $\mathcal{K}_0$ of least elements of K-classes (with a different join) to which the three element semilattice $\mathbb{N}_3^*$ has been adjoined as a $\wedge$-ideal. The isomorphism is given in an explicit form both ways, from and to $[\mathcal{S}, \mathcal{CR}]$, so that it can be, in sufficiently simple cases, concretely evaluated. The consequences of this theorem are far-reaching and would probably be very difficult to obtain otherwise. It is almost certain that further applications of this theorem await discovery and that some of these may further revolutionize our approach to the study of varieties of completely regular semigroups.

The main limitation of this theorem is that it is based on the lattice $\mathcal{K}_0$, a complete lower subsemilattice of $[\mathcal{S}, \mathcal{CR}]$, about which it says nothing for it takes it as its basic building block. Even though the theorem has been used without an explicit knowledge of $\mathcal{K}_0$, for a concrete determination of the structure of varieties and their lattices, it becomes necessary to know more about $\mathcal{K}_0$. This is amply illustrated by the fact that up to the varieties of orthodox or locally orthodox varieties, we can say much more since their corresponding parts of $\mathcal{K}_0$ fall below groups and completely simple semigroups, respectively. Further progress in the direction of complete and concrete determination of parts of the lattice $[\mathcal{S}, \mathcal{CR}]$, probably modulo varieties of groups or completely simple semigroups, evidently depends upon the amount of knowledge we shall have about the corresponding part of $\mathcal{K}_0$.

The second impediment, of a technical nature, is the appearance of the semilattice $\mathbb{N}_3^*$, the markers added to $\mathcal{K}_0$ to form $\mathcal{K}$. For we are dealing here with an appendix which serves its useful role of separating certain varieties, but it also seems both artificial and superfluous. A little practice is needed to efficiently overcome this inconvenience but this difficulty can be overcome to facilitate deeper investigations of the structure of $\mathcal{L}(\mathcal{CR})$. We will take up this challenge once more in [PR3]. The structure of $\mathcal{L}(\mathcal{B})$ is embedded in the Polák theorem and it could be argued that it is responsible for this difficulty. So it is not surprising, but is still disappointing, that there is no analogue of the Polák theorem for varieties of completely simple semigroups for which, as we have seen in [PR1, Chapter VIII], we are totally dependent on certain subgroups of a free group.

Conclusion

Our journey began in Chap. 1 with a discussion of the free completely regular semigroup on a nonempty set. We discussed Green's relations which gave us the broad picture. We then turned our attention to fully invariant congruences on the free object and saw that each fully invariant congruence has a halo of congruences associated with it via Green's relations. This enabled us to consider networks of congruences in the lattice of fully invariant congruences. We also saw that the lattice of fully invariant congruences satisfies a strong lattice identity (the Arguesian identity) which implies that lattice is modular. We then transferred all the machinery and results concerning congruences over to the lattice of varieties of completely regular semigroups thereby shedding new light on the relationships between known varieties and providing tools with which to dig deeper into the structure of the lattice of all completely regular semigroup varieties. Then we applied these tools to determine the lattice of varieties of bands. Since bands contain only trivial groups they are, in a sense, the antithesis of groups. The approach used is to view bands as special examples of completely regular semigroups. This way we explain the clearly visible relationships in the generally accepted depiction of the lattice of varieties of bands and can also investigate the very important $\mathbf{B}$-relation which imposes an important structure on the whole lattice of varieties of completely regular semigroups. It also introduces the very important class of varieties of the form $\mathcal{V}^B$, scattered deep into the lattice of all varieties. In the final chapter we present the remarkable theorem due to Polák. This is the most far reaching result. It provides a subdirect decomposition of the whole lattice of varieties of completely regular semigroups. The components of the decomposition are, in general, complicated but nevertheless permit us to give a complete description of certain special sublattices and to shed light on others. Our approach to the study of bands makes it possible to expose how critical the $\mathbf{B}$-relation is for the formulation of Polák's theorem.

In the sequel [PR3] to this text, we will explore further applications of the results presented here and also determine some limitations to the applications of Polák's theorem. In addition we will take advantage of the modularity of the lattice of all varieties of completely regular semigroups to determine the structure of some local pieces. We also consider more free objects and word problems. Our presentation in [PR3] will conclude with a study of the sublattice generated by the maximum elements in the classes of the $\mathbf{B}$-relation, ending with a conjecture regarding its structure.

References

Since all citations are made at the end of the relevant sections, each reference below is followed by the section number(s) of the section(s) in which it is cited.

Books

Petrich, M. [P1] *Lectures in Semigroups*, Akademie-Verlag, Berlin, 1977. Section 3.7.
Petrich, M. and N. R. Reilly
[PR1] *Completely Regular Semigroups*, Wiley–Interscience Publication, Wiley, New York, 1999. Introduction.
[PR2] *Completely Regular Semigroup Varieties: A Comprehensive Study of Advanced Techniques*, manuscript.
[PR3] *Completely Regular Semigroup Varieties: Applications and Modern Insights*, manuscript.

Papers

Ajan, K. S.
[1] *On some presentations of completely regular semigroups*, Semigroup Forum 45 (1992), 121–123. Section 1.1.1.
Ajan, K. S. and F. J. Pastijn
[1] *Finitely Presented groups and completely regular semigroups*, J. Pure Appl. Algebra 108 (1996), 219–230. Section 1.1.1.
AlDhamri, N.
[1] *Dualities for quasi-varieties of bands* (English summary), Semigroup Forum 88 (2014), no. 2, 417–432. Section 3.3.
Araujo, J. and J. Koniezny
[1] *Automorphisms of endomorphism monoids of relatively free bands*, Proc. Edinb. Math. Soc. 50 (2007), 1–21. Section 13.7.

© The Editor(s) (if applicable) and The Author(s), under exclusive license to Springer Nature Switzerland AG 2024
M. Petrich and N. R. Reilly, *Completely Regular Semigroup Varieties*, Synthesis Lectures on Mathematics & Statistics, https://doi.org/10.1007/978-3-031-42891-3

Auinger, K.

[1] *A method for the construction of complete congruences on lattices of pseudovarieties*, J. Pure Appl. Algebra 126 (1998), no. 1–3, 1–17. Section 2.3.

Auinger, K., T. E. Hall, N. R. Reilly and S. Zhang

[1] *Congruences on the lattice of pseudovarieties of finite semigroups*, Int. J. Algebra Comput. 7 (1997), 433–455. Section 2.3.

Birjukov, A. P.

[1] *Varieties of idempotent semigroups*, Algebra i Logika 9 (1970), 255–273 (in Russian). Sections 13.3, 3.5.

Bonzini, C. and Cherubini A.

[1] *Permutable completely regular semigroups; in: Bonzini, C. (ed.) et al., Semigroups: algebraic theory and applications to formal languages and codes*, Proc. Internat. Conf. on Semigroups, Luino, Italy, 22–27 June, 1992. World Scientific, Singapore, 36–41 (1993). Section 1.7.

Brown, T. C.

[1] *On the finiteness of semigroups in which $x^r = x$*, Proc. Camb. Philos. Soc. 60 (1964), 1028–1029. Section 3.7.

Casimiro, A. and E. Skapinakis

[1] *Basis reduction for cryptogroups and orthogroups*, Semigroup Forum 101, (2020), 779–785. Section 1.1.

Clifford, A. H.

[1] *The free completely regular semigroup on a set*, J. Algebra 59 (1979), 434–451. Section 1.1.

Ćirić, M. and S. Bogdanović

[1] *The lattice of varieties of bands*, in: J. M. Howie and N. Ruskuc (Eds), *Semigroups and Applications. Proc. Conf. St. Andrews, UK, July 2–9, 1997*, World Scientific, Singapore, 47–61 (1998). Section 3.3.

Dean, R. A. and R. H. Oehmke

[1] *Idempotent semigroups with distributive right congruence lattices*, Pac. J. Math. 14 (1964), 1187–1209. Section 3.3.

Demlova M. and V. Koubek

[1] *Endomorphism monoids of bands*, Semigroup Forum 38 (1989), 305–329. Section 3.3.

[2] *Endomorphism monoids in small varieties of bands*, Acta Sci. Math. (Szeged) 55 (1991), 9–20. Section 3.3.

[3] *Endomorphism monoids in varieties of bands*, Acta Sci. Math. (Szeged) 66 (2000), no. 3–4, 477–516. Section 3.3.

Dolinka, I.

[1] *All varieties of normal bands with involution*, Period. Math. Hung. 40 (2000), 109–122. Section 3.5

[2] *Remarks on varieties with involution bands*, Comm. Algebra 28 (2000), 2837–2852. Section 3.5.

[3] *Subdirectly irreducible bands with involution*, Acta Sci. Math. (Szeged) 67 (2001), 535–554. Section 3.5.

[4] *Finite bands are finitely related*, Semigroup Forum (2018) 97, 115–130. Section 3.5.

Fennemore, C. F.

[1] *All varieties of bands*, Semigroup Forum 1 (1970), 172–179. Sections 3.3, 3.5.

[2] *All varieties of bands, I*, Math. Nachr. 48 (1971), 237–252. Section 3.3.

[3] *All varieties of bands, II*, Math. Nachr. 48 (1971), 253–262. Section 3.3.

[4] *Characterization of bands satisfying no. non-trivial identity*, Semigroup Forum 2 (1971), 371–375. Section 3.5.

[5] *A subdirect irreducible band accepting a = a*, Semigroup Forum 9 (1974), 271–174. Section 3.5.

Fountain, J. B. and P. Lockley

[1] *Bands with distributive congruence lattice*, Proc. R. Soc. Edinb., Sect. A, Math. 84 (1979) 235–247. Section 3.3.

Freese, R. and J. B. Nation

[1] *Congruence lattices of semilattices*, Pac. J. Math. 49 (1973), 51–58. Section 3.3

Gabovich, E. Ya.

[1] *On bands in which every finitely generated subband is an endomorphic image*, Semigroup Forum 4 (1972), 335–340. Section 3.3.

Gerhard, J. A.

[1] *The lattice of equational classes of idempotent semigroups*, J. Algebra 15 (1970), 195–224. Sections 3.3, 3.5.

[2] *The number of polynomials of idempotent semigroups*, J. Algebra 18 (1971), 366–376. Section 3.3.

[3] *Subdirectly irreducible idempotent semigroups*, Pac. J. Math. 39 1971), 669–676. Section 3.3.

[4] *Some subdirectly irreducible idempotent semigroups*, Semigroup Forum 5 (1973), 362–369. Section 3.3.

[5] *Injectives in equational classes of idempotent semigroups*, Semigroup Forum 9 (1974), 36–53. Section 3.3.

[6] *The word problem for semigroups satisfying $x^3 = x$*, Math. Proc. Camb. Phil. Soc. 84 (1978), 11–19. Section 3.3.

[7] *Free completely regular semigroups I: Representation*, J. Algebra 82 (1983), 135-142. Section 1.1.

[8] *Free completely regular semigroups II: Word problem*, J. Algebra 82 (1983), 143–156. Section 1.1.

[9] *Free bands and free *-bands*, in: *Semigroups and their applications (Chico, Calif., 1986)*, 47–50, Reidel, Dordrecht, 1987. Section 3.7.

Gerhard, J. A. and M. Petrich

[I] *The word problem for orthogroups*, Canad. J. Math. 33 (1981), 893–900. Section 1.4.

[2] *Free bands and free *-bands*, Glasg. Math. J. 28 (1986), 161–179. Section 3.7.

[3] *Certain characterizations of varieties of bands*, Proc. Edinb. Math. Soc. (1988) 31, 301–319. Section 3.5.

[4] *Varieties of bands revisited*, Proc. Lond. Math. Soc. (3) 58 (1989), no. 2, 323–350. Sections 3.3, 3.5.

Green, J. A., and D. Rees

[1] *On semigroups in which $x^r = x$*, Proc. Camb. Philos. Soc. 48 (1952), 35–40. Section 3.7.

Hall, T. E.

[1] *On bands, free bands and identities*, Semigroup Forum 2 (1971), 83–84. Section 3.5.

[2] *On the lattice of congruences on a semilattice*, J. Aust. Math. Soc. 12 (1971), 456–460. Section 3.3.

Hall, T.E, and P. R. Jones

[1] *On the lattice of varieties of bands of groups*, Pac. J. Math. 91 (1980), 327–337. Section 1.7.

Jones, P. R.

[1] *Universal aspects of completely simple semigroups*, in: T. E. Hall et al. (Eds), *Semigroups, Proc. Conf., Clayton/Aust. 1979*; Academic Press, New York (1980), 27–46. Section 2.1.

[2] *Monoid varieties defined by $x^{n+1} = x$ are local*, Semigroup Forum 51 (1995), 357–377. Section 4.4.

[3] *A band whose congruence lattice has ACC or DCC is finite*, J. Algebra 64 (1980), 336–339. Section 13.3.

[4] *On free products of bands*, Semigroup Forum 43 (1991), 53–62. Section 13.7.

Jones, P. R. and M. B. Szendrei

[1] *Local varieties of completely regular monoids*, J. Algera 150 (1992), 1–27. Section 4.4.

Jones, P. R. and P. G. Trotter

[1] *The Howson property in var ieties of completely regular semigroups*, Simon Stevin 63 (1989), 285–294. Section 1.1.

Kadóurek, J.

[1] *On the word problem for free bands of groups and for free objects in some other varieties of completely regular semigroups*, Semigroup Forum 38 (1989), 1–55. Sections 2.4, 2.5.

Koryakov, I. O.

[1] *On basis ranks of varieties of idempotent semigroups*, Mat. Zap. Ural. Univ. 12 (1981), 48–52 (in Russian). Sections 3.3, 3.5.

Lee, E. W. H., J. Rhodes, and B. Steinberg

[1] *Join irreducible semigroups*, Int. J. Algebra Comput. 29 (2018), 1249–1310. Sections 2.6, 3.3.

Liu, G. X., H. Yan, and Z. Wang

[1] *A note on $\mathcal{D}_n$-testability for varieties of bands*, Algebra Colloq. 21 (2014), no. 3, 511-516. Section 3.3.

McAlister, D. B.

[1] *A homomorphism theorem for semigroups*, J. Lond. Math. Soc. 43 (1968), 355–366. Section 1.1.

McLean, D.

[1] *Idempotent semigroups*, Amer. Math. Monthly 61 (1954), 110–113. Section 3.7.

Nordahl, T. E.

[1] *Free products of normal bands*, Semigroup Forum 15 (1977), 87–88. Section 3.7.

Olin, P.

[1] *Varietal free products of bands*, Semigroup Forum 21 (1980), 83–87. Section 3.7.

Ol'shanskiĭ, A. Yu.

[1] *On the problem of a finite basis for the identities of groups*, Izv. Akad. Nauk SSFR 34(1970), 376–384 (in Russian); English translation: Math. USSR, Izv. 4 (1970), 381–389. Section 4.6.

Pastijn, F.

[1] *On completely regular inverse semigroups*, Simon Stevin 43 (1973–74), 135–138. Section 1.5.

[2] *The lattice of completely regular semigroup varieties*, J. Aust. Math. Soc. 49A (1990), 24–42. Sections 1.4, 1.5, 2.1, 4.4.

[3] *Pseudovarieties of completely regular semigroups*, Semigroup Forum 42 (1991), 1–46. Section 4.4.

[4] *Commuting fully invariant congruences on free completely regular semigroups*, Trans. Amer. Math. Soc. 323 (1991), 79–92. Section 1.7.

[5] *Pseudovarieties of completely regular monoids*, in: John Rhodes (ed.), *Monoids and semigroups with applications. Proc. Berkeley workshop on monoids, Berkeley, CA, USA, 31 July – 5 August 1989*, World Scientific, Singapore etc., 1991, 171–183. Section 4.4.

Pastijn, F. and J. Albert

[1] *Free split bands*, Semigroup Forum, 90 (2015) 3, 753–762. Section 3.7

[2] *Semilattice transversals of regular bands I*, Comm. Algebra 45 (2017), no. 11, 4979–4991. Section 3.3.

[3] *Semilattice transversals of regular bands II*, Semigroup Forum 95 (2017), no. 3, 423–440. Section 3.3.

Pastijn, F. G. and P. G. Trotter

[1] *Lattices of completely regular semigroup varieties*, Pac. J. Math. 119 (1985), 191–214. Sections 1.2, 1.3, 2.4, 2.5.

[2] *Complete congruences on lattices of varieties and pseudovarieties*, Int. J. Algebra Comput. 8 (1998), no. 2, 171–201. Section 2.3.

Petrich, M.

[1] *A construction and a classification of bands*, Math. Nachr. 48 (1971) 263–274. Section 3.7.

[2] *Identities without the star for *-bands*, Algebra Univers. 36 (1996), 46–65. Section 3.5.

[3] *New bases for band varieties*, Semigroup Forum 59 (1999), 141–151. Section 3.5

[4] *Varieties of bands*, Proc. Conf. Braga, World Scientific, (2000), 146–160. Section 3.5.

[5] *Two sided networks for completely simple semigroups*, Comm. Algebra 28 (2000), 3535–3553. Section 3.5.

[6] *Some relations on the lattice of varieties of completely regular semigroups*, Boll. Unione Mat. Ital. Sez. B Artic. Ric. Mat. (8) 5 (2002), no. 2, 265–278, Section 4.4.

[7] *The lattice of varieties of completely regular semigroups*, Results Math. 48 (2005), no. 1–2, 131–157. Sections 4.2, 4.4.

[8] *Completely regular monoids with two generators*, J. Aust. Math. Soc. 90 (2011), no. 2, 271–287. Sections 1.1, 3.6.

[9] *Some relations on a semilattice of varieties of completely regular semigroups*, Semigroup Forum 93 (2016), no. 3, 607–628. Section 3.6.

Petrich, M. and N. R. Reilly

[1] *Varieties of groups and completely simple semigroups*, Bull. Aust. Math. Soc. 23 (1981), 339–359. Sections 1.1, 1.2.

[2] *Bands of groups with universal properties*, Monatsh. Math. 94 (1982), 45–67. Section 2.4.

[3] *Semigroups generated by certain operators on varieties of completely regular semigroups*, Pac. J. Math. 132 (1988), 151–175. Section 2.5.

[4] *Operators related to E-disjunctive and fundamental completely regular semigroups*, J. Algebra 134 (1990), 1–27. Section 2.5.

[5] *The modularity of the lattice of varieties of completely regular semigroups*, Glasg. Math. J. 15 (1974), 109–120. Section 1.7.

Petrich, M. and P. V. Silva

[1] *Relatively free bands*, Comm. Algebra 28 (2000), no. 5, 2615–2631. Section 3.7.

[2] *Relatively free *-bands*, Beitr. Algebra Geom. 41 (2000), no. 2, 569–588. Section 3.7.

[3] *Structure of relatively free bands*, Comm. Algebra 30 (2002), no. 4, 303–322. Section 3.7.

[4] *On *-bands and their varieties*, Rocky Mt. J. Math. 33 (2003), no. 1, 217–252. Section 3.7.

Polák. L.

[1] *On varieties of completely regular semigroups I*, Semigroup Forum 32 (10985), 97–123. Sections 1.2, 1.4, 1.5, 3.

[2] *On varieties of completely regular semigroups II*, Semigroup Forum 36 (1987), 253–284, Sections 1.2, 1.4, 1.5.

[3] *On varieties of completely regular semigroups III*, Semigroup Forum 37 (1988), 1–30. Sections 1.2, 1.6, 3.

Reilly, N. R.

[1] *Varieties of completely regular semigroups*, J. Aust. Math. Soc. A38 (1985), 372–393. Section 1.7.

[2] *The Rhodes expansion and free objects in varieties of completely regular semigroups*, J. Pure Appl. Algebra, 69 (1990), 89–109. Sections 1.1, 1.2, 1.3.

[3] *Completely regular semigroups*, in: *Lattices, Semigroups, and Universal Algebra*, Proc. Int. Conf., Lisbon/Port. 1988 (1990) 225–242. Sections 1.7, 4.3, 4.4.

[4] *Kernel classes of varieties of completely regular semigroups I*, Semigroup Forum (2019), 814–839. Sections 4.3, 4.4,

[5] *Kernel classes of varieties of completely regular semigroups II*, Semigroup Forum (2019), 840–869. Section 4.4,

Reilly, N. R. and S. Zhang

[1] *Congruence relations on the lattice of existence varieties of regular semigroups*, J. Algebra 178 (1995), 733–759. Sections 2.3, 2.7.

[2] *Complete endomorphism of the lattice of pseudovarieties of finite semigroups*, Bull. Aust. Math. Soc. 55 (1997), no. 2, 207–218. Sections 13.2. 3.2, 3.4, 3.6.

[3] *Commutativity of operators on the lattice of existence varieties of finite semigroups*, Monatsh. Math. 123 (1997), 337–364, Sections 2.3 2.7.

[4] *Varieties and pseudovarieties generated by $\mathcal{D}$-chains*, in: J. M. Howie, (ed.) et al., *Semigroups and Applications. Proc. Conf. in St. Andrews, UK, July 2–9, 1997*, World Scientific, Singapore, 1998, 179–193, Section 3.3

[5] *Operators on the lattice of pseudo varieties of finite semigroups*, Semigroup Forum 57 (1998), 208–239. Section 2.7.

[6] *Decomposition of the lattice of pseudovarieties of finite semigroups induced by bands*, Algebra Univers. 44 (2000), no. 3–4, 217–239. Sections 3.2, 3.6.

Sapir, M. V.

[1] *On the lattice of quasivarieties of idempotent semigroups*, Mat. Zap. Ural. Univ. 11 (1979), 158–169 (Russian). Section 3.5.

Sapir, O.

[1] *The variety of idempotent semigroups is inherently non-finitely generated*, Semigroup Forum 71 (2005), 140–146. Section 3.3.

Scheiblich, H. E.

[1] *On epics and dimensions of bands*, Semigroup Forum 13 (1976–7), 103–114. Section 3.3.

Siekmann, J. and P. Szabó

[1] *A noetherian and confluent rewrite system for idempotent semigroups*, Semigroup Forum 25 (1982), 83–110. Section 3.5.

Sizer, W. S.

[1] *Representations of semigroups of idempotents*, Czech. Math. J. 30 (1980), 369–375. Section 3.2.

Tamura, T.

[1] *Note on attainability of identities on bands*, J. Alegbra 28 (1974), 1–9. Section 3.5.

Trotter, P. G.

[1] *Subdirect decompositions of the lattice of varieties of completely regular semigroups*, Bull. Aust. Math. Soc. 39 (1989), 343–351. Section 3.2

Trotter, P. G. and P. Weil

[1] *The lattice of pseudovarieties of idempotent semigroups and a nonregular analogue*, Algebra Univers. 37 (1997), 491–526. Section 3.6.

Vachuska, C. A.

[1] *On the lattice of completely regular monoid varieties*, Semigroup Forum 46 (1993), 168–186. Section 4.4

Vernikov, B. M.

[1] *Completely regular semigroup varieties whose free objects have weakly permutable fully invariant congruences*, Semigroup Forum 68 (2004), 154–158. Section 1.7.

[2] *Lower-modular elements of the lattice of semigroup varieties. II*, Acta Sci. Math. (Szeged) 74 (2008), no. 3–4, 539–556. Section 1.7.

Wang, Z. P., J. Leng and H. Y. Yu

[1] *On subdirectly irreducible regular bands*, Turkish J. Math. 41 (2017), no. 5, 1337–1343. Section 3.3.

Wismath, S. L.

[1] *The lattices of varieties and pseudo varieties of band monoids*, Semigroup Forum 33 (1986), 187–198. Section 3.8.

[2] *Hypersubstitutions for the variety of *-bands*, Sci. Math. Jpn 4 (2001), 785–795. Section 3.5.

Yamada, M.

[1] *Note on idempotent semigroups V. Implications in two variables*, Proc. Japan Acad. 37 (1958), 668–671. Section 3.3.

[2] *Certain congruences and the structure of some special bands*, Proc. Japan Acad. 36 (1960), 408–410. Section 3.

[3] *A note on subdirect decompositions of idempotent semigroups*, Proc. Japan Acad. 36 (1960), 411–414. Section 3.3.

[4] *On a certain class of idempotent semigroups*, Bull. Shimane Univ. (1960), 153–161. Section 3.3.

Papers containing solutions to problems from Petrich and Reilly [PR1]

Guo, C., G. Liu and Y. Guo

[1] *The congruence $\mathcal{Y}^*$ on completely regular semigroups*, Commun. Algebra 39 (2011), no. 6, 2082–2096. Problem II.5.10.

Hu, L. and G. X. Liu

[1] *A proof of $\mathcal{OBG} \vee \mathcal{BA} = C\mathcal{BG}$*, Southeast Asian Bull. Math. 35 (2011) no. 2, 229–236. Problem II.8.9.

Liu, G. and J. Zhang

[1] *A problem on central cryptogroups*, Semigroup Forum 73 (2006), no. 2, 261–266. Problem II.8.9.

Liu, Jing-guo and Xiang-zhi Kong

[1] *Completely regular semigroups for which $|C(S)/L| < 2$*, J. Math. Study 38 (2005) no. 2, 227–230. Problem VII.6.10(i).

[2] *On a problem of Petrich and Reilly*, J. Math. (Wuhan) 27 (2007), no. 5, 499–502. Problem VII.7.12.

Pan, X. J. and Y. Shao

[1] *P-completely regular semigroups*, Pure Appl. Math. (Xi'an) 24 (2008), no. 4, 793–795. VII.9.8.

Wang, L. M., Y. Y. Feng and H. H. Chen

[1] *Some special congruences on completely regular semigroups*, Commun. Algebra 47 (2019) no. 7, 2941–2953. Problems VII.3.13.

Wang, X. and G. X. Liu

[1] *A problem on bands*, Semigroup Forum 81 (2010), no. 3, 548–550. Problem VIII.2.13(ii).

Yu, H., Z. Wang, T. Wu and M. Ye

[1] *Classification of some r-congruence-free completely regular semigroups*, Semigroup Forum 84 (2012), no. 2, 308–322. Problems VI.1.23 and VII.6.10.

Index of Symbols

Upper Case Greek

$\Gamma(S)$	the lattice of fully invariant (unary) congruences on S, Sect. 1.2
$\Gamma(G)$	the lattice of fully invariant subgroups on G, Sect. 1.2
Γ	$\Gamma(CR_X)$, Notation 1.1.2
Δ	the interval $[\zeta, \omega]$, Notation 1.1.2
Θ	a relation on a completely regular semigroup, [PR1, Definition VI.2.5]
Θ	a set of words over T_ℓ, T_r, Definition 3.1.1
Θ_ℓ	T_ℓ-chain, Sect. 3.4
Θ_r	a relation on a completely regular semigroup, [PR1, Definition VI.2.5]
Θ_r	T_r-chain, Sect. 3.4
Λ	an equivalence relation on $\mathcal{L}(\mathcal{CR})$, Notation 2.2.2
Ξ	an equivalence relation on $\mathcal{L}(\mathcal{CR})$, Notation 2.2.2
Π	$\Lambda \cap \Xi$, Notation 2.2.2
Π	$\{s, e\}^+$, Definition 3.1.2
$\Pi_{\mathcal{V}}$	$\{S \in \mathcal{CR} \mid S/\pi_S \in \mathcal{V}\}$, Definition 2.3.7
Σ	$\{K, K_\ell, K_r, T, T_\ell, T_r\}$, Notation 1.3.2
Υ	$\{\mathcal{V}^B \mid \mathcal{V} \in \mathcal{L}(\mathcal{CR})\}$, Definition 3.6.2
Φ	a set of mappings, Notation 4.3.6
Φ_B	a set of mappings, Notation 4.4.6
Φ_D	set of mappings $\Theta^1 \to D$, Definition 3.2.1
$\Phi_{L\mathcal{O}}$	ideal of Φ, Notation 4.6.1
$\Phi_{\mathcal{LRO}}$	ideal of Φ, Notation 4.6.5
$\Phi_{\mathcal{O}}$	ideal of Φ, Notation 4.6.3
$\Phi_{\mathcal{OHU}}$	ideal of Φ, Notation 4.6.3
Ψ	subset of $(\mathcal{K}_1)^{\Theta^1} \times [\mathcal{S}, \mathcal{B}]$, Notation 4.2.3.

© The Editor(s) (if applicable) and The Author(s), under exclusive license to Springer
Nature Switzerland AG 2024
M. Petrich and N. R. Reilly, *Completely Regular Semigroup Varieties*, Synthesis Lectures
on Mathematics & Statistics, https://doi.org/10.1007/978-3-031-42891-3

Lower Case Greek

β	congruence induced by mapping b, Remark 3.7.9
γ	mapping from $[\mathcal{S}, \mathcal{CR}]$ to Ψ, Theorem 4.2.5
$\gamma_{\mathcal{V}}$	mapping from Θ^1 to $[\mathcal{S}, \mathcal{CR}]$, Theorem 4.2.5
$\gamma(u)$	$\overline{w}$, where w is the shortest initial segment of u with $c(u) = c(w)$, Sect. 1.1
δ	mapping from $[\mathcal{S}, \mathcal{CR}]$ to Φ_D, Theorem 3.2.8
$\delta(u)$	$\overline{w}$, where w is the shortest final segment of u with $c(u) = c(w)$, Sect. 1.1
δ^*	restriction of δ to $[\mathcal{S}, \mathcal{B}]$, Theorem 3.2.8
$\delta_{\mathcal{V}}$	a mapping $\Theta^1 \to D$, Theorem 3.2.8
$\epsilon(u)$	the first variable from X to appear in u for the last time, Sect. 1.1
ζ	least completely regular (unary) congruence on U, Notation 1.1.3
$\zeta_{\mathcal{V}}$	fully invariant congruence on U corresponding to $\mathcal{V}$, Sect. 1.3
η	the least semilattice congruence on U, Notation 1.1.12
$\theta_{\mathcal{B}}$	the mapping $\mathcal{V} \to (\mathcal{V} \cap \mathcal{B}, \mathcal{V} \vee \mathcal{B})$, Proposition 4.5.1
$\theta_{\mathcal{BM}}$	the mapping $\mathcal{V} \to \mathcal{V} \cap \mathcal{BM}$, Theorem 3.8.3
$\theta_{\mathcal{F}}$	the mapping $\mathcal{V} \to \mathcal{V} \cap \mathcal{F}$, Notation 2.4.1
$\theta_{\mathcal{RD}}$	the mapping $\mathcal{V} \to \mathcal{V} \cap \mathcal{RD}$, Theorem 2.6.2(vii)
$\theta_{\mathcal{RF}}$	the mapping $\mathcal{V} \to \mathcal{V} \cap \mathcal{RF}$, Notation 2.5.1
ι	identity congruence or mapping, Sect. 1.2
$\mu_{\mathcal{B}}$	the mapping $\mathcal{V} \to \mathcal{V} \cap \mathcal{B}$, Theorem 3.2.8
ξ	a mapping $\Phi \to [\mathcal{S}, \mathcal{CR}]$, Theorem 4.4.1
ξ_D	a mapping $\Phi_D \to [\mathcal{S}, \mathcal{B}]$, Theorem 3.2.8
$\xi_{\mathcal{K}_1}$	a mapping $\Psi \to [\mathcal{S}, \mathcal{CR}]$, Theorem 4.2.5
$\pi_{\mathcal{X}}$	a mapping of the form $\mathcal{V} \to \mathcal{V} \cap \mathcal{X}$, Corollary 4.2.10
π_S	a congruence on S, Definition 2.3.1
π^*	$\pi\varphi_{\Pi\Theta}$, Definition 3.1.2
ρ^*	least congruence containing ρ, Sect. 1.2
ρ^0	largest congruence contained in the relation ρ, Sect. 1.2
ρ_e	a relation on U, Notation 1.6.1
ρ_s	a relation on U, Notation 1.6.1
ρ_P	least element in P-class of ρ, Notation 1.3.5
ρ^P	greatest element in P-class of ρ, Notation 1.3.5
$\rho_{\mathcal{V}}$	fully invariant congruence on $F\mathcal{CR}(X)$ corresponding to $\mathcal{V}$, Sect. 1.3.
ρ^e	relation on U, Notation 1.5.1
ρ^s	relation on U, Notation 1.5.1
$\overline{\rho}$	a relation on U related to ρ, Notation 1.4.1
ρ	variety corresponding to $\rho \in \Delta$ or $\rho \in \Gamma$, Sect. 1.3
$\hat{\rho}$	least fully invariant congruence containing ρ, Notation 1.5.1
$\sigma(u)$	the last variable from X to appear in u for the first time, Sect. 1.1
$\sigma^{k+1}(w)$	$\sigma(s^k(w))$, Definition 3.7.2

$\sigma^A(w)$	the last variable in $c(w) \setminus A$ to appear for the first time, Definition 3.7.2
τ^d	word obtained from τ by replacing T_ℓ by T_r and T_r by T_ℓ, Notation 3.2.6
$|\tau|$	the length of $= n$, where $\tau = P_1 \cdots P_n$, Sect. 3.1
$\overline{\tau}$	the mirror image $P_n \cdots P_1$ of $\tau = P_1 \cdots P_n$, Sect. 3.1
$\tau(K_\ell, K_r)$	word obtained from τ by replacing T_ℓ, T_r by K_ℓ, K_r, Notation 3.2.4
$\varphi_{\Pi\Theta}$	mapping from Π to Θ, Definition 3.1.2
χ	mapping from $[\mathcal{S}, \mathcal{CR}]$ to Φ, Theorem 4.4.1
$\chi_\mathcal{V}$	ladder of $\mathcal{V}$, Theorem 4.4.1
ω	universal congruence, Sect. 1.2.

Upper Case Roman

B	the B-relation, Definition 3.6.1
B	$b(X^+)$ free band on X, Notation 3.7.7
CR_X	U/ζ, Notation 1.1.3
CR_n	$CR_{\{x_1,\ldots,x_n\}}$, Notation 1.1.3
D	$\{1, L, R, T\}$, Definition 3.2.1
D_n	fragment of $[\mathcal{S}, \mathcal{B}]$, Sect. 3.3
D'_n	fragment of $[\mathcal{S}, \mathcal{B}]$, Sect. 3.3
E	E-disjunctive relation, Sect. 2.2
E_k	boundary layer pattern, Sect. 3.3
$E_{n,k}$	ladder with pattern k at level n, Sect. 3.3
G_n	a system of words in X^+, Definition 3.5.1
H_n	a system of words in X^+, Definition 3.5.1
I_n	a system of words in X^+, Definition 3.5.1
P	$X \cup \{(,)^{-1}\}$, Chap. 1 preamble
P_n	a system of words in X^+, Notation 3.5.8
Q_n	a system of words in X^+, Notation 3.5.8
R_n	a system of words in X^+, Notation 3.5.8
$\mathrm{Soc}(L)$	part of a ladder, Definition 3.6.5
U	U_X, free unary semigroup on X, Preamble to Chap. 1
U_n	$\{u \in U \mid \#(u) = n\}$, Notation 1.4.7
U_n^*	$\{u \in U \mid \#(u) \leq n\}$, Notation 1.4.7
X	countably infinite set, Preamble to Chap. 1.

Lower Case Roman

Script

$\mathcal{K}_1$	$\{\mathcal{V} \vee \mathcal{S} \mid \mathcal{V} \in \mathcal{K}_0\}$, Notation 4.2.1
$\mathcal{K}_{\mathcal{CS}}$	$\mathcal{K}_0 \cap \mathcal{L}(\mathcal{CS})$, Notation 4.6.1
$\mathcal{L}(\mathcal{BM})$	the lattice of subvarieties of $\mathcal{BM}$, Definition 3.8.1
$\mathcal{L}_{\mathbb{P}}(\mathcal{CR})$	class of all $\mathbb{P}$-varieties, Definition 2.3.7
$\mathcal{M}$	the class of completely regular monoids, Definition 3.8.1
$\mathcal{MV}$	the L-class of the variety $\mathcal{V}$ of band monoids, Definition 3.8.1
$\mathcal{P}$	an integral class, Definition 2.3.3
$\mathcal{PD}$	$(\mathcal{CS} \cup (\mathcal{CR} \setminus \mathcal{LO})) \cap \mathcal{E}$, Notation 2.1.1
$\mathcal{PE}$	$\mathcal{G} \cup (\mathcal{CS} \setminus \mathcal{ReG}) \cup ((\mathcal{CR} \setminus \mathcal{LO}) \cap \mathcal{E})$, Notation 2.1.1
$\mathcal{P}(i), \mathcal{P}(ii)$	conditions for a $\mathbb{P}$-variety, Definition 2.3.7
$\mathcal{RD}$	class of right E-disjunctive completely regular semigroups, Sect. 2.6
$\mathcal{RF}$	class of right fundamental completely regular semigroups, Notation 2.5.1
$\mathcal{V}_P$	least element in the P-class of $\mathcal{V}$, Notation 1.3.5
$\mathcal{V}^P$	greatest element in the P-class of $\mathcal{V}$, Notation 1.3.5
$\mathcal{V}_B$	least element in B-class, Sect. 3.6
$\mathcal{V}^B$	greatest element in B-class, Sect. 3.6
$\mathcal{V}_{K^*}, \mathcal{V}^{K^*}$	operator on $\mathcal{L}(\mathcal{CR}) \cup \mathbb{N}_3^*$, Notation 4.3.3
$\mathcal{V}_\tau$	action of $\tau \in \Theta^1$ on $\mathcal{L}(\mathcal{CR})$, Definition 3.1.2
$\mathcal{V}^\tau$	action of $\tau \in \Theta^1$ on $\mathcal{L}(\mathcal{CR})$, Definition 3.1.2.

Open Back

$\mathbb{G}_n$	a sequence of words in U, Notation 3.6.7
$\mathbb{H}_n$	a sequence of words in U, Notation 3.6.7
$\mathbb{I}_n$	a sequence of words in U, Notation 3.6.7
$\mathbb{M}$	$\{\mu_S \mid S \in \mathcal{CR}\}$, Notation 2.4.1
$\mathbb{N}_3$	$\{\mathcal{LNB}, \mathcal{S}, \mathcal{RNB}\}$, Notation 3.1.5
$\mathbb{N}_3^*$	$\{L^*, T^*, R^*\}$, Notation 4.3.1
$\mathbb{N}_6$	$\{\mathcal{LZ}, \mathcal{T}, \mathcal{RZ}\} \cup \mathbb{N}_3$, Notation 4.3.1
$\mathbb{P}$	a congruence family, Definition 2.3.1
$\mathbb{P}_n$	a sequence of words in U, Notation 3.6.7
$\mathbb{P}(i)-\mathbb{P}(iv)$	conditions for a congruence system, Definition 2.3.1
$\mathbb{R}$	$\{\mathcal{R}_S^0 \mid S \in \mathcal{CR}\}$, Notation 2.5.1
$\mathbb{RD}$	$\{\tau_S \cap \mathcal{R}_S^0 \mid S \in \mathcal{CR}\}$, Notation 2.6.1
$\mathbb{T}$	$\{\tau_S \mid S \in \mathcal{CR}\}$, Sect. 3.6.

Simple Symbols

$u \cdot v$	$b(uv)$, Notation 3.7.7.

Index